A CONCISE ACCOUNT OF THE SEVERAL FOREIGN ORDERS OF KNIGHTHOOD, AND OTHER MARKS OF HONOURABLE DISTINCTION

A CONCISE ACCOUNT

OF

THE SEVERAL FOREIGN ORDERS OF KNIGHTHOOD,

AND

OTHER MARKS OF HONOURABLE DISTINCTION,

ESPECIALLY OF SUCH AS HAVE BEEN CONFERRED UPON

BRITISH SUBJECTS.

TOGETHER WITH

THE NAMES AND ACHIEVEMENTS OF THOSE GALLANT MEN

WHO HAVE BEEN PRESENTED WITH

HONORARY SWORDS, OR PLATE,

BY

THE PATRIOTIC FUND INSTITUTION.

BY NICHOLAS CARLISLE, ESQUIRE,

KNIGHT OF THE ROYAL HANOVERIAN GUELPHIC ORDER,
KNIGHT OF THE IMPERIAL AUSTRIAN ORDER OF THE IRON CROWN,
KNIGHT OF THE DANISH ORDER OF THE DANNEBROG,
GENTLEMAN OF HER MAJESTY'S MOST HONOURABLE PRIVY CHAMBER,
HONORARY DOCTOR OF CIVIL LAW IN THE UNIVERSITY OF OXFORD,
FELLOW OF THE ROYAL SOCIETY OF LONDON,
MEMBER OF THE ROYAL IRISH ACADEMY, DUBLIN,
MEMBER OF THE ROYAL SOCIETY OF NORTHERN ANTIQUARIES, COPENHAGEN,
MEMBER OF THE ROYAL ACADEMY OF SWEDEN, STOCKHOLM,
MEMBER OF THE LITERARY SOCIETY OF ICELAND,
MEMBER OF THE SOCIETY OF LITERATURE AND ART OF COURLAND, MITAU,
MEMBER OF THE AMERICAN PHILOSOPHICAL SOCIETY, PHILADELPHIA,
CORRESPONDING MEMBER OF THE ACADEMY OF SCIENCES AND BELLES LETTRES, PALERMO,
CORRESPONDING MEMBER OF THE ARCHÆOLOGICAL INSTITUTE, ROME,
HONORARY MEMBER OF THE ANTIQUARIAN SOCIETY OF NEWCASTLE-UPON-TYNE,
AND FELLOW AND SECRETARY OF THE SOCIETY OF ANTIQUARIES OF LONDON.

LONDON, MDCCCXXXIX.

REPRINTED BY
THE NAVAL AND MILITARY PRESS

THIS EDITION OF

A CONCISE ACCOUNT OF THE SEVERAL
FOREIGN ORDERS OF KNIGHTHOOD, &c., &c.

FIRST PUBLISHED 1992 BY

THE NAVAL & MILITARY PRESS,
DALLINGTON,
EAST SUSSEX.

PUBLISHING HISTORY

FIRST PUBLISHED 1839 BY JOHN HEARNE, LONDON.

THIS EDITION IS NOW REPRODUCED EXACTLY AS THE ORIGINAL VOLUME, COMPLETE AND UNABRIDGED.

TO

THE RIGHT HONOURABLE

SIR ROBERT PEEL, BARONET,

&c. &c. &c.

SIR,

The renown of those glorious Achievements which I have endeavoured to perpetuate, is the meed not of Individuals only, but of that also of the Nation, as the Parent of such gallant Offspring. Hence it becomes a respectful duty in me to introduce their heroick Actions under the Auspices of a Publick Character standing high in National esteem, — a Character, intimately associated with Honour, Ability, and Patriotism.

In being permitted, then, to dedicate these pages to Sir ROBERT PEEL, I gladly avail myself of the favourable opportunity of connecting the VALOUR of my Countrymen with the Name of a STATESMAN, educated in, and a most distinguished Minister under, that System of Policy, which led to the splendid VICTORIES herein recorded.

With sentiments of the highest respect, I have the honour to subscribe myself,

SIR,

Your most obedient,

and humble Servant,

NICHOLAS CARLISLE.

SOMERSET PLACE,
5th of April, 1839.

INTRODUCTION.

SPEAK TRULY, ON THY KNIGHTHOOD, AND THY OATH ;
AND SO DEFEND THEE HEAVEN, AND THY VALOUR !

SHAKSPEARE.

IT has been the endeavour of the greatest Writers in almost every Age, to expose with all the force that Wit and Argument could supply, the vanity of Ancestral Pride, and to show that true Nobility consists in VIRTUE, not in BIRTH.

Nevertheless, some distinguished Moralists have been of opinion that this rule, like every other, has it's exceptions, and may be carried beyond the boundaries of Reason. Not only Gratitude, but Policy, bids us to honour the Progeny of those who have exalted the character or improved the condition of their Country,— who, by their Valour or their Talents, have made it more respected, and happy. As, in truth, whatever may be his conduct, he who has had an honourable Name transmitted to him, will at least have an additional and powerful inducement to preserve it unsullied, and to convey it to his Posterity unimpaired, although he may not be able to impart greater Dignity to it.

For these reasons, ADDISON and Dr. JOHNSON agree in thinking, that he who to the claim of Merit adds that of illustrious Birth, is justly to be regarded more than one of equal Merit, who has no Hereditary pretensions.

As true worth, then, must ever be grounded upon the conduct of the individual, as a GENTLEMAN,—let it always be remembered, that much as this Title has been perverted from it's original meaning, it still includes in it's correct and proper sense, NOBILITY,—for any Individual, who distinguishes himself in Learning, in Arms, in Science or in Art, is self-ennobled.

That a Nobleman may be *made*, but that a Gentleman must be *born*, is an old and established maxim. FRANCIS the first, who boasted of the proud title of "*The King of Gentlemen*," declared his inability to make one. And JAMES the first, who was not over scrupulous as to what Titles he conferred, or on whom he bestowed them, when asked by his Nurse to confer this Honour upon her son, candidly told her,—"My good woman, a *Gentleman* I could never make him, though I could create him a *Lord*."

As true Merit, then, is generally speaking the surest road to Honours, let us contemplate it's reward and the DECORATIONS, which have been so frequently conferred upon the high-spirited race of our Nation,—a Nation, which, it is said, "may boast, beyond any other people in the World, of a kind of epidemick bravery, diffused equally through all it's Ranks."

It is by most persons acknowledged, that no maxim in Politicks is more indisputable, than that a Nation should have many Honours to bestow on those, who perform National services. Honorary Distinctions excite Emulation, cherish Publick spirit, and create an Ambition highly conducive to the good of the Country. The less burdensome these Honours are to the Treasury, the more productive are they of National advantage. Such Honours may, by an allowable adaptation of the language of BURKE, be called the "unbought graces of Life, the cheap defence of Nations, the promoters of manly Sentiments and Enterprise."

The Greeks and Romans well knew the Value of Honorary rewards,—rewards, which conferred Distinction and gave Precedence, uncontaminated by any mercenary association.

The multiplicity of ORDERS in The Northern States of Europe has frequently been condemned,—but it must be allowed, that it was a happy contrivance of the Sovereigns of Countries not remarkable for their Wealth, thus to provide means of rewarding eminent services, at the expense of a few STARS and RIBANDS,—since more substantial marks of Approbation would be attended with serious inconvenience. Besides, ORDERS and HONOURABLE EMPLOYMENTS inspire a higher degree of Emulation than pecuniary Recompenses,—and he who chiefly looks to *Gain*, as the meed of his efforts, will seldom entitle himself to the character of a true *Patriot*.

Those who deem the conferring of the various ORDERS of Nations a mere interchange of Trinkets, may also think that THE ORDERS themselves savour too much of the Chivalrous Age so long gone by!—and, that it is gone by, is to be deeply lamented,—It was the Age of unblemished Honour, the period when the *word* often bound stronger than *oaths* of the present day,—and when THE INSIGNIA of Virtue and of Valour were prized beyond all the Wealth that the Earth could furnish. To laugh at the Titles and Ceremonies of Honour, is to laugh at Honour itself,—to deride external Decorations is to treat spotless Virtue with contempt. It has been said, that some vulgar minds in our own Island turned the nobly earned Distinctions of our Immortal NELSON into ridicule,—they jested upon his Crosses and his Stars, when, had their Envy suffered them to look deeper, they would have seen *Wounds* under every Ornament, heroically received in making *his* breast the shield of *their* safety.

The necessity of THE BRITISH GENTRY asserting their Nobility increases, as the connection of GREAT BRITAIN with THE CONTINENT increases. Those, who deliver Passports for The Continent, ought to give the quality of *Gentleman* to those only, who are entitled to it,—but those, who really are entitled to it, should not suffer it to be omitted. The disuse of the word may be of the greatest disadvantage. When arrived at the place of his destination, his letters of recommendation may indeed prove who and what he is,—but he may be induced to change his Route, or may be involved in accidents or disputes, and then the omission of his real Title might occasionally prevent his receiving those

Civilities and that Hospitality, which he otherwise would experience.

A *Foreigner* in England usually passes for a person of greater Dignity than he really is, while the *Englishman* abroad often loses some of his importance. And this because our Gentry bear more modest Titles, and seem to have forgotten that they are Nobles,—

> " Stern o'er each bosom Reason holds her state
> With daring aims irregularly great;
> Pride in their port, defiance in their eye,
> I see the lords of human kind pass by."
>
> GOLDSMITH'S TRAVELLER.

The re-establishment of The Visitations of THE HERALDS would in the eye of the world replace the English Gentry, who, in no real respect, are at all below them, on a footing with *The Noblesse* of The Continent. If only those who are entitled to bear Arms, were allowed to use them, Heraldick Achievements would be more respected, and the true Rank of their possessors would again be recognised.

These observations may, possibly, be thought to be unnecessary,—but in a Work professedly written with a view to perpetuate the renown of Wisdom and of Valour, it would have been blameable not to recommend those attentions to personal consequence, which either the Authorities at home or abroad might inadvertently omit. For how can he claim a due respect from Foreigners, who neglects to let them know, what is his Station in Society in his own Country ?

A perfect knowledge of the various ORDERS of KNIGHTHOOD, which have been established at different times, is said to be a very necessary and important branch of the Science of Heraldry,—and many are the Writers, who have employed their pens upon the fascinating subject of Chivalry.

In the 50th Letter of Lord CHESTERFIELD to his son, dated in 1752, after commending him for going to *Versailles* to see the Ceremony of creating The Prince *de* CONDÉ, "*Chevalier de l'Ordre*,"—he observes, "as you will have been a great while in France, people will expect that you should be well informed of all these sort of things, relative to that Country. But the History of all THE ORDERS of all Countries is well worth your knowledge,—the subject occurs often, and one should not be ignorant of it, for fear of some such accident as happened to a solid Dane, at Paris, who, upon seeing *l'Ordre du St. Esprit,* said, *Nôtre St. Esprit chez nous c'est un Elephant.* Almost all the Princes in Germany have their ORDERS too, not dated, indeed, from any important events, or directed to any great object,—but because they will have ORDERS, to show that they may,—as some of them, who have the *Jus cudendæ monetæ,* borrow ten shillings worth of gold to coin a Ducat. However, wherever you meet with them, inform yourself, and minute down a short account of them,—they take in all the colours of Sir ISAAC NEWTON'S prisms."

But Honorary Distinctions are estimable, not on account of the Title they confer, but as a signal testimony of that esteem of Foreign States, which it should ever be the Ambition of an Individual to deserve, because respect is thereby reflected on his own Country.

And, as many of these Orders have been conferred upon our greatest Warriors and Statesmen, the honour may be considered as eminently increased by the Rank and Character of those, who are enrolled in the same Fraternity.

A subject more agreeable to an Englishman, as it is so immediately connected with History, and particularly so, with respect to The British Empire, cannot but be interesting to every Native of The United Kingdom, whose Relatives have gained imperishable honours by such exertions as won the Trophies, in the olden time, in the memorable fields of *Agincourt*, *Poitiers*, and *Cressy*.

The Institution of the early Orders of Knighthood may be traced to a high sense of Honour, combined with a prudent regard to the advancement of Military affairs,—whereby Virtue might be chiefly excited and promoted, and Valour accompanied by that moral Excellence, be decorated with an appropriate Reward.

It is a matter of History, that the Monarchs of Europe were in the practice of cementing their Alliances and Friendships, by a mutual exchange of Orders of Knighthood. But the Reformation, for obvious reasons, checked that kind of intercourse,—And the great Rebellion which raged in England in the Seventeenth Century,—together with the memorable War of *Thirty years*, that distracted The Continent,—seemed to have put an end to the gallant Civilities between Sovereigns, whereby their Subjects received those personal Distinctions, ever the most encouraging and gratifying to generous minds. But the late War in Europe, in which the legitimate Monarchs united for one common and laudable End, restored that intercourse which had seemed to be terminated, and revived with increased friendliness the interchange of Honours so productive of the best effects to all Parties.

Although some of these ORDERS of KNIGHTHOOD are obscure, and the descriptions of them are viewed as romantick, if not actually fabulous, while others may be thought inconsiderable,— yet, many of them have, nevertheless, been judged worthy of description here, equally with the most famous and splendid, " seeing, that once they had such in the Register of Honour."

There appears, however, to have been, at a very early period, some little jealousy in the minds of Sovereigns, as to the reception of FOREIGN ORDERS by their Subjects.

Mons. de WICQUEFORT (in *The Embassador and his Functions, p.* 354) informs us, that HENRY the Fourth of France having been assisted with a choice body of Forces by Queen ELIZABETH, invested NICHOLAS CLIFFORD, and ANTHONY SHIRLEY, two of the bravest and most gallant Officers of the English Auxiliaries, with THE COLLAR of THE ORDER of SAINT MICHAEL, for the services they had done him in the War. But, on their return to England, they were committed to prison for daring to accept this Honour, without having previously obtained the permission of The Queen's Grace, and were peremptorily ordered to send back THE COLLAR which they had received from The French Monarch,—Her Majesty proudly observing, "that as a virtuous Woman ought to look on none but her Husband, so a Subject ought not to cast his eyes on any other Sovereign than him, whom God had set over him. I will not," said she, "have my sheep marked with a strange brand, nor suffer them to follow the Pipe of a strange Shepherd."

CHRISTINA, Queen of Sweden, would not permit THE PRINCE PALATINE to receive THE ORDER of THE GARTER, nor THE COUNT *de la* GARDE to be made a PRINCE of THE EMPIRE.

These two Queens, *Mons. de* WICQUEFORT contends, were in the right to prohibit their Subjects from entering into engagements with Foreign Princes,—who cannot share out their affection, or their zeal, without robbing their Sovereign of all that portion, which they so bestow.

As *Mons. de* WICQUEFORT had just been remarking on the *Pensions* that had been paid to the Noblemen of the Court and Council, by Foreign Potentates, in preceding reigns, it is probable, that ELIZABETH might consider this Decoration of The French Monarch, as likely to operate as a corrupt influence on her Officers, and, therefore, acted in that peremptory and disdainful manner. It must, however, be acknowledged, that ELIZABETH often exhibited in her conduct a most unnecessary degree of Arbitrariness.

As to the corruption of Publick Ministers, it was made a jest of by JAMES the first,—who, with a baseness of sentiment, revolting and cowardly in any one, but especially so in a Sovereign, declared, that " he wished the King of *Spain would give them* ten times as much,—because this unprofitable expense *would render him less able to make war against him.*"

The Principle is clear, that no person can properly accept and wear THE INSIGNIA of a FOREIGN ORDER, without the previous consent of his own SOVEREIGN,—and for the obvious reason, that it might produce a division of his Allegiance. THE STATUTES of all ORDERS requiring Fealty from the Knights to the Chiefs, and imposing certain Obligations on them, which might prove incompatible with their superior duties.

The Sovereign, being the Fountain of Honour, can only regard with a jealous eye any Foreign sources from which Honourable Distinctions may flow, without his full consent, upon his own Subjects.

Previously to the year 1800, the instances of FOREIGN ORDERS of KNIGHTHOOD being conferred upon English Subjects were of such rare occurrence, as to attract little notice, either from The Publick or The Government. The permission of the British Sovereign was, of course, always necessary to legitimatize the acceptance of a *Foreign Order*, or authorize the wearing it's *Insignia* in this Country,—but, even for many years subsequent to the Accession of GEORGE the Third, there does not appear to have been any established Rule or Etiquette, as to the mode in which that permission was to be announced or recorded.

The *first* Royal Licence which appears on record, to enable a British Subject to accept *a Foreign Baronial Honour,* is dated the 21st of July, 1684,—and grants to JOHN, Earl of BATH, permission to assume the Dignity of *Count of The Holy Roman Empire,* from the Emperor LEOPOLD.

The *first* Royal Licence on record, to enable a British Subject to accept *a Foreign Order of Knighthood,* is dated the 16th of July, 1789,—and grants to SAMUEL BENTHAM, Esq., permission to accept and wear, in his own Country, the *Insignia* of THE ORDER of ST. GEORGE, of *Russia.*

Even FOREIGN POWERS seem to have doubted their ability to confer their DECORATIONS, or to dispense with their STATUTES, — for, although the brilliant and important service achieved by the Detachment of His Majesty's 15th Regiment of Light Dragoons in the Action of *Villers en Couché,* near Cambray, on the 24th of April, 1794, was deemed at the time worthy of THE CROSS of THE ORDER of MARIA THERESA, yet a doubt was then entertained at *Vienna,* whether this ORDER could be conferred upon *Foreigners.* It appears, however, that the difficulty being removed, HIS IMPERIAL MAJESTY availed himself with pleasure of the occasion to manifest his high esteem for the Regiment, as well as his regard for the Individuals, by investing with this distinguished ORDER of MERIT, Gentlemen who had proved themselves so worthy to wear it,—And Eight

Crosses were accordingly transmitted by His Excellency Lord Minto, then His Majesty's Envoy Extraordinary and Plenipotentiary at the Court of Vienna, on the 7th of November, 1800, to Lieutenant Colonel William Aylett, to be distributed to the Officers, for whom they were destined.

Every Country, since the first Institution of The Knights of Saint John of Jerusalem, has founded new Equestrian Fraternities,—Every Sovereign has carefully maintained the Splendour of most of those Orders, which he found existing,—And, since the year 1800, every Monarch has gratefully augmented their number.

Since the date of *The French Revolution,* the instances of British Subjects invested with Foreign Orders, have been very numerous,—and the frequency of these instances is, doubtless, to be attributed to the Political Alliances which England formed in the Progress of the severe Contest with the preponderating Power of France, and the consequent extension and success of The British Arms.

It appears that the genius of The Russian Government, in the year 1799, aimed at gaining a firm footing in *The Mediterranean,*—and that The Emperor Paul, among his caprices and cravings, had fixed his eyes and heart on *Malta,*—and it would seem, that some hopes had been holden out to him, or, at least that he was allowed, without being undeceived, to entertain a sanguine expectation of being presented with that important Island. A Fleet, with troops, sailed from *The Black Sea,* in August, 1800, for the express purpose of taking possession of the place, when it should surrender. No remonstrances were made, when PAUL haughtily assumed the Title of GRAND MASTER of THE SOVEREIGN ORDER of ST. JOHN of JERUSALEM,—and, even, when he conferred upon Captain POPHAM *The Cross of Malta*, as a mark of his particular esteem, his right to do so was recognised in THE LONDON GAZETTE.

We are, however, told, that, when it was communicated to GEORGE the Third, that Captain POPHAM was made *a Knight of Malta*, His Majesty declared in language similar to that of

—a fair Vestal, throned by the West,—

" that none of his Flock should wear any other mark than his own." And, as it is said, that Captain POPHAM was the only Knight of *that Order*, whose Promotion was formally recognised at The Court of St. James, it is probable that The King's repugnance to any of His Subjects being decorated with FOREIGN ORDERS might ultimately have occasioned the following REGULATIONS,—

Regulations of His Majesty and of His Royal Highness the Prince Regent, respecting Foreign Orders.

1st. That no British Subjects shall accept a Foreign Order, or wear its Insignia, without having previously obtained a Warrant under the Royal Sign Manual, (directed to the Earl Marshal of England,) granting them His Majesty's permission to accept and wear the same.

2d. That the intention of a Foreign Sovereign to decorate a British Subject with the Insignia of such Order, shall be notified to His Majesty's Principal Secretary of State for Foreign Affairs, either through the King's Minister abroad, or through the accredited Minister of the Foreign Sovereign resident at this Court.

3d. That when His Majesty's Principal Secretary of State for Foreign Affairs shall have taken His Majesty's pleasure, and obtained His consent upon the occasion, he shall then signify the same to His Majesty's Principal Secretary of State for the Home Department, in order that he may cause the usual Warrant to be prepared for the Royal Sign Manual, and take such other steps as may be necessary for carrying His Majesty's pleasure into effect accordingly.

4th. That when the Warrant is signed by the King, it shall be announced in the Gazette in the usual manner, and registered in His Majesty's College of Arms.

In addition to the above regulations of the King, His Royal Highness The Prince Regent deemed it expedient to announce, in the year 1812,

5th. "THAT no Subject of His Majesty could be allowed to accept the Insignia of a Foreign Order from any Sovereign of a Foreign State, except they shall be so conferred in consequence of active and distinguished services before the Enemy, either at Sea or in the Field; or unless he shall have been actually employed in the service of such Foreign Sovereign."

And in March 1813, His Royal Highness was pleased to command, that the following proviso should be thereafter inserted in all Royal Warrants for the acceptance of Foreign Orders:

6th. "That His Majesty's Licence and Permission doth not authorize, and shall not be deemed or construed to authorize, the assumption of any style, appellation, rank, precedence, or privilege appertaining unto a Knight Bachelor of these Realms."

N.B. Before the Principal Secretary of State for Foreign Affairs takes the Prince Regent's pleasure, on any application for an Officer in the Army to be permitted to accept a Foreign Order, he causes the same to be referred to the Commander-in-Chief, through his Royal Highness's Secretary, to know whether or not His Royal Highness sees any objection to the Prince Regent's pleasure being taken thereupon.

And if the application be in favour of a Naval Officer, the Principal Secretary of State for Foreign Affairs communicates with the First Lord of the Admiralty, to the same effect, previous to the Prince Regent's pleasure being taken thereon. (Signed) CASTLEREAGH.

The Policy which dictated these REGULATIONS, is still observed, inasmuch as no British Subject is allowed to wear *The Insignia* of any FOREIGN ORDER of KNIGHTHOOD in The Royal Presence, without first obtaining His Sovereign's permission,—and having the same duly registered in THE COLLEGE of ARMS,—nor does any permission granted subsequently to March 1813, authorize the assumption of any style, appellation, rank, precedence, or privilege appertaining to that of a Knight-Bachelor of The United Kingdom of Great Britain and Ireland.

It is true, that the REGULATIONS have occasionally been departed from,—but the instances are few, and these only in favour of Persons of High Rank, or through Official Influence,—while members, both *Civil* and *Military*, are left to deplore with CYMBELINE, that

——" Knighthoods and honours, borne
As I wear mine, are titles but of scorn."

The Practice of Government is, therefore, to allow such persons only to accept FOREIGN ORDERS, as shall have obtained the Distinction, *for services in the Field*, —And, in consequence, several applications of CIVILIANS for Royal Licences have been refused,—nevertheless, those CIVILIANS who have received from Foreign Potentates the Decorations of the various ORDERS of KNIGHTHOOD, cannot be deprived of the Celebrity which led to the acquisition of such Honours, although they be denied the Privilege of wearing them.

It is neither my intention, nor my inclination, to

censure REGULATIONS emanating from THE CROWN,—but it may not be disrespectful to state, what has been the practice in other Kingdoms, in the distribution of Honours, and in the appropriation of suitable proofs of acknowledgement of every degree of Merit.

FOREIGN GOVERNMENTS have studied to reward Merit of every kind.—To the *Military* are assigned special ORDERS, according to their Rank and Service,—To the *Civilians*, are DECORATIONS and MEDALS, adapted to every occasion,—And, with a noble and wise Policy, they are open to persons of all Religious tenets, and neither interest, rank, birth, or patronage, have any influence in the distribution of them,—but true MERIT is alone the recommendation to such Distinction. Why, then, is Great Britain so sparing in Civil Honours?—Why is this Illustrious Empire the only one in Europe, where there is no such dignified and generous mode of publickly distinguishing superior acquirements in Science, in Learning, and in liberal Art?

If the Laws of the Country are found to be unjust or imperfect, a remedy is provided by THE SENATE. If by some oversight extraordinary Knowledge, rare Talents, and exemplary Industry, be treated with indifference, why permit the reproach to continue? Is any encouragement of that spirit of emulation, which animates the Civilian in his highly important but peaceful labours, to be found in the preceding REGULATIONS?—If not, then let so injurious a *Partiality* no longer exercise it's chilling influence,—let every Class of the Community be incited, by the hope of National Honours, to direct their best energies for the benefit and glory of a Country capable and willing, to reward most bountifully and gratefully CIVIL as well as MILITARY Services.

I may safely predict, that The Institution of a CIVIL ORDER, would obtain the unanimous approbation of The British People,—and that every National expression of Gratitude would be offered at the foot of the Throne, for such an act of wise and liberal Policy,—an act the more auspicious, as gracefully marking the commencement of the reign of a young and lovely Queen.

When the Royal Licence for accepting a Foreign Order had been obtained, the claimant assumed the style of *Sir*,—But GEORGE the Third was always averse to the grant of such Licences, and first stopped their issue, in the case of KER PORTER, who had obtained THE ORDER of ST. JOACHIM, in or about the year 1807. Shortly after, a strong effort was made to obtain for Sir SAMUEL EGERTON BRYDGES permission to accept the like ORDER of ST. JOACHIM,—but in vain.

Thus, up to the year 1810, it appears to have been generally considered, and acted upon, that the acceptance of a *Foreign Order of Knighthood*, with the consent of The British Sovereign, gave a title to the appellation of *Sir*, and to the rank and other privileges of a Knight-Bachelor of these Realms.

But, in that year, the question as to the propriety of this practice, was raised, in the case of Sir HENRY CLEMENTS THOMPSON, Captain in The Royal Navy, who had received, with his Sovereign's consent, THE ORDER of THE SWORD, of *Sweden*, but was appointed by The Lords of The Admiralty to the command of a Ship, by the description of "HENRY CLEMENTS THOMPSON, *Esquire*." The Case was referred for the consideration of THE HERALDS' COLLEGE, where the opinions were much divided,—but the majority decided to report, that the acceptance of a *Foreign Order*, under such circumstances, did *not* confer the Title or Rank so claimed.

The direct *contrary* had, however, been previously *ruled* by the highest legal authority in the Kingdom, though the Case in which it occurred, being one of *Assault* only, had doubtless not come to the knowledge of either the Parties or the Referees.

In a Pamphlet privately printed, in 1809,—detailing a Prosecution against WILLIAM DEARSLEY, for an assault on The Rev. SIR ROBERT PEAT, it appears, that the Counsel for the Defendant in that Case, took an objection to the description of the Plaintiff, arguing, that as he had not been *Knighted* by The King, he had no right to the appellation, *Sir*.

Lord ELLENBOROUGH overruled this objection, observing, "That the Order of Knighthood having been confirmed, by Patent, from The King of England, no doubt whatever could be entertained respecting it's validity.—The King is the Fountain of Honour,—and no one ever doubted the *Knighthood* of *Sir* SYDNEY SMITH, with many others, whose *Rank* had been confirmed by The King. Had it been written *Baronet*, the objection would then have been fatal."

In the year 1813, Sir ISAAC HEARD, Garter King of Arms, suggested the expediency of restraining, in some measure, this practice of assuming the style of an English Knight-Bachelor, in consequence of the reception of Foreign Orders,—and a clause, providing against such construction of the Licence, was first introduced into The Royal Warrant, allowing John Milley Doyle to accept The Portuguese Order of The Tower and Sword, on the 20th of March, 1813.

It may now, therefore, be laid down as a principle, that *previous to the Regulation of* 1812, the Investiture with a Foreign Order, sanctioned by His Majesty, entitled the Party to the rank, appellation, and privileges, of a Knight-Bachelor.

But the spirit of Enterprise gained strength as the late War advanced,—and almost numberless are the British Subjects who, in the bright career of honourable and distinguished service, made their Valour and Intelligence conducive to the interests of other Countries, as well as their own.

In conferring The Orders gained by such bravery and ability, the Sovereigns aimed at bestowing the strongest Publick marks of their approbation and esteem, —and wished, that these should be regarded, as the best proofs they could offer of the sentiments which they entertained of services so signal and important.

In all civilized and polished States, these incentives to deeds of Glory,—these rewards destined to crown Military Valour and Intellectual Superiority, as well as to recompense high Personal Merit of every other description, must ever exist.

Even in the gloomy days of Cromwell, which displayed any spirit, save that of courteous Gallantry, WHITELOCK, the Ambassador from THE PROTECTOR to the Court of Sweden, accepted THE ORDER of AMARANTA, from the renowned CHRISTINA. Her Majesty invested him with THE ENSIGNS of THE ORDER, upon it's first Institution,—nor did THE PROTECTOR express any disapprobation.

After the Battle of *Leipsic*, in 1813, EUROPE might be said to have at length approached her deliverance,—and ENGLAND confidently looked forward to repose under those Laurels, which her steady and unexampled efforts in the common Cause of Nations had entitled her to demand.

Then came, to crown the valour, the sacrifices, and the perseverance through a period of nearly a quarter of a Century, THE VICTORY of WATERLOO, which made triumphant the Cause espoused by BRITAIN, and laid all EUROPE under everlasting obligations to the Country and the Genius of WELLINGTON!

ENGLAND may well be proud of all her sons,—for there is scarcely a Promontory overlooking the Ocean, or a Plain sufficiently spacious for Conquest, that does not bear testimony to their Prowess,—our Fleets and our Armies have given to Localities immortal Names, and Victors and Places will together be recorded in the brightest pages of unfading History!

It is a feature of the late War, not perhaps generally known, though peculiarly honourable to THE BRITISH NATION, that the Achievements of her Naval and Military Officers not only obtained them admission into the already existing ORDERS of almost every Country in Europe, but even gave rise to the Establishment of NEW ORDERS of KNIGHTHOOD, principally, if not solely, for the purpose of rewarding their Services.

THE SUBLIME PORTE, before the year 1800, possessed no ORDER of CHIVALRY,—And THE ORDERS already established in the stricter Catholick Countries, "being united with Religious Ceremonies and Institutions," could not correctly be bestowed upon Persons belonging to a different Communion.

But the Navy and Army of GREAT BRITAIN having expelled *The French* from *Naples,* in 1798,—and from *Egypt,* in 1801,—and having protected THE KING of PORTUGAL in his removal to *The Brazils,* in 1808,—The gratitude of THE SULTAN was shown by the Establishment of THE ORDER of THE CRESCENT,—and that of THE KINGS of NAPLES and of PORTUGAL, by their respectively founding and reviving THE ORDERS of ST. FERDINAND and of MERIT, and of THE TOWER and SWORD.

That the amount of FOREIGN ORDERS bestowed on our Officers, is so considerable as to afford a sufficient reason for debarring their indiscriminate admission at THE BRITISH COURT, cannot be granted,—indeed, the estimated number of such ORDERS must be founded on erroneous data,—they are by no means so numerous, as is generally supposed,—the whole of them that have come under my notice, do not exceed 795.—Of this number, many of THE KNIGHTS are dead,—many received the honour, prior to the Regulations of 1812,—

many have not had their applications and claims complied with,—and many have hesitated at the expense attending The Royal Licence to accept and wear the Decorations.

I have been induced, at the suggestion of a brave Officer,—Lieut. Colonel CHARLES MENZIES,—to deviate a little from my original design, in order to do justice to ENGLAND herself,—by introducing THE HONOURABLE REWARDS "*for gallant deeds in Arms*," which have been so nobly bestowed upon many distinguished Officers by The high-spirited, impartial, and discriminating COMMITTEE of THE PATRIOTIC FUND INSTITUTION.

It is of great importance to the interests of Society, that generous labours and sacrifices in the cause of Humanity, should not remain unrewarded by The Publick. Open and recorded demonstrations of Gratitude for noble Actions will inflame the Youth of future times with a holy zeal, and inspire them with a desire to emulate the glorious deeds which History, in glowing colours, shall have handed down to them.

In pursuit of my object, I have consulted the most esteemed works of Heraldry, together with the Writings of such Authors, both British and Foreign, as have expressly treated on the Subject which, in the following pages, I have attempted more fully to describe.

The kindness of those whom I esteem, is always to me the most agreeable topick,—and I rejoice in the opportunity of publickly expressing my obligations and my gratitude to Sir WILLIAM WOODS, K. H., GARTER KING of ARMS,—who, not waiting to be invited to contribute to the Work, at once and with the liberality of an old and steady friend, threw open all the Treasures of THE COLLEGE of ARMS to my unrestrained perusal, and free use. From the conduct of a Parent so kindly disposed, I was naturally led to anticipate similar kindness in his Son,—I am, therefore, not a little gratified in having to express my obligations to ALBERT WILLIAM WOODS, Esq., PORTCULLIS, for many valuable observations, which he frankly suggested to me.

But, that no omission or inaccuracy should occur, at least as far as Prudence could prevent,—I inserted a letter, in the month of October, 1838, in those widely extended Publications, THE GENTLEMAN'S MAGAZINE, and THE UNITED SERVICE JOURNAL, stating my intention, and requesting information, applicable to the Work, as well from CIVILIANS as MILITARY MEN. From THE EDITORS of those well-conducted Journals, the most prompt and obliging courtesy was received,—especially by the gallant EDITOR of the latter, who not only complied with my request, but added such recommendation of his own, as gave vigour to study, and diminished the labour by inducing the kind communications of my Correspondents.

By the politeness of ROBERT GEORGE CLARKE, Esq., The London Gazettes have been examined with the greatest convenience. Here we find recorded the Wisdom of our Statesmen, and the brilliant Achievements of our Warriors,—Here we behold the Sovereigns of the greatest Empires hastening to testify their admiration of our Countrymen in the most animating language,—Here, one Monarch describes his *rescue,* from the Legions of a sanguinary Enemy,—there, another actually ascribing the Resumption of his Throne to irresistible valour,—nay, even a barbarous Government is compelled to renounce the savage practice of making Christian Slaves, and to obey the dictates of Humanity, at the voice of The British Lion. What other Nation can boast of such proud Principles, and such glorious Results!

> This happy breed of men, this little world;
> This precious stone set in the silver Sea,
> Which serves it in the office of a wall,
> Or as a moat defensive to a house,
> Against the envy of less happier lands;
> This blessed plot, this earth, this realm, this England!
>
> RICHARD II.

With a view to obtain the most accurate information, I ventured to address each of THE FOREIGN AMBASSADORS, now resident at THE BRITISH COURT,—and I acknowledge with pleasure and gratitude, their immediate compliance, as well as that of THE CONSULS GENERAL,—all of whom, with a politeness and alacrity, in accordance with the Dignity of their Stations, not only supplied me with Books and Official Documents, but rendered the service doubly valuable, by the affable manner in which the obligations were conferred.

But I have still a further and not less agreeable duty to perform, in the expression of my most sincere Thanks to that Courtly Establishment, THE COLLEGE of ARMS,—where every Gentleman to whom my object and wishes were made known, hastened to supply me with information, and to relieve my inquiries from their well-stored Collections.

Yet I linger on the gratification of the pleasures of Friendship, in my acknowledgement of the zealous assistance which I have received from Mr. JONES, in the various Translations,—a Gentleman, alike conspicuous in his skill of modern Languages, and in the faithful discharge of his Publick duties at THE BRITISH MUSEUM.

I AM sensible that THIS WORK is imperfect, as relates to the enumeration of the several KNIGHTS,—but I have used my best endeavours to obtain the most accurate statements. Those who have neglected to comply with my Publick request, may possibly have done injury to their Parental feelings,—for, here the Child may learn in infancy to lisp the noble Deeds of his Forefathers, and, by times, to direct his most ardent aspirations and energies to the Prosperity, the Happiness, the Greatness, and the Glory of his NATIVE LAND.

NICHOLAS CARLISLE.

SOMERSET PLACE,
29th of January, 1839.

THE EMPIRE OF AUSTRIA.

THE ORDER OF THE GOLDEN FLEECE.

THIS ORDER, which is justly esteemed one of the most Honourable and Illustrious of the many Orders of Chivalry existing in Europe,—being Second only to THE MOST NOBLE BRITISH Order of THE GARTER,—was instituted, on the 10th of January, 1429, at *Bruges,* in Flanders, by PHILIP *the Good,* Duke of Burgundy and Brabant, and Earl of Flanders, the most Puissant Prince of his age,—that being the day of his Marriage with his third wife, ISABELLA, of Portugal.

It was founded in honour of THE HOLY VIRGIN and THE APOSTLE SAINT ANDREW, the Duke being incited so to do, by his great love of Chivalry, the honour and zeal of which he particularly wished to promote, in order that the Catholick Faith, the tranquillity and welfare of the State, should be defended, protected, and maintained, as much as possible,—and for the purpose of diffusing and strengthening Religion, the Church, Virtue, and good Manners.

The occasion of its Institution is differently related,—but it appears most probable, that, having determined to institute an ORDER of KNIGHTHOOD, PHILIP chose for the Badge of it, the material of the staple manufactures of his Country, which was *The Fleece*,—and this emblem might have been the more agreeable to him, from the figure which it made in the Heroick ages of the World, when *The Argonautic* expedition was undertaken to obtain it under the guidance of JASON, as is so beautifully described in the Grecian Mythology.

PHILIP is likewise stated to have chosen SAINT ANDREW for the Patron of THE ORDER, because a *Cross* of that Saint had been in the possession of the ancient Dukes of Burgundy,—and hence the reason, that even now the Cross of *Burgundy*, and the Cross of *St. Andrew*, are alike.

The City of *Bruges* was the Staple of the Wool trade, which was carried on with all the other Countries of Europe, and especially with the Realm of England.

And when THE ORDER was first established, THE GRAND MASTERSHIP was inalienably attached to the Earldom of *Flanders*, that once flourishing Country, the birth-place and seat of European commerce.

At the second Festival of the Order, which was celebrated at *Ryssel*, on the 30th of November, 1431, THE STATUTES, which had been promulgated on the 27th of the same month, at *Lille*, contained the following Clause—

It was therein provided, that if the Male Line of the House of Burgundy should become extinct, the husband of the daughter and heiress of the last Sovereign, should be Chief of THE ORDER. In this manner it passed to THE HOUSE of AUSTRIA, together with THE LOW COUNTRIES, when the only daughter and sole Heiress of CHARLES *le Téméraire*,—(killed in the Battle of *Nancy*, on the 2d of January, 1477),—married, some months afterwards, MAXIMILIAN, Archduke of AUSTRIA, and ultimately EMPEROR of *Germany*.

After the Abdication of his Grandson, CHARLES the fifth, in 1556, the Spanish and Austrian Line of Succession remained in possession of THE ORDER, until its failure by the death of King CHARLES the Second, on the 1st of November, 1700, when the war of Succession commenced in Spain,—in which CHARLES the Third, (afterwards the Emperor CHARLES the Sixth), and PHILIP the Fifth contested the Sovereignty of that Crown.

CHARLES having lost the Dominions of Spain, preserved only *The Low Countries*,—but, as the Ancient Master of those Provinces had been THE FOUNDER of THE ORDER, he declared himself to be the sole and legitimate Head of it,—carried away with him the Archives of THE ORDER on leaving Spain, and celebrated with great Pomp at *Vienna* the re-establishment of THE ORDER, in the year 1713.

PHILIP the Fifth of Spain, on his part, proclaimed also that he was the only possessor of the dignity of GRAND MASTER of THE ORDER, and protested at The Congress of *Cambray*, in 1721, against the declaration of CHARLES.

Both Sovereigns, however, at the Peace of *Vienna*, in 1725, agreed, that each of them, until his death, should retain the Titles he had assumed and borne up to that day,—thereby tacitly compromising on either side, the Dignity of Grand Master,—but that their heirs should only take the Titles of those Countries, which they actually possessed as Sovereigns.

After the death of CHARLES, MARIA THERESA transmitted the Dignity of Grand Master to her Husband, FRANCIS the First, which produced on the part of PHILIP the Fifth, before The Electoral Assembly at *Vienna* and at *Franckfort*, a new Protest, whereby he summoned AUSTRIA to lay down her pretensions, and to concede entirely THIS ORDER to Spain.

At the Peace of *Aix-la-Chapelle*, in 1748, France, England, and Holland, offered their mediation between the litigating Powers,—but FERDINAND the Sixth caused his Ambassador to declare, that no conciliatory steps were necessary, as the Dignity of GRAND MASTER of THE GOLDEN FLEECE was and should remain inseparable from the Crown of Spain.

MARIA THERESA protested in her turn, and vindicated that Dignity in favour of her Husband. Neither of the two Parties being willing to yield, the dispute remained undecided, and the Sovereigns of both Kingdoms have ever since that time indiscriminately conferred the Dignity of Knight of THE GOLDEN FLEECE,—adding, however, as a mark of distinction, the words *Austrian* or *Spanish*, and using almost similar Insignia.

On account of the much more remote Antiquity of

the other Spanish Orders, The Golden Fleece obtained but little consideration in Spain at the time of its Institution,—but it subsequently acquired a degree of great importance, under the reigns of Charles the Fifth and Philip the Fifth. Ever since that period, The Order has been borne, not only by many Kings and Princes, but by all The Emperors who have sprung from The Illustrious House of Austria,—and it has only been conferred upon Personages of the highest Birth, and for the most eminent Services rendered to The Illustrious Houses of Austria and of Spain.

The Knights received a Rank above all other persons of The Court, except only The Princes of The Blood of Crowned Heads,—they were exempted from all Taxes,—Philip the Fourth permitted them to *cover* themselves in presence of the King, like Grandees of the Kingdom, to enter without molestation all the Chambers of the Palace, &c.

The number of Knights is limited to Fifty-one, being the number appointed by Charles the Fifth.

In all other respects the Statutes and Collars of both Branches, under the respective Sovereigns of Austria and of Spain, are exactly the same.

The Statutes of The Order, in which it is styled, "*L' Ordre de la Toison d'or*," have by degrees undergone various alterations. The most essential, and one which is still in force, is, that The Chief of The Order appoints the Knights entirely from his own free choice, and that they must all be of *The Roman Catholick* Religion,—for it can only be conferred upon a *Pro-*

testant, by means of a dispensation from the Pope.

THE BADGE consists of a Golden Lamb or Fleece, with a gold flint stone, enamelled blue, on which is engraven the MOTTO, from CLAUDIAN, "*Pretium non vile laborum.*"

The Knights usually wear the Badge, pendent to a broad plain red riband, round their necks,—But, on days of Ceremony, they wear THE COLLAR of THE ORDER, which is composed of double steels, interwoven with flint stones, emitting sparks of fire,—the whole enamelled in their proper colours, at the end whereof hangs a GOLDEN FLEECE. The fusils are joined two and two together, as if they were double B B's, the cyphers of *Burgundy,* and the flint stones, the ancient Arms of the Sovereigns of Burgundy of the first Race,—with their Motto, "*Ante ferit quam flamma micet.*"

On days of Ceremony The Knights wear an exceedingly splendid COSTUME, consisting of a Robe of deep red velvet, lined with white silk, and of a long purple velvet Mantle, lined with white satin, with a border richly embroidered in gold,—and this sentence, in old French, "*Je l' ay empris,*" is embroidered in gold, several times, on the exterior white satin border of the mantle. The CAP is of purple velvet, with gold embroidery, from which falls behind a piece of the same material, which is attached to the shoulder,—a plain band is suspended from the left of the Cap. The shoes and stockings are red.

THE ANNIVERSARY of THE ORDER is celebrated at *Vienna* on *St. Andrew's* day, and if that day falls on a week day, on the following Sunday,—when The Emperor, and all the Knights who happen to be present, repair in grand Costume, in procession, to the Church of the Court, hear Divine Service, and then return to the Palace, where they dine in The Hall of The Chapter.

On the day of Epiphany, a Chapter is holden in presence of The Emperor. At it's conclusion The Emperor preceded by his Court, and accompanied by The Knights, proceeds to The Hall of Knights, and takes his seat under the Canopy, while the Knights also occupy their accustomed places. Here the Candidates, who have been waiting in The Council Chamber in the Dress of their Order, are led by the oldest Knight, preceded by The King at Arms, into The Hall of Knights before the Throne. After having received from The Grand Master the Investiture by three blows of the Sword, and taken the prescribed Oath, in which they promise Fidelity and Obedience to The Sovereign, and Submission to The Statutes, The Emperor, with his own hands, puts THE COLLAR round their necks, and embraces them,—which ceremony is then repeated by the whole of the Knights. On the completion of the Ceremony, the Procession returns in a similar manner, and the newly received Knights occupy their allotted places.

There are four Great Officers, *viz.*, a Chancellor, Treasurer, Registrar or Secretary, and a King of Arms, called "*Toison d'or.*"

THE IMPERIAL MILITARY ORDER OF MARIA THERESA.

The stormy times which assailed Her Imperial Majesty, MARIA THERESA, at her Accession to the Throne, form an interesting part of History, as well as the ardent devotedness and loyalty with which her armies fought for the rights of their Queen.

The Peace of *Aix-la-Chapelle*, in 1748, at last restored tranquillity to The Austrian dominions,—but The Empress had not forgotten the important services of her brave Champions, and having in the first instance seized every opportunity to evince her gratitude towards them, she resolved afterwards to confer upon them a Publick recompense, which should at the same time incite them to fresh proofs of heroick valour.

In 1756, a war with Prussia having just broken out, opened to the warriors of MARIA THERESA a new field for gathering Laurels. The Empress then determined upon the foundation of an ORDER, with which she might reward those of her Officers, who should have distinguished themselves by their courage and talents. The Emperor FRANCIS the First, that great Protector of true merit, gave his entire assent to this noble idea, and the creation of THE ORDER took place, on the 13th of May, 1757. The Austrian Army, however, having obtained the victory of *Collin*, on the 18th of June, 1757, that memorable day was considered as the day of Foundation of the NEW ORDER, and it was announced to The

Imperial Armies, that The Order should be called, "The Military Order of Maria Theresa,"—and that The Emperor, Francis the First, had assumed the dignity of Grand Master of it.

His Highness Duke Charles of Lorraine, the Emperor's brother, and Field Marshal The Count of Daun, were the First and Second Grand Knights, who were received into The Order.

The Statutes appeared on the 11th of October, 1758,—but they have been changed and modified by a Supplementary decree, dated the 12th of December, 1810, by The Emperor Francis the Second.

The Emperor is always The Grand Master of The Order,—which formerly consisted only of Grand Crosses and Knights,—but which is now composed of three classes,—*viz.*,

Grand Crosses,
Commanders, and
Knights.

The third Class having been added by The Emperor Joseph the Second, on the 16th of October, 1765.

This Order is only granted to true Military Merit,—and neither interest, birth, religion, antiquity of family, rank, nor other circumstances, can in anywise influence the appointment. It is indispensably necessary, that The Chapter should regard it as an invariable standard, "that, all such Deeds as might have been omitted and left undone, without responsibility, but nevertheless were performed, are entitled to The Order." The distribution is to be made, conformably to a previous decision of The Chapter, under the observance of certain solemn and impressive formalities,—or, *on the very spot*, immediately by The Grand Master of The Order.

In the first case, The Chapter is to examine with the utmost impartiality and secrecy, the value of the Exploit which is to be rewarded. When the Decoration is conferred by The Grand Master himself, the following Ceremonies are observed,—The Chancellor of The Order gives notice to the Candidates of their nomination, and of the day and hour of their reception,—and invites at the same time, all the Generals and Superior Officers to be present at the Ceremony. At the hour appointed, The Emperor, preceded by the Dignitaries of The Order, Chamberlains, Privy Counsellors and other persons of Rank, enters the Hall of reception, immediately preceded by The Great Marshal of The Court, holding the Sword of the Empire out of its scabbard, and takes his seat on the Throne. On the right hand of The Emperor stands a table, covered with red velvet, on which are laid cushions of the same fabrick,—upon these, the Diplomas and Decorations are placed. The Treasurer and Secretary stand near to this table. When the Personages present have taken the places assigned to them, the Members to be received are then introduced, and take their places opposite the Throne. The Chancellor of The Order approaches the Throne, kneels down, and receives The Order from The Emperor, withdraws a few paces, and addresses the Assembly in a short discourse, in which he mentions the foundation of The Order, and the purpose for which it was created, and alludes to the merit of the Candidates, who are to be admitted. After this, those persons approach the Throne one after the other, and kneel on a red velvet cushion placed at the foot of it. The Treasurer gives the Decoration to The Chancellor, by whom it is presented to The Emperor, who then decorates the new Knight, pronouncing these words,—"Receive from our hands the Decoration of The Military Order of Maria Theresa,—it is to be an acknowledgement of your achievements, and a sign of your reception into this Order, which is devoted alone to valour and ability. Use it to the Glory of God, in the service of Our Illustrious House, and in the defence of your Country." After this speech, The Emperor embraces the Knight, who is now received, and the Procession quits the Hall in the same order, as that observed at their entrance.

When the Army is in Campaign, the reception takes place at The Head Quarters, in presence of all the Officers. The place of The Emperor is then supplied by The President of The Chapter of The Order, who is usually The Commander in Chief of the Army, and a Grand Cross,—or the Decoration is sent, accompanied by a Letter from The President of The Chapter.

All the Patents of Promotion are delivered, free of expense.

The Badge of The Order is a Cross, of eight points, of gold, enamelled white, edged with gold,—in the centre, bordered with gold, are the Arms of Austria, encircled with a white margin, on which is the word, "*Fortitudini,*" in gold letters,—and, on the reverse, on a white ground, is a cypher of the letters M. T. F., (*Maria Theresa Fundator*), in gold, on an enamelled green ground, and a wreath of laurel. It is worn, pendent to a white, watered riband, edged with red, of one hand's breadth, sashwise over the right shoulder,—and the Star, the obverse of the Cross, which was added by Joseph the Second, in 1765, on the left breast of the apparel, embroidered in silver.

The Commanders wear the Decoration suspended from a narrower riband, round the neck, and without the Star.

The Knights wear the Decoration, of a smaller size, at the left button-hole, and pendent to a riband two fingers in breadth.

The number of Members of The Order is unlimited. Eight Pensions of 1,500 Vienna florins are assigned to the Class of The Grand Crosses,—Sixteen Pensions of 800 florins, to that of Commanders,—One hundred Pensions of 600 florins, to the first division of the class of Knights,—and One hundred Pensions of 400 florins, to the first division. The possessors of these Pensions succeed each other, and are replaced by non-pensioned Members. The Widows of Members having a pension, receive one half of it until their death. A Mass for the Roman Catholick members is solemnized in the Church of The Augustine Friars.

The Knights obtain the Title of Barons, if they wish it, and are breveted as such, free of all expense. They may ask for, and obtain an Audience of The Emperor, without announcing themselves through the Chief Chamberlain, — and have free access to the Grand Festivals of The Court, as well as to the Entertainments in the Private Apartments.

The Anniversary of The Order is celebrated on the 15th of October, which is *Saint Theresa's* Festival, as well as the Name-day of The Illustrious Foundress of The Order,—or the next Sunday, at Court in time of Peace, and at the Head Quarters in time of War.

The Military Order of Maria Theresa has been liberally granted to the Military of The Allied Powers of Austria, in recompense of their brilliant exploits. The Duke of Wellington, after the battle of *Waterloo*, received a number of Crosses, for his own distribution, which is the only circumstance of the kind on record.

The value of this Order, being united with singularly able conduct, has been preserved in an exalted state, and continued in the very spirit of its Illustrious Foundress.

And the Emperor Francis, in the 40th article of the Statutes, declares, that, since no other Order can be preferred in his esteem before one which can alone be acquired by extraordinary bravery and merit, he had found it expedient, in favour of The Military Order of Maria Theresa only, to make an exception to that rule regarding The Order of The Golden Fleece, which requires that no other Order be worn at the same time with it,—and he thereby graciously permits, that this Badge of Honour may be worn at the same time with The Golden Fleece,—but that no Order of Knighthood of a *Foreign Power* can or shall be worn with The Military Order of Maria Theresa, as in the case of The Golden Fleece.

On the death of a Grand Cross or Knight, his *Insignia* of THE ORDER shall be given or sent to The Chancellor of The Order by his heirs, or by whomsoever happens to take charge of the same.

GRAND CROSS.

ARTHUR, Marquess of WELLINGTON, K. G., Field-Marshal in The Army, Commander of The Allied Forces in Spain and Portugal, to accept and wear the Insignia of a Knight Grand Cross of The Imperial Military Order of Maria Theresa, and of a Knight Grand Cross of The Royal Swedish Military Order of The Sword, which their Majesties The Emperor of AUSTRIA and The King of SWEDEN have respectively been pleased to confer upon His Lordship, "in testimony of the high sense which those Sovereigns entertain of his distinguished merits and brilliant achievements."—Royal Licence, 4th of March, 1814.

COMMANDERS.

SIR ROBERT THOMAS WILSON, Knight of The Imperial Military Order of Maria Theresa, Commander of The Royal Portuguese Military Order of The Tower and Sword, Knight of The Imperial Russian Military Order of St. George, of the Third Class, and a Major General in The Army, to accept and wear the Cross of a Commander of the said Imperial Military Order of Maria Theresa, and the Insignia of The Grand Order of The Red Eagle of Prussia, which their Majesties The Emperor of AUSTRIA, and The King of PRUSSIA, have respectively been pleased to confer upon him, "in testimony of their esteem and the high sense which they entertain of his merits and his services in the common cause, and of the valour displayed by him upon every occasion."—Royal Licence, 4th of March, 1814.

THOMAS FRANCIS FREMANTLE, Esq., Rear Admiral of The White Squadron of His Majesty's Fleet, "in testimony of the high sense which The Emperor of AUSTRIA entertains of the distinguished gallantry and conduct displayed by him (in co-operation with The Imperial Troops), on the Coast of *The Adriatic.*"—Royal Licence, 23d of May, 1814.—Created a Baronet, 14th of August, 1821.—*See*, THE TWO SICILIES.

KNIGHTS.

JAMES LOCKHART WISHART, Esq., Count of The Holy Roman Empire. —14th of May, 1785.

SIR JAMES WILLIAMS, Colonel in The Austrian Service, and Commander of The Imperial Flotillas on the *Adriatic*, the *Italian Lakes*, and the *Danube*, in 1796 and 1799.

Major WILLIAM AYLETT, then Lieut. Colonel in the Army,—

Captain ROBERT POCKLINGTON, late Major of the 15th Dragoons,—

Captain EDWARD MICHAEL RYAN, then Major in the Army,—

Lieutenant THOMAS GRANBY CALCRAFT, Lieutenant Colonel of the 3d Dragoon Guards,—

Lieutenant WILLIAM KEIR (now GRANT), Major of the 6th Dragoon Guards,—

Lieutenant CHARLES BURRELL BLOUNT, late Captain of the 15th Light Dragoons,—

Cornet EDWARD GERALD BUTLER, then Major of the 87th Regt. of Foot,—

Cornet ROBERT THOMAS WILSON, then Major in *Hompesch's* Regt. of mounted Riflemen,—

Officers, who served in the Detachment of His Majesty's 15th Regt. of Light Dragoons in the Action of *Villers en Couché*, near Cambray, on the 24th of April, 1794,—to each of whom

His Majesty has been graciously pleased to grant His Royal Licence and Permission to accept the Rank of Knight of The Imperial Military Order of Maria Theresa, and bear the Insignia thereof,—the same having been conferred upon them by The Emperor of Germany, in testimony of the high sense which His Imperial Majesty entertained of their distinguished Conduct upon that occasion.—*Downing Street*, 2d of June, 1801.

As so signal an Exploit was performed in the presence of His Imperial Majesty, and the same having been most explicitly ascertained by the Generals in Chief, and Superior Officers, The Emperor was most graciously pleased to command His Excellency The Baron DE THUGUT, to communicate to Lieut. Colonel AYLETT and the other Officers concerned therein, His Princely resolutions upon their conduct,—this was done by the following letter from the Baron DE THUGUT to Lieut. Colonel AYLETT.—

" Sir,

The Emperor remembers with satisfaction the distinguished proofs of valour that you, and the other Officers of the 15th Light Dragoons, manifested on the 24th of April, 1794. near *Cambray.* His Majesty regrets, that the Statutes of THE ORDER of MARIA THERESA, confirmed by a constant custom, forbids the Cross of this Order, strictly National, being conferred on Officers so worthy of being decorated with it,—but wishing to give you, as also your honourable Companions, a Publick mark of his particular esteem, His Majesty has commanded a MEDAL to be struck, to perpetuate the remembrance of this brilliant Action, and has ordered me to offer to them *the only Impressions* which have been struck, except *one*, which is placed in The Imperial Cabinet of Vienna. In fulfilling the intentions of His Imperial Majesty, I beg you, consequently, to receive for yourself, Sir, and to distribute to the other Officers who, on the memorable 24th of April, 1794, fought under your orders, the Medals which I have delivered to Captain RYAN.

I have the honour to be, &c.

Vienna, LE BARON THUGUT.'

5th of March, 1798.

It appears that a doubt had been entertained, at the time of this gallant exploit, whether THE ORDER of MARIA THERESA could be conferred upon *Foreigners*,—but that doubt having been dispelled, His Imperial Majesty's grateful intentions were fulfilled by granting to each Officer the *Medal* and *Riband* of MARIA THERESA.—And the Royal Licence was officially notified in The London Gazette, on the 2d of June, 1801.

In a sketch of this brave action, when the Emperor *was rescued* from the French Cavalry, it is stated, that the 15th Light Dragoons had as many men opposed to them (in comparison to their numbers), as the renowned BLACK PRINCE at the hard fought battles of *Cressy* and *Poictiers*.

This rare and distinguished MEDAL which is of Silver, bears,

On the obverse,—The Bust of the Emperor to the right, hair long, laureated, neck bare, with the legend, IMP. CÆS. FRANCISCUS II. P. F. AUG.,—I. M. WIRT, F.

On the reverse,—This inscription, FORTI BRITTANNO IN EXERCITU FOED. AD CAMERACUM XXIV. Apr. MDCCXCIV,—underneath, two sprigs of laurel.

Sir THOMAS LOUIS, Rear Admiral.—Services in the re-conquest of *Naples* from the French, in 1799.—Created a Baronet, in 1806. Died, 17th of May, 1807.

CHARLES ROWLEY, Esq., late Captain of His Majesty's Ship *Eagle*, "in testimony of the high sense which The Emperor of AUSTRIA entertains of his distinguished gallantry and services (in co-operation with the Imperial Troops), on the Coast of *The Adriatic*."—Royal Licence, 23d of May, 1814.

WILLIAM HOSTE, Esq., late Captain of His Majesty's Ship *Bacchante*, "in testimony of the high sense which The Emperor of AUSTRIA entertains of his distinguished gallantry and services (in co-operation with the Imperial Troops), on the Coast of *The Adriatic*."—Royal Licence, 23d of May, 1814.—Created a Baronet, 3d of July, 1814.

FAIRFAX MORESBY, Esq., Commander of His Majesty's Sloop *Weazle*, "in testimony of the high sense which The Emperor of AUSTRIA entertains of his distinguished gallantry and services (in co-operation with the Imperial Troops), on the Coast of *The Adriatic*."—Royal Licence, 23d of May, 1814.

John Fane, Esq., commonly called Lord Burghersh (son and heir Apparent of John, Earl of Westmorland, K. G.), a Colonel in The Army, Aid-de-Camp to His Royal Highness the Prince Regent, and Lieut. Colonel of the 63d (*or*, West Suffolk) Regt. of Foot, lately employed on a particular service at the Head Quarters of the Imperial Austrian Armies, " as a signal testimony of the high sense which His Majesty The Emperor of Austria entertains of the services rendered by him to the cause of Europe, while employed with his Armies."—Royal Licence, 7th of June, 1814.—*See*, Tuscany.

The Honourable George Cadogan, a Post Captain in the Royal Navy, " in testimony of the high sense which the Emperor of Austria entertains of the distinguished gallantry displayed by him (in co-operation with the Imperial Troops), on the Coast of *The Adriatic*."—Royal Licence, 22d of July, 1814.

Sir Neil Douglas, K. C. H., and C. B., Colonel in The Army, Aid-de-Camp to His Majesty, Lieut. Colonel of the 79th Regt. of Foot (*or*, Cameron Highlanders),—Royal Licence, dated 7th of November, 1831, to accept and wear The Insignia of a Knight of The Imperial Austrian Order of Maria Theresa,—and of The Imperial Russian Order of St. Wladimir, of the Fourth Class, which their Imperial Majesties The Emperor of Austria and the late Emperor of Russia were pleased to confer upon him, " in testimony of the high sense which those Sovereigns entertained of his Services before the Enemy in the Field in *The Netherlands*, in the year 1815."

Sir Granby Thomas Calcraft,—*See*, Portugal, Order of The Tower and Sword.

Sir Colin Campbell, K. C. B., Colonel in the Army, Captain in The Coldstream Regt. of Foot Guards, Assistant Quarter-Master-General of His Majesty's Forces, and Aid-de-Camp to Field Marshal His Grace The Duke of Wellington,—

Lord Fitzroy Somerset, K. C. B., Colonel in the Army, Extra Aid-de-Camp to His Royal Highness The Prince Regent, Captain in the 1st *or* Grenadier Regt. of Foot Guards, and Military Secretary to His Grace The Duke of Wellington.—Royal Licence and permission, dated the 4th of April, 1816, that they may respectively accept and wear the Insignia of a Knight of The Imperial Military Orders of Maria Theresa of Austria,—of St. George of Russia, of The Fourth Class,—of The Royal Bavarian Order of Maximilian Joseph,—and of an Honorary Knight Commander of The Royal Portuguese Military Order of The Tower and Sword,—conferred upon them by their Imperial Majesties The Emperors of Austria and Russia, His Majesty The King of Bavaria, and The Prince Regent of Portugal, " in testimony of their approbation of the distinguished conduct and intrepidity displayed by these Officers throughout the recent arduous Campaigns in *The Peninsula,* and at the memorable Battle of *Waterloo.*"

Sir George Adam Wood, Knt., Colonel in The Army, C. B., and K. C. H., to accept and wear the Insignia of a Knight of The Imperial Military Order of Maria Theresa, of The Imperial Russian Order of St. Wladimir, of the Fourth Class, and of the Royal Military Order of William of The Netherlands, of the Third Class,—which their Imperial Majesties The Emperors of Austria and Russia, and His Majesty The King of The Netherlands have been respectively pleased to honour him, " in testimony of the high sense entertained by these Sovereigns of the signal courage and intrepidity displayed by him in the memorable engagements with the Enemy in *The Netherlands,* in the year 1815."—Royal Licence, 25th of May, 1820.

Lieutenant-General Sir Frederick Adam,—*See,* Russia, Order of St. Anne, of The First Class.

Captain James Black,—Died, in 1835.

Lieut. General John, Lord Strafford.

A Dispatch from Field-Marshal The Duke of Wellington, dated, *Head-Quarters, Paris ! ! !* — the 2d of

August, 1815, contains the following List of Officers, upon whom The Emperor of Austria had conferred The Cross of a Commander, and of a Knight, respectively of The Order of Maria Theresa, "in testimony of His Imperial Majesty's approbation of their services and conduct, particularly in the late Battles in *The Netherlands,*"—

To be Commanders of
The Order of Maria Theresa.

Lieut. General The Marquess of Anglesey, G. C. B.
Lieut. General Lord Hill, G. C. B.

To be Knights of
The Order of Maria Theresa.

Lieut. General Sir Henry Clinton, G. C. B.
Major General Sir James Kempt, K. C. B.
Major General Sir Edward Barnes, K. C. B.
Major General Lord Edward Somerset, K. C. B.
Colonel Sir John Elley, K. C. B., Royal Horse Guards.
Colonel Thomas Reynell, 71st Regt.
Colonel Sir Andrew Barnard, K. C. B., 95th Regt.
Colonel The Honble. Alexander Abercromby, Coldstream Guards.
Colonel Sir George Wood, Royal Artillery.
Colonel Sir Colin Campbell, Coldstream Guards.
Colonel Sir John Colborne, K. C. B., 52d Regt.
Colonel Alexander Woodford, Coldstream Guards.
Colonel The Honble. Frederick Ponsonby, 12th Light Dragoons.
Colonel Felton B. Harvey, 14th Light Dragoons.
Colonel Carmichael Smith, Royal Engineers.
Lieut. Colonel James Mc Donnell, Coldstream Guards.
Lieut. Colonel Sir Robert Hill, Knt., Royal Horse Guards.

Lieut. Colonel Lord FITZROY SOMERSET, K. C. B., 1st Guards.
Lieut. Colonel ROBERT DICK, 42d Regt.
Lieut. Colonel NEIL DOUGLAS, 79th Regt.
Lieut. Colonel LORD SALTOUN, 1st Guards.

Another Dispatch from Field-Marshal The Duke of WELLINGTON, dated, *Head-Quarters, Paris,* the 8th of October, 1815, contains the following further List of General Officers, upon whom His Majesty THE EMPEROR OF AUSTRIA had conferred THE ORDER OF MARIA THERESA, "in testimony of His Majesty's approbation of their services and conduct, particularly in the late Battles fought in *The Netherlands.*"—

THE ORDER OF MARIA THERESA.

Major General Sir JOHN BYNG.
Major General Sir FREDERICK ADAM.
Major General Sir DENIS PACK.
Major General Sir HUSSEY VIVIAN.

SILVER MEDALS were given to every Officer and Soldier, who were present at WATERLOO, on the memorable days of the 16th, 17th, and 18th of June, 1815,—

OBVERSE,—A laureated head of The Prince Regent, with the words, " GEORGE P. REGENT,"—beneath the neck, T. WYON, *Jun., s.*

REVERSE,—Victory, seated, with expanded wings, on a base, inscribed WATERLOO, holding in her right hand a Palm branch, and in her left, a branch of Laurel,—above, WELLINGTON.— In the Exergue, June 18, 1815.

The Order of Saint Stephen.

This Imperial and Royal Order was instituted by The Empress Queen, Maria Theresa, on the 5th of May, 1761,—that being the day of the Coronation, as Roman Emperor, of her Successor The Archduke Joseph, afterwards The Emperor Joseph the Second,—and it was named, "The Order of The Holy and Apostolick King Stephen," in memory and veneration of The King of that name who was canonized, and who founded the Kingdom of Hungary.

Her Majesty wishing to give to Civil Merit and Talents, such an ostensible mark of her gratitude, as she had already done in favour of Military Merit.

In conformity with The Statutes of the 6th of May, 1764, the Dignity of Grand Master of The Order is inseparable from the Crown of Hungary,—when, however, the Imperial Throne is occupied by a *Princess,* her Successor is The Grand Master.

The Emperor Joseph the Second, when King of The Romans, was the first Knight who was invested with the Insignia. This Ceremony was performed with great Solemnity by the August Foundress, in The Chapel of The Imperial Palace at *Vienna,*—and in the Castle at *Inspruck,* which is styled "*The Residence,*" is a fine historical Picture representing that event.

The Order consisted formerly of Twenty Grand Crosses,—Thirty Commanders,—and Fifty Knights of the third class,—exclusive of the Ecclesiastical Members,—but now the number of Knights is more considerable, and unlimited.

One of the principal conditions for the attainment of this Order is, to be of Noble descent,—and, indeed, only ancient and high Nobility can belong to the first two Classes,—the third class is appropriated to inferior Nobility. Noblemen are obliged to prove their nobility up to the fourth generation. Austrian Counts, Barons and Chamberlains, are exempted from this formality, as their Antiquity is previously known and acknowledged.

When Natives obtain the Grand Crosses they become Actual Privy Counsellors,—and with the Cross of Commanders, Decretal Privy Counsellors,—and Knights of the third class are, if they wish, raised, *without expense*, to the Dignity of Counts and Barons.

The Grand Crosses are addressed by the Title of "*Excellency*."

The possessor of The Order of St. Stephen has the faculty to wear it in his Arms. And, with regard to *Foreigners*, it is conferred as the first Order of The Royal Family and the Court.

The Badge of The Order is a Cross, of eight points, of gold, enamelled green, and surmounted with the antique Regal crown, which is peculiar to the Kingdom of Hungary. On one side, is the effigy of the Patron, enamelled, proper,—and in the centre, on a field *gules*, appears, *argent*, a Patriarchal cross, being the Arms of Hungary, between the letters, M. T. (*Maria Theresia*), *argent*,—and around the whole, are the words, "*Publicum Meritorum Præmium.*" On the Reverse, on a field, *argent*, is a garland, or crown of Oak leaves, proper, which surround the letters STO. ST. RI. AP. (*Sancto Stephano, Regi Apostolico*). It is worn, pendent to a broad, richly watered riband, composed of three stripes,—the one in the middle being crimson, and the two lateral, dark green, sashwise from the right shoulder to the left side.

The Ecclesiastics who have The Grand Cross, wear it suspended from their neck.

The Grand Crosses wear upon the left side of their upper apparel, a

STAR of Silver embroidery, in the centre of which, on a field *purpure*, is a Patriarchal cross, *argent*, surrounded with a crown of Oak leaves, proper. It is in the form of a radiated sun.

THE COLLAR, of gold, is composed of double S. S. (*Sancto Stephano*), surmounted with Hungarian Crowns, and of the cypher $^{T}_{M}$ (*Maria Theresia*), surmounted with the Imperial diadem. These Monograms are so disposed and connected together, as to form THE COLLAR,—and from a Rose placed in the centre, upon which is a golden eagle, with outspread wings, and the words, "*Stringit amore,*" the Grand Cross of The Order is supended.

COMMANDERS wear the same Cross, pendent from their neck by a rather narrower riband.

And KNIGHTS of the third class wear the Small Cross, at the button-hole.

THE ROBES OF THE ORDER are of green velvet, richly embroidered, and lined with crimson satin, and made up after the old Hungarian fashion.

THE CAP is furred with ermine, and ornamented with Heron plumes, which issue from a crimson and green enamelled sheath.

THE COMMANDERS have, instead of a garland of golden oak leaves on the border of their Cloaks, a single border, embroidered in gold,—and which is not so wide, on the Cloak of The KNIGHTS.

The reception of the Knights is performed with great solemnity, in the presence of The Grand Master of The Order, and with a suitable admonition.

THE DIPLOMAS, which are sent to every Member, are in the shape of a Book, for Grand Crosses,—and that of a Patent, for the two other classes. They all bear the Signatures of the Grand Master, Chancellor, and Secretary.

The Grand Crosses receive from The Grand Master the title of "COUSIN," and dine at the same table with him on the Anniversary of The Order, *St. Stephen's* day. Another table is served at Court, for the two other classes.

Every Member can obtain an Audience of The Grand Master, without the previous sanction of The Grand Chamberlain. The first two classes have a free access to The Privy Council Chamber,—and the third class, only on the Festivals of the Order, and on their arrival or taking leave, when they are admitted to kiss hands.

THE DIGNITARIES and FUNCTIONARIES of THE ORDER are,—

The Prelate, who is always Archbishop of *Gran*.

The Chancellor, a function which is always united with the Chancellorship of The Kingdom of Hungary.

A Secretary,

A Treasurer,—and an Officer, in the Chancellory.

THE EMPEROR sometimes bestows this Order, ornamented with *diamonds*,—which is then esteemed an extraordinary mark of Imperial favour.

When the Empress MARIA THERESA visited her Hungarian Dominions, in the midst of her troubles, with which she was overwhelmed, soon after her accession to the Hereditary States, and painted her distressed situation to The General Diet, THE MAGNATES all drew their swords, and unanimously and heroically exclaimed,

Moriamur pro rege nostro Maria Theresia!

GRAND CROSS.

GEORGE, Earl of ABERDEEN, K. T., His Majesty's Plenipotentiary at The Court of Vienna,—4th September, 1813.

The Imperial Austrian Order of Leopold.

The Emperor Francis the First, ever eager to obtain a new opportunity so agreeable to his heart, of giving a fresh proof to his faithful Subjects of the favour and affection with which he was so entirely devoted to them, —and that those who particularly distinguished themselves in the service of His Majesty and of their Country, might be assured of it by a Publick token,—Having had laid before him the Statutes and Ordinances, which were to form the foundation of a New Order,—and as these had been approved of by him in the following manner, it was his Royal will and command, that these Statutes and Ordinances should be strictly observed by all The Knights of the newly created Order, both then and thereafter, and serve to them as a constant Rule,—

1. To celebrate the memory of his late Illustrious Lord and Father, His Majesty The Emperor Leopold the Second, he appointed to the New Order the name of "The Imperial Austrian Order of Leopold,"—and all its Members should thereafter be styled, "*Knights of The Order of Leopold.*"

2. The dignity of Grand Master is constantly and inseparably attached to The Imperial Throne of Austria.

3. The Order consists of three Degrees, according to Merit,—the first comprehending The Knights Grand Cross,—the second, The Knights Commanders,—and the third, The Knights.

4. The Knights Grand Cross, on account of their Dignity, take

precedency of all the others, and the Knights Commanders of the Knights.

In all the three Orders the Rank is arranged, according to the time of reception,—and in this way the Members of each degree take their place, according to the period of their reception into The Order.

When, on one and the same day, several Knights are received into The Order, the precedency is assigned to him, who first received The Badge of The Order.

5. The chief purpose of The Order is to signalize, by Publick acknowledgement and reward, the services rendered to The State, and to The Imperial House. Those only, therefore, can receive The Order, who have given distinguished proofs of their attachment to their Country, and to their Native Princes, by strenuous and successful efforts to promote the Welfare of The State,—who, by distinguished Learning, have directed it to the best interests of the Community, and the Glory of the Country,—or, who have distinguished themselves by other great Undertakings for the Common benefit,—while, at the same time, it is expressly to be understood, that perfectly honourable Deportment and a blameless Reputation are indispensable conditions of Admission.

6. Admission to The Order is open to all, without distinction of Rank, whether in the *Civil* or *Military* Service, or not,—provided the individual possesses the necessary qualifications.

7. The bestowal of The Order is the exclusive right of the Grand Master, and is withholden from motives entirely personal,—so that Petitions to be admitted into it can never be received.

8. The following is The Badge of the Order, which is granted to The Knights, as a mark of the Dignity conferred upon them,—

An enamelled golden Cross, of an octagonal form outwardly,—the exterior line of each part of the Cross is bent inwardly, in the form of a plain circular line. The Cross itself is red, edged with a white stripe,—it has a circular shield in the middle, on which, in front, are the three letters, F. I. A. in cypher, "*Franciscus Imperator Austriæ.*" This shield is also red, and is surrounded with a white stripe, in which are the words, "*Integritati et Merito.*" The reverse of the Cross is formed as above, with the exception of the shield in the middle, which is also circular, but white, surrounded with a golden oak-wreath, and bearing for inscription the Motto of the late Emperor Leopold the Second, "*Opes Regum, Corda Subditorum.*" Between each of the four parts of the Cross are three oak-leaves, with acorns, the latter being placed near the two sides. Over the Cross the Imperial Crown is suspended, as a loop-ring.

The Knights Grand Cross wear The Badge of their Order by a red riband with a white stripe at the edge, hanging down from the right shoulder towards the left side near the sword. Together with this, The Knights Grand Cross wear on their breast an octagonal Star, adorned with silver, in the middle of which is the Cross of The Order, as described above, having the letters F. I. A. in cypher, and the words, "*Integritati et Merito.*" On extraordinary occasions The Knights of The Grand Cross wear the above described Cross of The Order, on the breast, by a gold Chain round the neck. On the gold Chain of The Order the letters F. and L., *Franciscus Leopoldus,* first are seen, in cypher,—above them is the Imperial Crown of Austria, then a wreath of oak, then the above letters, and so on to the end.

The Knights Commanders wear their Cross, which differs from that of The Knights Grand Cross only in size, by a red riband two inches in breadth, having a white stripe at the edges, and suspended round the neck on the breast.

The smaller Cross is worn by the Knights on the left breast, in the button-hole or in a loop, by a red riband three-fourths of an inch in width, and edged with white.

The Cross of The Herald, and of the other Officers of The Order, having the Motto, "*Opes Regum, Corda Subditorum,*" is displayed on a large Gold Medal. It is worn round the neck by a red riband, edged with white, three-fourths of an inch in width, and holden together at the top by The Imperial Crown of Austria.

9. No Knight is permitted to adorn the Badge of his Order with precious stones, as he would then appear to be particularly distinguished by such a mark.

This manner of adorning the Ornaments of The Order is exclusively reserved for The Grand Master, and The Crown Prince, as Successor to the Throne.

10. On the other hand, every Knight is permitted to adorn his

ARMORIAL BEARINGS with the *Insignia* of THE ORDER, and to employ his ARMS so adorned on every occasion.

11. Prescribes the Dress of The Order, which is of the colour of The Austrian Arms, red and white,—the edging, and other ornaments being of gold.

The Under-dress is the same for all the three ranks.

The mantle which is white, as emblematical of the moral purity of the Knights, is of velvet.

12. The following Officers are appointed for the various occasions, on which they may be required by The Order,—*viz.*, A Prelate, Chancellor, Registrar, Treasurer, Herald, and Chancery-Clerk.

The appointment of these Officers is exclusively committed to The Grand Master.

13. Directs the Ceremony and usage to be observed on the solemn reception of a Knight of The Order,—which are nearly the same as those prescribed for The Order of Maria Theresa,—and the form of the Oath,—which having been taken, The Grand Master again exhorts the Knight Elect to the fulfilment of his Oath, and delivers over to him The Badge of The Order, with the following discourse, printed in German or in Latin,—

"We are convinced that what you have now sworn to, you will perform at all times like a brave and honourable Knight. Receive, therefore, THE BADGE of THE ORDER of LEOPOLD, as a reward for your merits, which you have continually to wear in order that by this honourable token, you may always have before you, what you owe to God, Ourselves, Our House, and the Dignity of The Order."

The dispensation of The Oath of The Order to a Knight Elect is committed to The Grand Master.

14. As soon as the Reception is completed in the form prescribed, The Grand Cross, as a mark of distinguished grace, is then embraced by The Grand Master, and in the same manner the Members of The Order present, assure him of their friendship.

15. The Diploma of Reception is prepared for The Grand Crosses in the form of a Book,—for The Commanders and Small Crosses in the form of a Patent, and signed by the Grand Master, the Chancellor of The Order, and the Registrar of The Order. The Commanders

receive the same with a pendent, the small Crosses with a fastened Seal.

16. In order that The Knights may be made acquainted with their duties and constant in their performance, each is furnished with a copy of THE STATUTES of THE ORDER.

17. Each Knight of The Order may supply himself, at his own expense, with several Badges of The Order, but he must previously announce it to The Chancellor of The Order.

18. The Knights as they have vowed, have always to wear The Badge of The Order, and none shall appear publickly without it,—neither must any *Foreign Badge of Honour* be worn next to it, unless the express sanction of The Grand Master be obtained for that purpose.

19. After the decease of a Grand Cross, THE CHAIN of The Order must be delivered to The Grand Master,—but THE BADGE of The Order of the other Members must be returned to the Registrar of The Order.

20. On the first day after Epiphany, a Festival of The Order is holden annually in The Royal Chapel, at which the whole of the Members of The Order here present must appear unless through any valid cause prevented, with their Badges of The Order, and in the Costume of The Order.

21. At the Festivals of The Order, the Small Crosses have admission to the Private Council Chamber, where Grand Crosses and Commanders have always the privilege of being present.

22. Likewise, all those who are honoured with The Order, may, without any distinction of Class, be received at all the Court Festivals and Levees.

23. If any Knight, favoured with the Grand Cross, does not yet possess the dignity of a PRIVY COUNSELLOR, the same shall be bestowed upon him, *gratuitously*. The Commanders, upon their application, will be raised to the Rank of BARONS,—and The Knights of the Small Cross will be promoted to the Dignity of HEREDITARY KNIGHTHOOD, free of expense.

24. To all our Hereditary-domain Officers, We ordain, that whenever a grant is made by them to The Knights of The Order, to the title of such Grant shall he added, besides the ordinary proper one, also the title of The Order.

25. Likewise, We ordain, that whenever Decrees made in the name of The Grand Master, are sent to The Grand Crosses, they, as a greater mark of our favour and inclination, shall be honoured with the appellation of "*Our Cousin.*"

26. In conclusion, it is Our will, that although The Order of St. Stephen, regarded as a united body, takes rank before The Order of Leopold, and when on the same day Orders of both descriptions are presented, The Knight of The St. Stephen's Order takes precedence of The Knight of The Leopold Order, bearing the same rank,—yet, in all other respects, both Orders shall be holden equal, and between those Grand Crosses, Commanders, and Small Crosses, appointed at unequal periods, only the date of Reception shall make the exception.

These are the Laws and Regulations upon which We have founded this Noble Order of Knighthood.

As We Ourselves will not fail equally to observe the same, so shall all Our Successors to the Throne be obliged always to observe them.

Retains the right of explaining and removing doubts or objections, and of extending, altering, and of adding to the Statutes and Regulations.

Given at *Vienna*, the 14th of July, 1806.

His Imperial Majesty was graciously pleased, by Order of the 2d of January, 1812, to command that, in the Crosses of this Order, the Oak leaves branch should be removed,—And this, in consequence of a Royal Decree of the 17th of January, 1818, was then added to the Statutes.

Grand Cross.

The Right Honourable Sir George Murray, G. C. B., G. C. H., Knight Commander of The Royal Portuguese Military Order of The Tower and Sword, Lieut. General of His Majesty's Forces, Colonel of the 42d (Royal Highland) Regt. of Foot, and Commander of the Forces in Ireland, in compliance with the earnest desire expressed by their Majesties The Emperor of Austria, The Emperor of Russia, The King of Prussia, The King of Bavaria, and The King of Saxony, to accept and wear the Insignia of the undermentioned Orders of those Sovereigns respectively, viz., The Insignia of a Knight Grand Cross of The Imperial Austrian Order of Leopold, the Insignia of a Knight Grand Cross of The Imperial Russian Order of St. Alexander Newski, the Insignia of a Knight Grand Cross of The Royal Prussian Order of The Red Eagle, the Insignia of a Knight Commander of The Royal Bavarian Order of Maximilian Joseph, and the Insignia of a Commander of The Royal Saxon Order of St. Henry, with which their said Majesties have been pleased to honour him, " in consideration of his highly distinguished and important services against the Enemy in the Field, during the last war."—Royal Licence, 16th of September, 1825.

KNIGHTS.

WILLIAM HAMLEY, Esq., a Commander in The Royal Navy, " as a testimony of the high sense which The Emperor of Austria entertains of the services rendered by him, as First Lieutenant of His Majesty's Ship *Havannah*, at the siege of *Zara*, in the late Campaign."—Royal Licence, 1st of June, 1815.

Lieutenant WILLIAM KINGDOM RAINS, of The Royal Artillery, " in testimony of the high sense which The Emperor of Austria entertains of his distinguished conduct and services in the field, during the recent operations in *Italy*."—Royal Licence, 9th of October, 1815.

Captain RICHARD SAUMAREZ, of The Royal Navy, " in testimony of the approbation of His Imperial Majesty The Emperor of Austria of the distinguished services performed by him before the Enemy, during the Campaign of 1813."—Royal Licence, 28th of April, 1821.

GEORGE DUNCAN ROBERTSON, Esq., Colonel in The Army, Lieut. Colonel in The Sicilian Regiment of Foot, and C. B., " in testimony of the approbation of His Imperial Majesty The Emperor of AUSTRIA, of the distinguished Services rendered by him, while commanding THE BRITISH TROOPS acting in co-operation with THE IMPERIAL FORCES, at the Siege and Capture of *Trieste*, and other operations in *Italy*, during the Campaigns of 1813 and 1814."—Royal Licence, 5th of March, 1816.

JOHN DUFF MARKLAND, Esq., Post Captain in The Royal Navy, and C. B. "in testimony of the approbation of His Imperial Majesty The Emperor of AUSTRIA, of the distinguished Services rendered by him at the Siege and Capture of *Trieste*, and the other operations in *Italy*, during the Campaigns of 1813 and 1814."—Royal Licence, 5th of March, 1816.

WESTBY PERCEVAL, Esq., Post Captain in The Royal Navy, "in testimony of the approbation of His Imperial Majesty The Emperor of AUSTRIA, of the distinguished services rendered by him at the Siege and Capture of *Trieste*, and the other operations in *Italy*, during the Campaigns of 1813 and 1814."—Royal Licence, 5th of March, 1816.

Captain (now Lieutenant-Colonel) EDWARD ANTHONY ANGELO, conferred expressly upon him, for his *personal* exertions and intrepidity in leading the Austrians, on storming a Redoubt, which ensured the capture of The Castle of *Trieste*, in 1813,—Captain ANGELO had previously served with the Austrian Army in the Campaign against The Vice-Roy of Italy, as well afterwards at the Sieges of *Cattaro*, *Ragusa*, and other Fortresses, which were given up to The Austrian Government, when captured by the British arms.—Permission to accept and wear the same by Warrant and Sign Manual of King WILLIAM the Fourth.

Lieut. Colonel JOHN HOBART CARADOC.

The Imperial Austrian Order of The Iron Crown.

His Imperial Majesty, Francis the First, Emperor of Austria, in order to perpetuate the Epoch of the happy Reunion of His Italian Provinces under His Sceptre by a continual Memorial of His particular favour and grace, felt himself induced to found an Order, under the appellation of "The Iron Crown," and to place it in the number of His other Orders of His Crown and House. To this end, His Majesty submitted the following plan of the fundamental System of this Order, in accordance with His intention, and having approved of the same, He commanded, that the annexed Regulations for all the Knights of this Order should be then, and in all future times strictly observed, and serve to them as a constant guide.—

1. All Members of this Order shall be styled Knights of The Iron Crown.

2. The Dignity of Grand Master remains always and inseparably connected with The Austrian Empire, and consequently with himself, and after him, restricted to His legitimate Successors to The Imperial Austrian States.

3. The Order consists of three degrees, according to Merit, viz., Knights of the first, the second, and the third Classes.

4. The Knights of the First Class take precedence of those of the second, and the latter of those of the third class. The Knights of the

same degree take their places according to the period of their reception into The Order. When on one and the same day, several Knights are received into The Order, they take their places according to the Regulation in which they have received The Badge of The Order.

5. Admission into The Order is granted to every one without difference of Rank, whether or not he hold immediate Office in Civil or Military Service, provided he possesses the necessary qualifications. As such qualifications will be considered, distinguished proofs of attachment to the reigning Princes and to the State,—successful endeavours to promote the welfare of the Monarchy—finally, being distinguished by other great and generally beneficial Undertakings.

6. The bestowal of The Order is reserved to The Grand Master, and no person shall ever be permitted to petition for it.

7. The Badge of Honour, which is conferred upon the Knights as a proof of their acquired Honour, is, as follows,—

The Iron Crown, on which the doubly crowned Imperial Austrian Eagle rests,—the same has at both sides on the breast, a dark blue enamelled shield in the middle, on which, in front, is a plain golden F., —on the reverse side appears the date of the year 1815.

The Knights of the first Class wear The Badge of The Order by a broad yellow riband, having a dark blue stripe at the edges, suspended round the neck on the breast. They wear, besides this, on their breast, a four cornered Silver Star, in the middle of which is the Iron Crown on a gold ground, and surrounded by a dark blue enamelled wreath, with the inscription, "*Avita et Aucta.*"

On extraordinary occasions The Knights of the first class wear the above described Badge of The Order by a Gold Chain, round the neck, —on which are first introduced, entwined within each other, the cyphers F. and P.,—then follow in succession, The Iron Crown, a wreath of Oak leaves, again the aforementioned letters F. and P., and so continued on, in alternate variation, to the end.

The Badge of The Order of the second class is only distinguished from that of the first by its size,—it being hung round the neck by a yellow riband, with blue stripes, two inches wide.

The Knights of the third class wear The Badge of rather reduced proportions, at the button-hole, on the left breast, with a similar riband, nine lines in breadth.

The distinguishing marks worn by The Herald and the other Officers of The Order, when attending at the Festivals of The Order, consist of The Badge inclosed in a large gold Medal, which is attached to the riband worn by The Knights of the third class,—besides which, the Herald is distinguished by his Staff.

8. The Decoration of The Badge with precious stones is left to the exclusive consideration and decision of The Grand Master,—and no Knight is allowed to wear The Badge in such form, unless he has been graciously favoured with the sanction of The Grand Master.

9. On the other hand, The Knight is permitted to adorn his Armorial Bearings with the *Insignia* of The Order, and to make use of the same on all occasions during life.

10. Prescribes the Costume, whereby to distinguish the Members of The Order, in their different gradations.

The colours of The Order are yellow, blue, and white,—the edging and all other ornaments being of silver.

11. The number of Knights is fixed at One hundred,—*viz.*, *Twenty* of the first, *Thirty* of the second, and *Fifty* of the third class. The Princes of The Imperial House are not included in this number.

12. The Officers appointed for the various occasions on which they may be required to attend, are,—The Prelate, Chancellor, Treasurer, Secretary, Herald, and Chancery Clerk.

The appointment of these Officers is exclusively committed to The Grand Master.

13. Prescribes the Ceremony and Oath to be observed, on the occasion of a solemn reception of a Knight of The Order.

The dispensation of The Oath of The Order to a Knight Elect is committed to The Grand Master.

14. As soon as the Reception is completed, and the form prescribed, The Knight of the first class, as a mark of distinguished grace, is then embraced by The Grand Master, and in the same manner the Members of The Order present, assure him of their friendship.

15. The Diploma of the Reception is prepared for The Knights of the first class, in the form of a *Book*,—for the Knights of the second

and third classes, in the form of a *Patent*, and signed by The Grand Master, The Chancellor of The Order, and The Secretary. The Knights of the first and second classes receive their Diploma with a pendent Seal,—those of the third class, with a fixed Seal.

16. The Knights are required always to wear The Badge of The Order, and none shall appear publickly without it. Neither must any *Foreign Badge* of Honour be worn next to it, unless the express sanction of The Grand Master be obtained for that purpose.

17. After the decease of a Knight of the first class, the Chain of The Order, with the Badge, together with the Book of the Statutes, must be returned to The Treasurer of The Order.

18. On the first Sunday after the 7th of April, as the day of The Foundation of The Lombardo-Venetian Kingdom, a Festival of The Order will be annually holden, at which the whole of the Members of The Order there present must appear,—unless through any valid cause prevented,—with their Badges of The Order, and in the prescribed Costume.

19. At the Festivals of The Order, the Knights of the third class have admission to the Privy Council Chamber,—where the Knights of the first and second classes have always the privilege of being present.

20. Likewise, all those who are honoured with THE ORDER may, without any distinction of class, be received at all the Court Festivals and Levees.

21. If any Knight, favoured with The Order of the first class, does not possess the Dignity of a PRIVY COUNSELLOR, the same shall be bestowed upon him, *gratuitously*,—The Knights of the second class, upon their application, will be raised to the rank of BARONS,—and The Knights of the third class will be promoted to the rank of KNIGHTHOOD.

22. To all Officials the command is given, that whenever a Grant is made by them to The Knights of The Order, to such shall be added, besides the ordinary proper title, also the title of The Order.

23. Likewise, that whenever Decrees made in the name of The Grand Master are sent to The Knights of the first class, they, as a greater mark of His favour and inclination, shall be honoured with the appellation of "*Our Cousin.*"

24. In conclusion, it is His Majesty's will, that although The Orders of St. Stephen and Leopold, regarded as united bodies, take rank before The Order of The Iron Crown, when on the same day Orders of the three descriptions are presented, The Knight of St. Stephen's Order takes precedence of The Knight of the Leopold's Order, and the latter takes precedence of The Knight of The Order of The Iron Crown, bearing the same rank,—yet, in all other respects, these three Orders shall be holden equal,—between those Grand Crosses or Knights of the first class, between the Commanders or Knights of the second class, and between the small Crosses or Knights of the third class, only the date of the Reception shall make the exception.

These are the Laws and Regulations upon which is founded The Order of Knighthood of The Iron Crown.

As His Imperial Majesty will not fail equally to observe the same, so shall all His Successors to the Throne be obliged to observe them, until His Majesty or His Successors, by the altered spirit of the times and circumstances, shall find it necessary to introduce alterations therein.

Given in His Royal City of *Milan*, the 1st of January, 1816, in the 24th year of His reign.

The Iron Crown which is that of the antient Kings of Lombardy, is preserved in the Treasury of the fine Church of *Saint John the Baptist* at *Monza*, about 10 English miles to the North of the City of Milan.

It is a Circlet of pure gold of four fingers in breadth, carved and adorned externally with enamel and precious stones, in the shape of an antient Diadem, and is $18\frac{1}{10}$th English inches in circumference. Within the golden Circlet is fixed a thin rim of *Iron*, a quarter of an inch in breadth, which, according to tradition, is made of *one of the Nails* of The Cross of Christ, and from thence the Crown takes its Name.

According to tradition, THEODOLINDA, Queen of LOMBARDY, received this Crown from Pope GREGORY the First, called, *The Great,*—but MURATORI shows, that it is not of such Antiquity. It was taken to *Bologna* to crown The Emperor CHARLES the Fifth,—and it was not until after his Coronation, *that the Iron rim was reported to be made of the Nail of The Holy Cross.* The story was founded upon the antient traditions, respecting the Crown or Helmet of CONSTANTINE.

Saint AMBROSE relates, in his Sermon on the death of THEODOSIUS, that *Saint* HELENA, the Mother of CONSTANTINE, sought for and found the *four Nails* with which CHRIST was crucified, and that she caused *one* of them to be formed into a Diadem set with Gems,—itself a more precious Gem,—and *another* to be made into the *Bit of a Bridle,* and sent them to her son CONSTANTINE. Other authors, contemporary with *Saint* AMBROSE, state, that the *Nail* was made to form a part of the *Helmet* of CONSTANTINE, that he might thereby be protected from danger in battle.

In what manner it descended from The Greek Emperor to THEODOLINDA does not clearly appear,—but the bequest of it by her to the venerable Cathedral of *Monza* is an historical fact.

This antient and venerated Crown is preserved with the most religious care in a Shrine contained in an enormous Cross, suspended over the Altar of *The Duomo,* or *Cathedral* of the Royal Ville of *Monza.* At ordinary periods, even, it is difficult to obtain permission to view the Crown,—an order signed by THE GRAND

DUKE, and countersigned by THE GOVERNOR of MILAN, being required for the purpose. In exhibiting it, much Ceremony is observed by the Canons and Priests, who appear in grand *Pontificalibus*. The Sacristan ascends a ladder which is placed against The Cross, opens the Shrine, and this most legitimate of Relicks is displayed to Publick view amidst the blaze of Torch light and fragrant Incense, from silver censers. Merely the Jewels are visible, owing to the glare and smoke,—the Sacristan then descends, and the Canons exhibit a *Model Crown*, in imitation of the real one, in order that the spectator may examine its ornaments, and form an idea of the size and value of the Gems.

SECOND CLASS.

EDWARD YOUNG, First Lieutenant.

EDWARD, Viscount KINGSBOROUGH, Author of the splendid Work on the "*Antiquities of Mexico,*" in 1830. He died on the 27th of February, 1837.

THIRD CLASS.

ANDREW PELLAT GREEN, Captain in The Royal Navy, conferred upon him by THE EMPEROR of AUSTRIA, when he went on board His Majesty's Ship, "*Rochford,*" then under his command, at *Naples*.

NICHOLAS CARLISLE, Esq., K. H., K. D., D. C. L., &c., Secretary of THE SOCIETY of ANTIQUARIES of LONDON.

THE ORDER OF ELISABETH THERESA.

THIS ORDER is the Second Military Order in Austria, and it is also called, "THE MILITARY FOUNDATION OF ELISABETH THERESA." It was instituted in the year 1750, by The Empress ELISABETH CHRISTINA, widow of The Emperor CHARLES the Fifth, for the purpose of rewarding with it, Twenty Officers, from the rank of Colonel up to that of General, who should have faithfully served THE HOUSE OF AUSTRIA, during a period of *Thirty years.* These Knights were divided into Three Classes, and endowed with an annual Revenue of 16,000 florins,—out of which, an annual pension of 1,000 florins was to be paid to every Member of the first Class,—of 800 to the second,—and of 300 to the third.

On the 16th of November, 1771, MARIA THERESA, daughter of The Foundress of The Order, renewed it,—modified it,—and fixed the number of Knights at *Twenty-one,*—and she decreed, that *six* of them should receive 1000 florins,—*eight* of them, 800 florins,—and *seven,* 500 florins, annually, as Pensions. These Regulations continue the same at the present day.

The Knights are named Members of THIS ORDER, by the Monarch himself, on the presentation of The Aulic Council of War,—and neither birth, country, religion, nor the previous possession of any other Military Orders, are grounds of exclusion.

THE BADGE is a Star, bordered with gold, of eight points, enamelled half red, half white,—in the middle is an oval, bordered with a gold margin, on which, beneath a gold Imperial crown, are the united cyphers E. C. and M. T. (ELISABETH CHRISTINA, and MARIA THERESA), and around them are the words, *M. Theresia Parentis gratiam perennem voluit.*

The whole of the Knights,—who form but one Class,—wear, without any farther distinction, this Cross on a black silk riband, which is pressed tight on the Cross by a gold Imperial Crown, on the left side in the button-hole.

On the 19th of November, which is *St. Elisabeth'*- day, and the Anniversary of The Empress ELISABETH, divine service is solemnized in the Church of The Augustine Friars in *Vienna,* in the presence of all the Knights, Generals, Superior and other Officers of the Garrison.

The possessors of THIS ORDER, who are of the Roman Catholick faith, are under the obligation of reciting daily, three *Ave Marias,* and three *Pater Nosters,* for THE FOUNDRESS of THE ORDER, and the reigning SOVEREIGNS. Knights, of any other Communion, pay annually three ducats to the funds of the Invalids.

The Order of Ladies of The Star of The Cross.

From very antient times The Imperial House of Austria is said to have been in possession of a small piece of The Cross of Christ.

The Emperors Maximilian and Ferdinand were accustomed to bear with them constantly in War and Peace this Relick, inserted in a Cross of gold.

After the death of Ferdinand, his Successor Leopold, presented it to the widowed Empress Eleanora, a daughter of Duke Charles, of Mantua, in order by it's means to soften the sorrows of her Widowhood.

By her it was preserved with great care in a small wooden box, adorned with crystal and enamel, over which was a silk covering. It happened, that, on the 2nd of February, 1668, a fire suddenly broke out in the night in The Imperial Palace at Vienna, beneath the Apartment of The Empress Eleanora, which spread with such rapidity, and so speedily reached the apartment of The Empress, that she saved herself with much difficulty, and the apartment in which the box containing The Cross was placed, was speedily consumed and fell to the ground. By careful search amid the ruins, on the 6th of February the Gold Cross was found almost wholly uninjured, being only melted in a trifling degree.

The Empress was so rejoiced at this remarkable preservation and recovery of The Cross, that she commanded a Religious Procession to take place in celebration, and The Prince Bishop of Vienna caused a formal declaration to be made of this singular event.

This occurrence induced The Empress to found a FEMALE ORDER, with the view not only of preserving the remembrance of this extraordinary circumstance, but also of inducing the Members of THIS ORDER to devote themselves to the service and honour of The Holy Cross, to pursue a virtuous course of life, and to practise religious actions and deeds of charity.

After Pope CLEMENT the ninth had confirmed this new Order, by a Bull, dated the 28th of July, 1668, with a grant of the accustomed privileges, and had confided to The Prince Bishop of Vienna the superintendence of its Ecclesiastical concerns, and The Emperor LEOPOLD the first had not only confirmed The Order and its Statutes, on the 9th of September, but had taken it under his care and protection,—The Empress ELEANORA declared herself, on the 18th of September, 1668, it's Foundress and Protectress.

THIS ORDER which is only conferred on Princesses, Countesses, and Noble Ladies, received upon its establishment the name of "*Community of Noble Ladies*," under the title of "*Star of The Cross*",—and the Members were styled, "*Wearers of The Cross*", or, "*Ladies of The Star of The Cross.*" It subsequently received the title of "*Order of Ladies of The Star of The Cross.*

The name of The Star of The Cross contains an allusion to a Constellation of four Stars at *The South Pole*, called "*The Cross Star*," or, "*Star of The Cross*,"— which, however, is not visible in our horizon.

The Empress or a Princess of The Imperial House of Austria is Chief Protectress of The Order. She appoints the Members, and chooses from them two, as her assistants, who must be in attendance at her side at the Ceremonies of The Order.

The number of Members is undetermined, and depends entirely on the pleasure of the Protectress.

The Festival of The Order, at which all resident Ladies of The Order are obliged to attend, is celebrated twice in the year, *viz.*, the day of the discovery of *The Cross*, the 3d of May, and the day of the elevation of *The Cross*, the 14th of September. Formerly a third Festival was holden on Good Friday, or on some other day appointed by The Protectress, when a solemn Procession took place from The Imperial Chapel at *Schönbrunn* to the Church of The Mother of God at *Hietzing*,—this, however, is no longer observed. The service for the Foundress and all deceased Members of The Order, which all resident Ladies of The Order are bound to attend, is usually holden on the 6th of February.

The Badge, which has undergone four alterations since the time of Maria Theresa, is an oval Medallion, with a broad blue enamelled border, inclosing a black enamelled Eagle with two heads, and claws, both of gold, on which lies a gold Cross, enamelled green, and bordered with brown wood. Over this, on an intwined wreath, in black letters, on a white ground, is the Motto of The Order, "*Salus et Gloria.*" It is worn, pendent to a strip of black riband, on the left breast.

On the Festival days of The Order, appointed for the admission of new Members,—the 3d of May, and the 14th of September,—The Badges are placed on the Altar, are then consecrated according to the Ritual, and are bestowed by The Chief Protectress on the newly appointed Ladies of The Order, kneeling before her.

The Officers of The Order are, a Treasurer and a Chancery Clerk.

The Order of Ladies Slaves to Virtue.

This Order was instituted at Vienna, in 1662, by The Empress Eleanora, Widow of the Emperor Ferdinand the Third,—who declared herself Sovereign of it.

The Ensign which is worn by the Ladies of this Order, is a golden Sun, encircled with a chaplet of laurel, enamelled green, with this Motto over it, "*Sola ubique triumphat.*" On solemn occasions it is worn, pendent at the breast to a small Chain of gold,—but on other days, they wear a smaller Medal, attached to a plain narrow black riband.

The number of The Ladies is limited to *Thirty*, and they must be of very ancient and noble Families. They promise to be faithful to The Empress, as their Grand Mistress, and to make Virtue the only guide of their conduct.

When a Lady dies, her Relations are bound to transmit the great Medal to The Empress,—but the smaller they are allowed to retain, as a Memorial that one of their Family has had the honour of being a Member of this Order.

Medals and Marks of Honour.

Bronze Cross for the Army,—On the 30th of May, 1814, The Emperor Francis the first, in order to perpetuate the memory of this remarkable Æra, and also of the great events in which The Imperial Austrian Army had taken so active a part, founded *at Paris!* this Medal, *cast in the metal of the Cannon taken from the Enemy,*—the Medal to be worn at the button-hole by Military persons of every Rank. On the death of the possessor, this Medal becomes the property of his heirs, but without the right of wearing it.

Cross of Honour for the Civil Service,—This Cross was instituted by The Emperor Francis the first, for those individuals of The Civil Service, who had rendered themselves meritorious by lending their aid towards the prosperous issue of the War of 1813 and 1814. This mark of Honour is composed of a large Cross of gold, and also of a Cross of gold and silver, and is worn at the button-hole. The first distribution was made by the Emperor Francis the first in person, on the 26th of May, 1815. This Cross becomes the property of the heirs of the possessor. It is highly esteemed.

Military Medal for Valour,—The Emperor Joseph the second, in order to reward Valour and Military Merit in those who were not eligible for the Military Order of *Maria Theresa,* instituted the Military Medal of Honour, in gold and silver, and appropriated it for distribution among Soldiers of inferior Rank, commencing with Sergeant Majors, as a reward for their brilliant Actions in the face of the enemy. Times and circumstances having rendered necessary various changes, in the form of this Institution, The Emperor Francis the first, who always paid particular attention to Military Merit in all ranks of the Army, renewed the Statutes of the Medal for Valour, and enlarged the number of its Prerogatives. A valourous Action, of decided importance, executed with intrepidity and extraordinary efforts, is recompensed with *The Golden Medal*, without regard to Rank. *The Silver Medal* is conferred as a reward for deeds of arms of inferior valour. The General-in-Chief alone possesses the right to award The Medal for Valour,—but he may delegate this power temporarily to Generals commanding the Corps of the Army. The person who has obtained the Medal of *gold*, receives double pay,—he who possesses the *silver* Medal, the moiety of what his pay would be on the Peace establishment. This addition is continued to them for life, whether they continue in the Army or obtain Government employment in the Civil Service. The Medal descends to the heirs of the possessor, who, if they desire it, may receive from The Minister of War an equivalent in money,—for a Golden Medal 25 florins 28 kreuzer. The Medal for Military Valour of Joseph the second, and of Francis the first, are worn at the button-hole,—the name of the possessor, the designation of his Regiment and the Rank he held in it, as also a succinct account of his action, are sent to The Aulic Council of War, by which they are inserted in the Journals.

Cross of Ecclesiastical Merit,—*Pro piis Meritis.*—This Cross which is of gold and silver, was instituted by Francis the first, on the 26th of November, 1801, and intended for Chaplains of Regiments, who had distinguished themselves. In order to obtain this mark of distinction it is necessary,—*First,* That the Ecclesiastic should have displayed ardent zeal in the performance of his Holy duties,—or, *Secondly,* That he should have distinguished himself by honourable Military actions, such as having led on or animated the Troops in combat, or having personally taken part therein. For this last class of deeds of valour, Medals of *gold* and *silver* are conferred, which are worn attached to the same riband, as the Cross of Ecclesiastical Merit.

Medal of Civic Honour,—*Large, Middling, Small,*—The Emperor Francis the first, in order to reward Civil Merit, conferred a Medal of Civic Honour. There are no rules for the distribution of this Medal,—it is granted from time to time to persons filling offices at

Court. These Medals are of gold and silver,—the former are divided into three sections, and are all worn attached to a riband of deep scarlet colour. The great Civic Medal of gold is sometimes given attached to a Chain of gold, as a yet more conspicuous reward for eminent Merit and distinguished Services rendered to the State,—in this case, it is worn suspended from the neck.

The Obverse of the great Medal represents the portrait of The Emperor FRANCIS the first, with the inscription "*Franciscus Austriæ Imperator,*"—on the Reverse, is a temple, with the words, "*Austriæ ad imperii dignitatem evecta.*"

The two other Medals represent on the Obverse, the portrait of The Emperor, with the inscription "*Franciscus Aust. Imp. Hun. Boh. Gal. Lod. Rex. A A.*",—and, on the Reverse, a balance of justice, a sceptre, and a caduceus, surmounted by a Crown, with the inscription, "*Justitia regnorum fundamentum.*"

CROSS of BOHEMIA,—granted in 1814, by the Emperor FRANCIS the first, to those of the Bohemian Nobility, who, during the War, had formed around him a particular Guard, and had accompanied him throughout The Campaign. The Cross, which is of gold and red enamel, bears on one side the White Lion of Bohemia,—and, on the other, the inscription, "*Nobilibus Bohemis bello Gallico fidis corporis custodibus Franciscus Augustus* MDCCCIV." It is worn, attached to the button-hole by a riband of three stripes,—the centre red, and the two others white.

MARK of DISTINCTION for VETERANS.—In order to confer a Mark of Distinction on Soldiers, from the rank of Serjeant-Major who have served a greater number of years than the original stipulation, and to encourage them to re-enter the Service, two escutcheons of distinction have been introduced, one of them hexagonal, and the other oval, which are worn on the left breast of the coat.

FOREIGNERS who, before being naturalized, engage themselves

permanently, after having completed the term of their final engagement, also obtain this Mark of Distinction.

Native Veterans who, after the completion of their term of Service, continue in the ranks until they are invalided, without receiving the price of re-engagement, may also obtain the escutcheon.

THE GRAND DUCHY OF BADEN.

THE ORDER OF FIDELITY.

CHARLES WILLIAM, Margrave of BADEN-DURLACH, after having laid the first stone of his new residence at *Carlsruhe*, instituted in honour of this event, on the 17th of June, 1715, THE ORDER of FIDELITY,—consisting only of Grand Crosses.

On the 8th of May, 1803, CHARLES FREDERICK, in commemoration of his acquiring The Electoral Dignity, renewed this ORDER, and divided it into Two Classes,—*viz.*,

GRAND CROSSES,—and

COMMANDERS, in which class a distinction is also made between The Knights nominated *before* or *after* the 26th of December, 1812, the day of the foundation of THE ORDER of THE LION of ZAEHRINGEN.

The reigning Prince is always The Grand Master of The Order, and nominates the Members. Princes of the Blood are Grand Crosses from the period of their birth.

THE BADGE is a gold Cross, of eight points, enamelled red, the points being tipped with gold balls, and in the middle is a white enamelled shield, upon which appear masses of stone and green hills, and thereon is a double C., and, around the same, the word "*Fidelitas.*" Upon the reverse, is an escutcheon of white enamel, with the arms of The Margrave, on a field of gold. The Cross is surmounted with a Prince's Cap, which is turned up with a border of ermine. In each of the four principal angles, is a double C. of gold, intertwined with each other.

This CROSS is suspended from a lemon coloured, watered riband of the breadth of three inches and a half, which is edged with a narrow border of silver, and passed across the right shoulder.

The Knights likewise wear an eight pointed STAR, embroidered in silver,—in the centre of which is represented the obverse of the Cross, on an orange coloured ground, and, in the four angles of the rays, are the double C's, in gold. This Star is worn upon the left breast of the upper apparel.

COMMANDERS of the First Class wear the JEWEL or STAR in the same manner, and the Decoration of The Order hanging from the neck by a riband of two inches broad.

Those of the Second Class wear the Decoration in the same manner, but without a Star.

Those persons who are received as KNIGHTS of THIS ORDER, must be of very antient and noble Families.

THE OFFICERS of THE ORDER are, a Secretary, and a Treasurer, who is also the Registrar. The former wears The Badge, like that of the Commanders,—the latter, a smaller Cross, on a gold chain,—but both of them, without the Star.

According to the STATUTES, which, however, have never appeared in print, *Twenty ducats* are paid on admission.

THE ORDER OF MILITARY MERIT OF CHARLES FREDERICK.

THIS ORDER was instituted on the 4th of April, 1807, by The Grand Duke CHARLES FREDERICK, to promote Valour, Fidelity, and Attachment to his person,—and to recompense the merit of Officers, but particularly that of Generals and Superior Officers, who had been in the presence of the Enemy.

THE ORDER is composed of three Classes,—*viz.*, GRAND CROSSES,—COMMANDERS,—and KNIGHTS.

The King is always the Grand Master. Princes of the Blood are Members by birth-right, without however having any claim to the *Pension*, which is granted with it.

GENERALS only can receive the Grand Cross,—the number of the KNIGHTS is unlimited. Officers of every Rank, and of all the branches of the Military Administration are admitted without regard to Religion, Birth, or any other circumstance.

A Pension of 400 florins is paid to the two oldest Members of THE ORDER,—one of 200 florins to the three oldest Commanders,—and eight of 100 to the eight oldest Knights.

Every military exploit, which, if not performed, would have in no wise affected the responsibility of the individual who undertook to execute it, or any other proof of valour, resolution, and sagacity, as illustrated by several examples in the 10th article of the STATUTES, confers a right to this ORDER. Claims may also be acquired by services rendered to THE STATE during a period of 25 years, or by proofs of attachment and zeal to the person of THE SOVEREIGN.

In order to decide upon these several rights, THE CHAPTER of THE ORDER assembles every other year, on the 20th of November, under the direction of THE GRAND MASTER,—or, in case of his absence, under that of the Senior Grand Cross. The right of decision, however, belongs to The Grand Master, who likewise enjoys the power of nominating at all times, without assembling The Chapter.

The Seniority of the Member received, and his right to the *Pension,* date from the day on which the Action to be rewarded, was performed, —and that day, and, if it be possible, *the very hour*, are recorded in the Diploma.

For those Members who are recompensed for their long and faithful services, or for any other cause, their Seniority dates from the day of their reception. They have the right to annex it to their ARMS.

The KNIGHTS suspend it to the bottom of their Escutcheon by a knot of Riband,—for COMMANDERS, the Riband surrounds the Escutcheon, —which, for the GRAND CROSSES, is placed on the Decoration of The Order.

Those who quit the Military service, preserve the right of wearing the Decoration,—but lose it, however, together with the Pension, if they enter into any *Foreign* Service without the consent of their Sovereign.

The honours paid to Members of The Order on their decease, are the same as those shown to the next superior Class.

THE GRAND CROSSES wear the DECORATION by a riband with three stripes,—the middle one being yellow, and the two others red, bordered with white. This riband hangs across the left shoulder, and the Jewel is placed on the left breast. The letters C. F., in the centre of the Decoration, are the initials of The Grand Duke CHARLES FREDERICK. The MOTTO of THE ORDER is, " *For the Honour of Baden.*"

COMMANDERS wear the Cross by a riband of one half the width. As soon as they have attained the Rank of GENERAL, they add to it the JEWEL, on the left breast.

THE KNIGHTS wear the SMALL CROSS, by a still narrower riband, at the left button-hole.

At their nomination The Grand Crosses pay 20 ducats,—Commanders 10,—and Knights 5., to the Treasury of THE ORDER,—the affairs of which are managed by a Chancellor and a Treasurer.

The Order of The Lion of Zaehringen.

The Order of The Lion of Zaehringen was instituted, on the 26th of December, 1812, by The Grand Duke Charles Louis Frederick, of Baden, being the Name-day of his Consort, The Grand Duchess Stephania Louisa Adriana. The Sovereign is always The Grand Master,—and Princes of his Family are Grand Crosses by right of birth.

The Ruins which are painted on the Medal, represent the Castle of the Dukes of *Zaehringen*, ancestors of the Dukes of *Baden*,—on the reverse are the Arms of Zaehringen,—and, in the centre, is a golden Lion, looking towards the left, on a red enamel.

Grand Crosses wear the Decoration suspended by a dark green, watered riband, edged with an orange colour, three inches and a-half wide, passing across the right shoulder.

The Jewel, which is a Star, of eight points, and on which the golden Lion of Zaehringen appears in a red field, with the Motto, "*For Honour and Truth,*" is to be worn on the left breast.

By the Commanders, the Cross is worn by a riband, two inches wide, around the neck. The Knights wear it by a riband of an inch and a quarter broad, attached to the first button-hole on the left.

The different Decorations of this Order are conferred on particular occasions, ornamented with *Oak-leaves* in gold, on which are the initials of the Sovereign, surmounted with a crown.

The affairs of The Order are managed by The Treasurer.

Medals, and Marks of Distinction.

Gold, and Silver, Medal for Military Merit,—The Grand Duke Charles, after having founded The Military Order of Charles Frederick, instituted on the 4th of April, 1807, this Medal, for rewarding Actions more or less noble of Subalterns and Soldiers against the Enemy.

This Medal represents on one side, a Griffin ready for combat, holdhing with his left claws the Arms of Baden, and with his right a Sword, and the inscription, "*For the Honour of Baden.*"—On the Reverse, are the words, "*To the brave,*" with the name of him, on whom the Medal has been bestowed,—the whole inclosed within a crown of laurel.

The Medals are worn at the button on the left, by the riband of The Order of Military Merit. The possessors of the *gold* Medal receive additional pay, equal to the pay itself. A sum equal to half the amount of pay, is granted to those who are decorated with the *silver* Medal.

Marks of Distinction for Service,—The Grand Duke Leopold founded on the 18th of February, 1831, certain Marks of Distinction, to honour, by an external sign, Services which had been rendered to the Country, during a certain period.

A Cross, of eight points, in the centre of which is the letter L., surmounted by a crown, and pendent to a red riband, edged with yellow, red, and white, is granted to Officers, for 25 years of service, comprising the years of service as a Private and Subaltern officer, —on the reverse, are the Roman numerals XXV.

A particular Mark of Distinction of Service of the first class, is granted to Under-Officers and Privates for 25 years,—another of the second class, for 18 years,—and that of the third class, for 12 years. One Campaign is reckoned two years.

The Cross is to be worn at the button-hole,—and the Marks of Distinction for Service, on the left breast.

THE KINGDOM OF BAVARIA.

The Order of Saint Hubert.

This Order was founded by Gerhard the Fifth, Duke of Jülich and Berg, in the year 1444, in commemoration of a Victory obtained on the 3d of November, *St. Hubert's* day, at *Ravensberg*, over Arnold *von* Egmont, who had assumed the title of Duke of *Geldern*, and had invaded the Territory of Jülich.

His son, William, Duke of Jülich and Berg, confirmed this Institution by giving to The Order, in 1476, the first Statutes.

The Order continued to flourish under his Successors to the Government, until the year 1609, when the Male line of The Founder became *extinct* with Duke William.

When John William, of The House of *Neuberg*, Elector of The Palatinate, obtained the Office of *Chief Carver*, together with the Upper Palatinate, he, as owner of part of the Country to which formerly THE ORDER appertained, renewed, on the 29th of September, 1708, the previous Institution by improved Regulations, and declared himself Grand Master of THE ORDER.

His Successor to the Government, The Elector CHARLES PHILIP, confirmed, on the 31st of January, 1718, the STATUTES,—and CHARLES THEODORE, Elector of The Palatinate, enlarged the same, at first on the 7th of April, 1744, and, afterwards, on the 30th of January, 1760.

After the happy Re-union of The Palatine-Bavarian Countries, under the Sceptre of the most Illustrious Elector, MAXIMILIAN JOSEPH the Fourth, subsequently KING of BAVARIA, the variance which had previously existed on the question, *as to which Province* THIS ORDER *originally belonged*, ceased of itself, and the present STATUTES of THE ORDER were, on the 30th of March, 1800, by Royal right and title re-established.

THE ORDER consists of *Twelve* COUNTS and BARON-CAPITULARS, and a GRAND COMMANDER.

Besides these, THE ORDER can only receive Crowned Heads, Sovereigns, or Members of antient Princely Houses, or such Foreigners, as His Majesty may deem especially worthy of the same. The number of Princely Knights is, therefore, indefinite.

After the re-establishment of The Royal Dignity an Ordinance, of the 18th of May, 1808, was enacted, determining that THE SAINT HUBERT'S ORDER, with the confirmed Laws and Privileges, should remain the *First* ORDER of The Kingdom, and be placed in such degree of proportion and connection with the newly founded Order of CIVIL MERIT of THE BAVARIAN CROWN, that the hitherto Capitulars of *The Order of St. Hubert*, when also Grand Crosses of *The Order of Civil*

Merit, shall take precedence in rank before the other Grand Crosses of *The Order of St. Hubert*,—The Order of Civil Merit, however, itself to be distinguished by the Small Cross or Riband at the button-hole.

The Badge of The Order, with the Motto "*In Trau vast*," *(in Fidelity firm)*, consisted at the period of the first foundation by The Duke Gerhard the Fifth, of a Chain, in the links of which gold Hunting-horns and decorations were placed in succession,—to which, at the commencement, a kind of Medal was attached, but, subsequently, was substituted by another green coloured mark in the form of a *Peea*, upon which the Legend of *St. Hubert* was represented in gold, and to which was fastened, with gold cords, a Hunting-horn, also of gold.

Of the Legend of *St. Hubert*, whose Name the Order bears, the following particulars may be related. *St. Hubert* was a son of Duke Bertrand of *Aquitania*, to whom, whilst he was still a *Heathen*, when on a Hunting excursion with his dogs and attendants, suddenly The Saviour of the World appeared, and that in the visible shape of a *Cross* above the head of a *Stag*, instructing him to go to Bishop Lambert. When he did this, he was received by the Bishop with great affection, converted and baptized. Afterwards he went to Rome, in order to visit the Holy places, and then ordained a Bishop, as successor to the deceased *St. Lambert*.

In what manner the history of the Conversion of *St. Hubert* is to be represented on the Chain of The Order, as well as on the *Cross*, subsequently introduced, is exactly appointed in the Statutes of The Order, of the 29th of September, 1708.

The Chain of this Order, which has always been holden in great honour, consists of 42 links, and of these 21 form oblong squares, on each of which appears the said history of the Conversion, as follows,—On the right side of a mountainous, woody country is seen, ascending a hill, the forepart of a *Stag*, bearing on his head, between the antlers, *the Cross of The Saviour*,—close by it, *St. Hubert* is seen kneeling, worshipping The Saviour of the World,—before him is his *dog*, and to the right side, his *servant*, holding the horse by the bridle. The other 21 links of the Chain, more oblong in form, and which are distinguished alternately by red and green colours, contain the twisted Gothick cyphers, T. V. *(Trau vast)*.

The present Cross of The Order to this Chain consists of a gold, eight pointed, white enamelled Cross. The circular middle shield represents again the Conversion scene on one side, and, on the opposite side, an Imperial *apple*, with the Cross in the form of a terrestrial globe, where the circumscription appears,—"*In memoriam recuperatæ Dignitatis avitæ*, 1798," which refers to the renewal and re-establishment of The Order. This Cross was afterwards ornamented with 12 gold points in the corners, and covered with a Royal crown.

The Knights wear besides this, on the left breast, a pointed, faint-rayed Silver Star, upon which is fastened a Cross of silver, inworked with gold and inlet with gold. In the middle of the Star is a scarlet velvet round surface, on which is seen the Motto of The Order, "*In Trau vast*," in the old Teutonick character.

With the Grand Cross and the Grand Chain, or the Riband on many occasions, is worn a very small Cross, similar in form with the larger one, and, at the button-hole, another Chain of equally small proportions. The Grand Cross is worn by a scarlet riband, of a hand's breadth, with a narrow green border, over the left shoulder to the right hip.

The Costume of Ceremony on solemn occasions is, a black Collet, a similar Scarf, tight breeches with knee-bands, ornamented with rosettes, a short black Mantle, Hat with feathers, and Sword, according to the old Spanish taste.

When His late Electoral Highness Charles Theodore visited Italy, he was received and treated at *Florence* with so much Pomp and Magnificence, by George Nassau, third Earl Cowper, that he requested His Lordship to accept of this, his own Order, as a small testimony of his regard, and of the due sense which he entertained of His Lordship's distinguished hospitality. This accomplished Nobleman was created *a Count of the Holy Roman Empire*, on the 31st of January, 1778. He died at Florence, on the 22d of December, 1789.

The Order of Saint George.

According to the authority of the Historians of Bavaria, the Origin of this Order is traced to the period of *The Crusades,* in which The Dukes Guelph the First, Otho the Third, Eckart the Third, and Otho the Fourth, of Bavaria, bore a share in person, and The Company of The Knights under the Shield of *St. George,* whose object had been to protect and defend the Christian faith, were transplanted into Germany, as early as the Twelfth Century.

It is further stated, that the Emperor Maximilian the First, before the war with the *Turks,* and with the object of defending the Catholick faith, founded, in the year 1494, an Order of Saint George, which circumstance is considered as the *first renewal* of this Order, but it soon afterwards fell into oblivion,—until the Elector Maximilian Emanuel, of Bavaria, animated by a pious zeal for the honour of his Church and the worship of her Saints, formed amidst his Campaigns against the *Turks,* the determination to re-establish this Order which had already existed for some Centuries previously, and to raise it to the rank of a Military Order. But overtaken by death, he could not accomplish his design,—the completion of which was left to his son Charles Albert, afterwards The Emperor Charles the Seventh, who, on the 28th of March,

1729, in the third year of his reign, undertook the *second renewal* of THIS ORDER, under the Title of "*Protector of The Holy Knight and Martyr George, and the Immaculate Conception of the Virgin Mary,*" for the honour and benefit of the antient Nobility of Bavaria. He endowed it with formal Statutes, great Priorships, Benefices, (and, subsequently, also Provostships), and Pope BENEDICT the Thirteenth confirmed THE ORDER by a Bull, of the 15th of April, 1828, and endowed it with the privileges, honours, indulgences and other advantages, which had been granted to *The German Orders* by all the preceding Popes.

After the extinction of The Bavarian Line of LOUIS, it was confirmed by the Elector CHARLES THEODORE, in the year 1778, as a Palatine-Bavarian Order,—by his Successor, The Elector MAXIMILIAN JOSEPH the Fourth, afterwards King, again confirmed, but it was made subordinate to THE ORDER of ST. HUBERT,—so that it holds at present the *Second* Rank.

His present Majesty, King LEWIS, has had its former STATUTES revised, and, with some alterations, has sanctioned them anew,—they are dated, at *Munich*, the 25th of February, 1827.

The King is Grand Master,—The Crown Prince, Grand Prior,—and the Prince immediately after him, Second Grand Prior.

THE ORDER is divided into Two Classes or Branches, The *German* and *Foreign*.

THE MEMBERS of THE ORDER are divided into three Classes,—*viz.*, GRAND COMMANDERS, COMMANDERS, and KNIGHTS. The number of the first, as Capitularies, is fixed at *Six*,—those of the second, as

Capitularies, at *Twelve*,—the Grand Master being at the same time allowed to nominate some of the Knights, whose number is not fixed, as Honorary Commanders or Grand Commanders.

Besides these, The Order has, since the year 1741, admitted as eligible to Knighthood, some of *The Clergy*, consisting of one Bishop, one Provost, four Deans, and Chaplains of The Order (with the distinction of Domestick Papal Prelates), agreeably to a Bull of Pope Benedict the Fourteenth, of the 6th of October, 1741, and of Pope Pius, of the 30th of April, 1782.

The Order elects as Members, only such as profess *The Roman Catholick faith*, and who can produce the strictest proofs of their *Noble* descent.

The Celebration of The Order takes place regularly twice a year,—*viz.*, on the 24th of April, and the 8th of December, on which occasion the Knights appear in the dress of The Order. This Dress consists of a Collet of white satin, richly embroidered with silver, a Scarf of the same, breeches of the same material with bows at the knees, and tassels hanging from the Sword,—a light blue Mantle, differing in length according to the degree of rank in The Order, richly embroidered with silver, and lined with white,—a Ruff round the neck, a Hat *à la Henri*, with white plumes,—the whole after the antient Costume of The Burgundian Knights.

Besides this Dress, since the year 1802, another is worn, consisting of a modern scarlet Uniform with white cuffs, and breeches, and silver Epaulettes.

The Cross of The Order is of gold, having eight points, and the Bavarian rhomboids in the corners, upon the points of which, as well as upon those of the Cross, there appear gold studs. The front is blue enamel, with white edges. In the centre, on a field of gold, the Virgin Mary is represented upon a semi-lune, standing in clouds,—her head is surrounded by five Stars,—beneath the Moon is a Serpent, lying at length, the head of which she is treading under foot. On the four rhomboids are the following letters, in gold, V. I. B. I., *Virgini Immaculatæ Bavaria Immaculata.*"

On the reverse it is red, with a white enamelled border, the centre of which is surrounded by a laurel crown, containing upon a gold ground the Knight of *St. George* on horseback, killing the Dragon with his sword. On the rhomboids are the following letters, in gold, the Motto of The Order, J. V. P. F., "*Justus ut Palma florebit.*" Over the Cross there appears, instead of a crown, a Lion's head, which is fixed to it.

The gold Chain of The Order consists of three different links,—the first forms an oblong square, being adorned on it's extended sides with gold flames, and on the shorter sides, with regal crowns. On the square itself are the following words, "*In fide, justitia, et fortitudine.*" The second link consists of two annexed Bavarian rhomboids, of blue and white enamel,—the third, of two Lions standing opposite to each other, on a ground of gold, bearing in one fore-paw a torch, in the other a naked sword, and encompassing a white enamelled column crowned with the Imperial Globe.

Besides the gold Cross of The Order, the Members wear, according to the class to which they belong, on the left breast, a blue *embroidered* Star of eight points, with silver binding and rhomboids in the corners, in the centre of which is a red cross, on a silver shield. This Star may also consist of *metal.*

The riband of The Order is of light blue, having a narrow white and dark blue edging. It is worn across the right shoulder to the left.

The Commanders of the Grand Cross are permitted to surround their Family Arms with the Chain of The Order.

The Shield of The Order, upon which the Candidates have to place their right hand, when they give their word of Honour to observe the Statutes of The Order, is that of Duke Christopher, after whose death, at *The Isle of Rhodes,* it was brought to *Bavaria.* It is of blue Steel, and decorated on the outside, which is in the form of a *heart,* with antient workmanship, in gold,—upon the inside, is the embroidered Cross of The Order.

Upon the Shield, which every Candidate wears the day he is admitted, is seen, upon a field of gold, the Cross of The Order, with the Motto, "*In fide, justitia, et fortitudine.*" On the upper part are represented his Coat of Arms, on the lower part the Candidate's Christian and Family name, as well as the year and day of his admission.

The Members of The Order receive the Chain, the Grand, Intermediate, and Small Cross from The Grand Master. These Crosses revert to The Order after the death or resignation of a Member, and they must be sent within two months of the time to the Office of The Chancellor of The Order, where a receipt will be given. In cases where a Knight is promoted to a Commander, or a Commander to a Grand Cross, he is under the same obligation to return the Cross, worn previous to his promotion.

Every Candidate for The Order is obliged, before his admission, to prove by written documents his own Extraction, as well as that of all the Ancestors of his Pedigree, five Generations back, to be *German*, and of *Chivalrick Nobility*.

THE ROYAL MILITARY ORDER OF MAXIMILIAN JOSEPH.

The Elector CHARLES THEODORE had already founded, on the 8th of June, 1797, a *Military Decoration* for Officers,—

When, on the 1st of March, 1806, His Majesty MAXIMILIAN JOSEPH the First graciously ordained, that, from the 1st of January of the same year, being the day whereon The Royal Dignity was re-instated in BAVARIA, this *Military Decoration* should be raised to the Rank of a ROYAL ORDER,—the Preamble to THE STATUTES declaring,—

That the attachment, fidelity, and valour shown by the whole Army during the late War, then brought to a close in so glorious a manner for the Country, had induced His Majesty The King to found a permanent monument of His high satisfaction, by changing the hitherto *Honorary Military Decoration* into an ORDER of MERIT,—and to endow it with a secure FUND, that the great merits of Officers should not merely be distinguished by a Publick Decoration, but that they should also ensure to them an increase of Pay.

His Majesty hoped, that the object which he had in view might not be mistaken,—but, that this *newly created* ORDER might be looked upon as a peculiar mark of Distinction for those, who, by distinguished actions, had acquired the most just claims to the favour of their King and the gratitude of their Country.

In consideration of these circumstances, His Majesty expected that it would be accepted and received with that respect, which He himself bestowed upon services exerted in the cause of the Country, and for which the heart of every true BAVARIAN so deeply felt.

It was, moreover, His Majesty's pleasure, that THIS ORDER should be considered as founded, commencing with the first of January, 1806, as the day on which *The Title of* KING was assumed,—and that the Generals rewarded with The Grand Cross, should take rank before The Aulic Privy Counsellors, such being also Chamberlains.

THE KING is Grand Master.

The assembled CHAPTER of THE ORDER examine, according to the Statutes, the testimonials which are to prove such Military exploits as have been executed with skill, presence of mind, and valour, from a free impulse and at the peril of life, for the benefit and glory of The Royal Service,—and lay before His Majesty, for his decision, the result of their votes, stating the motives of their resolution. The Rank is fixed from the day, on which such Officers distinguished themselves.

THE ORDER is divided into three Classes,—*viz.*, GRAND CROSSES, COMMANDERS, and KNIGHTS.

Particular Preferments and Privileges are attached to THIS ORDER.

Every Officer of whatever rank or branch of the Army, may be admitted into the same, without regard to Religion, birth, or other circumstances,—but Generals only can obtain the Grand Cross.

Six Commanders of The Grand Cross have an annual income of 1500 florins,—*eight* of the Commanders have 500 florins,—and *fifty* of the Knights, 300 florins.

TWELVE CHILDREN of Members have the prospect of receiving the contributions instituted for their support.

The granting of THIS ORDER to *Natives* confers at the same time the Rank of *Nobility* upon those, who did not already belong to the Rank of Nobility. This Rank of Nobility, however, is confined only to the individual who is thus favoured. Any Member of THE ORDER, whose Father and Grand-father had acquired this Distinction for Merit, can lay claim to the grant of *Hereditary Nobility*, free of expense.

THE DECORATION of THE ORDER is a gold Cross, enamelled white, and a gold crown,—in the centre, on a round shield of blue enamel, are the initials, in gold, of His Majesty, King MAXIMILIAN, as the Founder of The Order, on one side,—and, on the other, the MOTTO, "*Virtuti pro Patria.*" It is of different sizes, according to the Rank in The Order,—and is worn, pendent to a black riband, having a small white and blue stripe on each side, by the Knights, on the left breast,—by the Commanders, round the neck,—and by the Grand Crosses, on the left breast, the riband passing over the right shoulder to the left hip.

The Grand Crosses also wear a STAR, embroidered upon the left breast, surrounded with rays of silver, and with the MOTTO "*Virtuti pro Patria.*" They wear likewise the smaller Orders of their Rank, round the neck.

COMMANDERS.

Lieutenant-General Sir JOHN LAMBERT,—*See*, RUSSIA, ORDER of ST. WLADIMIR, of The Third Class.

The Right Hon. Sir GEORGE MURRAY,—*See*, AUSTRIA, ORDER of LEOPOLD.

KNIGHTS.

Sir COLIN CAMPBELL,—*See*, AUSTRIA, ORDER of MARIA THERESA.

LORD FITZROY SOMERSET,—*See*, AUSTRIA, ORDER of MARIA THERESA.

The Hon. ARTHUR PERCY UPTON, Colonel in The Army, a Brigadier-General on The Continent, and C. B., "in testimony of the approbation of His Majesty The King of BAVARIA of the distinguished services rendered by him in the Field, while attached to the Bavarian Army in the last Campaign, particularly in the Action at the passage of *The Saare*, on the 23d of June, 1815."—Royal Licence, 16th of October, 1816.

A Dispatch from Field-Marshal the Duke of WELLINGTON, dated, *Head Quarters, Paris!* the 24th of September, 1815, contains the following List of Officers, upon whom His Majesty THE KING of BAVARIA had conferred Decorations of different Classes of *The Order of Maximilian Joseph,*" in testimony of His Majesty's approbation of their services and conduct, particularly in the late Battles fought in *The Netherlands,*—

COMMANDERS.

Major General Sir Colin Halkett, K. C. B.
Major General Sir John Lambert, K. C. B.—*Gazetted.*
Major General Sir James Lyon, K. C. B.
Major General Sir John Vandeleur, K. C. B.

KNIGHTS.

Colonel Sir Colin Campbell, K. C. B.—*Gazetted.*
Colonel F. B. Hervey.
Colonel Lord Fitzroy Somerset, K. C. B.—*Gazetted.*
Lieut. Colonel Sir Noel Hill, K. C. B.
Lieut. Colonel Norcott.
Lieut. Colonel Fremantle.
Major Hon. G. Dawson.

The Order of Merit of the Bavarian Crown.

As it was the intention of His Majesty, King Maximilian Joseph, to institute an honourable Memorial to distinguished Merit in the Army, by the foundation of The Military Order of Merit,—in the same manner did his Majesty wish to afford an honourable distinction to deserving Servants of the State, and to eminent Virtues and Merits of Citizens of all Classes, by commanding, on the 19th of May, 1808, the foundation of this Order.

Admission into The Order is open to every one, who has rendered good service to the State, has distinguished himself by superior Civic virtues, or has deserved the gratitude of his Country by advancing its interests and glory.

The Statutes of that period appoint The Order to consist of four Classes,—viz., 12 Grand Crosses,—24 Commanders,—100 Knights, and an indefinite number of those, who may obtain the Gold or Silver Medal of Civil Merit.

The revised Statutes, of the 8th of October, 1817, limit the number of Grand Crosses to 24, exclusive of those who are Knights of The Order of St. Hubert,—of Commanders to 40,—and of the Knights to 160.

The granting of the Cross of Knighthood to *Natives*, who previously possess the rank of Collegiate or Spiritual Counsellor as servants of the Court, the State or the Church, includes the rank of *Nobility*, for those who did not previously belong to that Class. The rank of *Nobility* thus bestowed, is limited, however, to the persons thus favoured.

Any Member of THE ORDER, whose Father and Grand-father had likewise acquired this distinction for Merit, may lay claim to the bestowing of *Hereditary Nobility*, free of expense.

The DECORATION of THE ORDER is an octangular Cross, enamelled white, surrounded by a wreath of oak leaves, and surmounted with the Royal crown,—in the centre of the Cross, are the blue and white rhomboids, together with the golden crown, and the MOTTO, "*Virtus et Honos*,"—on the reverse, is the portrait of The Royal Founder, in gold, with the inscription, "*Max. Joseph. Rex Bojoariæ.*"

It is worn of different sizes, according to the rank of the possessor, by a blue, watered riband, the edges of which are bordered by a small white stripe. By the KNIGHTS, it is worn in the button-hole, —by the COMMANDERS, round the neck.—and by the GRAND CROSSES, over the left shoulder down to the right hip.

THE STAR which is worn on the left breast, consists of a SILVER CROSS,.of eight points, with bright rays,—in the centre whereof is the regal crown, on the blue and white rhomboids, with the MOTTO, "*Virtus et Honos*," on a ground of deep red, which is surrounded with a wreath of oak leaves.

In the middle of THE COLLAR of THE ORDER are the initials of it's august Founder,—the links form alternately, a crown and a gold oval, which are surrounded by green laurel wreaths, and upon them are alternately the letters, V. E. H., (*Virtus et Honos.*)

Every class shall be permitted to surround THE ARMS of his FAMILY with the *Insignia* of THE ORDER.

THE ORDER of Merit of THE PALATINE LION, founded by the Elector Palatine CHARLES THEODORE, in 1768,—*ceases* by the present foundation of THE NEW ORDER of MERIT, and will not, therefore, be granted in future,—but those who already possess it, both *Natives* and *Foreigners*, may continue to wear it under it's name.

The Council assembles once a year, on the 27th of May, to propose to His Majesty those who are considered worthy of the different classes of THE ORDER,—To this Council are likewise to be made, all applications for the MEDAL of MERIT.

A Fund was established by His Majesty MAXIMILIAN

JOSEPH, to be applied in yearly pensions for 20 Children of Members of the first three Classes,—which pious legacy was graciously confirmed by His Majesty, King LEWIS,—and, finally, by a Resolution, of the 12th of October, 1835, their number was increased from 30 to 38.

This Fund was created by the property, both in Land and Capital of the former ORDER of ST. RUPERT, for the Principality of *Salzburg*, which had reverted to His Majesty's disposal, *at the Dissolution of* THAT ORDER, —and was thus generously applied, with a view to commemorate by a new benefit, a prosperous reign of *Twenty-five* years.

The Order of Merit of Saint Michael.

The former Order of The Knights of Saint Michael was founded, on the 29th of September, 1693, by The Elector of Cologne, Joseph Clement, as Duke of Bavaria, when present at *Munich*,—being intended as an Order of Ancestry, for Members of *The Roman Catholick Faith*, and it was retained during the reign of His Majesty, King Maximilian Joseph, under The Grandmastership of His Royal Highness Duke William in Bavaria. The original intention of it was the maintenance of Religion and the Defence of the Honour of God,—subsequently, another object of it was, the Support of The Defenders of the Country.

After the death of the last mentioned Grand Master of this Order, it's Statutes were abolished on the 16th of February, 1837,—new Statutes were made, relating to the granting of it for the future,—and it was, by his present Majesty King Lewis, created into an Order of Merit, to be obtained without distinction of rank, birth, or religion.

The number of Members is fixed for *Natives* to 24 Commanders of The Grand Cross,—40 Commanders,—and 300 Knights,—but for *Foreigners*, it is unlimited.

The Decoration of all the Classes consists for the future of a gold Cross of azure blue enamel, having eight broad points, and on the upper part a regal crown,—upon the four wings of it, which are set in gold, are the gold letters, P. F. F. P., implying "*Principi Fidelis favere Patriæ.*"

In the centre of the Decorations of The Grand Crosses and Commanders is represented, in gold relief, *St. Michael* in martial attire, surrounded by refulgent rays. On his shield is the MOTTO, "*Quis ut Deus,*"—on the reverse, in the centre, upon a gold ground, is the word "*Virtuti,*" in blue enamel.

The Crosses of The Knights contain, instead of the image of *St. Michael*, the Motto, "*Quis ut Deus,*" on a gold ground in azure blue enamel,—and, on the reverse, in azure blue enamel of gold, is the word "*Virtuti.*"

Two-thirds of the width of the riband by which the Decoration is worn, are dark blue, one-third rose-colour, and the latter colour equally extends to both the extreme sides.

The Commanders of The Grand Cross wear the Decoration of the first class by the same kind of riband, of a hand's breadth, from the right shoulder down to the left side,—they also wear, on the left breast, a gold embroidered STAR with silver rays, upon which is again the Cross of The Order, with the Motto, "*Quis ut Deus.*"

The Commanders wear a smaller Decoration by a smaller riband, round the neck, hanging down upon the breast, without the embroidered Star.

The Knights wear a still smaller Cross, differing somewhat in form, at the same riband, only still narrower, and fastened to the coat.

The Knights of THE ORDER *of St. Hubert*, upon whom is conferred The Grand Cross of *St. Michael*, indicate this in the same manner as they do *The Civil Order of Merit*, by wearing only the Cross of Knighthood.

The Commanders of The Grand Cross of THE CIVIL ORDER of THE BAVARIAN CROWN, who, at the same time, as Commanders of The Grand Cross of *St. Michael*, wear both Stars together.

THIS ORDER does not confer the rank of *Nobility*.

Taxes, contributions and other payments of THE ORDER are no longer to be paid, either by the former or future Members of The Order.

No person is allowed to apply for THE ORDER of ST. MICHAEL by petition.

THE STATUTES which are comprised in thirteen articles, are dated at *Munich*, on the 16th of February, 1837.

KNIGHT.

AUGUSTUS BOZZI GRANVILLE, M. D., Knight of The Order of *The Crown* of WÜRTEMBERG, and of The Royal Order of *St. Michael* of BAVARIA,—" His Majesty THE KING of BAVARIA having been moved to such a gracious act of favour, in consequence of having greatly approved of his work, on "*The Spas of Germany,*" &c.

The Royal Order of Lewis.

This Order was founded, on the 25th of August, 1827, by His present Majesty King Lewis, for rewarding those Servants who had completed the *Fiftieth* year of their Service, after the Birth-day of His Majesty in that year.

According to the Statutes, which are dated at *Bruckenau*, on the 25th of August, 1827, it is ordained,—

That every Servant who has been during *Fifty years* a Servant of The Court, the State, the Army, or the Church, and has given satisfaction to His Majesty, may be admitted into This Order. To complete this number, those years of Service may be reckoned that have been rendered to Provinces, not formerly belonging to, but which have afterwards been incorporated with the Kingdom. Every year passed in a Campaign counts double,—whereas, on the other hand, the time during which any one may have been pensioned off, is not reckoned at all.

The Decorations of The Order consist,—For the Officers or those holding the rank of Officer in the Army, as also for Servants of The Court and State, and for Servants of the Church, who are of the rank of Counsellor, of a gold Cross, with the Royal crown, having the portrait of The Founder in gold, on a ground of white enamel, and, on it's four corners, the inscription, "*Lewis, King of Bavaria,*"—on the reverse, is a green enamelled wreath of oak leaves, which surrounds the inscription of gold, on a white ground, "*For Fifty years of Honourable Service.*" The four corners of the reversed side, indicate the day of it's foundation by the words, "*On the 25th of August,* 1827,"

For Members of inferior rank, The Decoration consists of a gold MEDAL, having on both sides the same inscriptions, as the Cross.

THE CROSS, as well as THE MEDAL, are both of them worn by a crimson and light blue riband, which is only wider with the former than the latter, and fastened in the button-hole.

THE BADGES of THE ORDER, after the decease of a Member, are to be returned to The Grand Chancellor of The Civil Order for Merit of The Bavarian Crown.

The Order of Saint Elizabeth.

This Order was instituted, at *Mannheim*, on the 18th of October, 1766, by Her Most Serene Electoral Highness Elizabeth-Augusta, Countess Palatine of The Rhine, *&c.*,—who, from a motive of particular devotion towards *Saint Elizabeth*, her Patroness, had been induced to establish an Order of Ladies, in honour of that Saint,—the principal object of which should be, *to afford help and assistance to the needy*,—and having obtained the consent of Her Consort, The Elector, to that effect, she had, therefore, established the said Order, and had thought proper to ordain certain Statutes for its better maintenance and government,—among which are,—

That this Order shall be conferred only upon such Ladies, *as profess The Catholick Faith*, and who are in a state of exhibiting and proving Sixteen Quarters of uninterrupted and unsullied *Nobility*, as well on behalf of themselves, as on that of their husbands, in case they are or have been married.

With an exception, so far as relates to Princesses of The Electoral House, and those of other antient and Princely Families, this Order is limited to the great Mistress of the Household,—to the Ladies of the Court, especially attached to the service of Her Highness's Person,—and to *Six* Ladies, who are married, or in a state of widowhood.

No Lady whatsoever, beyond this number, shall be received.

The distinctive Badge of this Order is a Cross, *pattée*, of white enamel. On the principal side, upon a field, *azure*, is a representation of *Saint Elizabeth* bestowing Alms, the figures enamelled, proper. The reverse is distinguished by Her Highness's cypher, in gold, upon a field, *argent*, and the same is surrounded with three circles,—that in the middle, is green, and the two lateral circles and the contours of the Cross are of burnished gold. The whole is surmounted with an Electoral crown.

The Ladies shall wear this Cross attached to their left breast, and it shall be suspended from a blue riband, with a small border of red.

No Lady shall appear in publick without this Cross,—if she does, she shall be fined *one Ducat* for every such omission.

Upon the demise of a Lady, THE CROSS must be transmitted to The Treasurer.

THE ORDER OF THERESA.

THIS ORDER was founded by Her Majesty the reigning Queen THERESA, of Bavaria, on the 12th of December, 1827, and confirmed by His Majesty The King.

The object of this Institution is, to afford to a certain number of unmarried Daughters of Noble Birth, together with The Badge of Honour, a certain sum *per annum* to improve their own Income.

The Nomination is made by The Queen, as GRAND MISTRESS of THE ORDER, with the sanction of The King.

THE BADGE of THE ORDER consists of a gold, light blue enamelled Cross, having at the top the Royal Crown,—in the centre of which, on the obverse, is the initial T., in gold, upon a white ground, surrounded with a wreath of *Rue,*—and, on the reverse, is the year 1827, the date of the foundation of The Order, surrounded with the words, "*Let our Life on Earth be Faith in Eternity,*" also in gold, upon a white ground.

The Order of St. Anna, at Munich.

This Order was founded by the Widow of The Elector Maximilian the Third, Maria Anna Sophia, in the year 1784, for the benefit of Ten Noble Ladies, who had attained the 15th year of their age, and able to prove their Descent from *sixteen Noble* Ancestors,—under the care and superintendence of an Abbess, to dwell and board in The Institution.

This Institution, however, underwent a complete change on the Succession of Maximilian Joseph the Fourth to the Throne, — the Building itself was abolished, but the Income was left to remain for each Lady, married or single, during life.

The office of Abbess was to be holden, either by the reigning Female Sovereign or a Princess of The Royal House.

The usual Dress of The Ladies is *black*.

The Badge of The Order consists of a gold, white enamelled Cross, rounded off at the extreme points, bordered with blue, and having gold rings in the corners of the points,—on the obverse, upon a white ground, in gold, is the figure of *The Virgin Mary*, and, on the points of the Cross, the Motto, "*Sub tuum præsidium*,"—upon the reverse, also on a white ground, in gold, is the figure of *St. Benno*, as the Patron Saint of Bavaria, and, on the points, the inscription, "*Patronus Noster*."

The Cross is worn by a light blue riband, bordered with white.

His late Majesty, MAXIMILIAN JOSEPH, was graciously pleased to order, on the 6th of May, 1809, that the daughters of all such officers as had fallen in battle, or had shown distinguished proofs of bravery and of skill, should have *a third share* of the Income arising from this, as well as all other Ladies' Institutions.

The Order of St. Anna, at Würzburg.

The Order of Saint Anna, at *Würzburg*, was founded by a Countess, Anna Maria, of *Dernbach*, who, by will, dated in 1683, devoted the whole of her fortune to the erection of this Institution for Noble Ladies of *The Franconian Nobility*,—and which was acted upon and completed in the year 1714, by the Bishop, John Philip.

It consists, besides The Abbess and Ladies of Honour, of 12 Members of the first class, and 20 of the second class,—whose appointment is sanctioned and directed by The King.

The Badge of The Order consists of a gold, white enamelled Cross, rounded off at the points, and bordered with red,—on the obverse, on a white ground, is the figure of *St. Anna*, in gold, and, on the points of the Cross, the Motto "*To her Noble daughters*,"—and, on the reverse, the Family Coat of Arms of The Founder.

The Cross is worn by a red riband, bordered with silver. The usual Dress of The Ladies of The Institution is *black*.

MEDALS.

THE GOLD and SILVER MEDALS for CIVIL MERIT.—This was added to the Under-division, or Fourth Class of The Institution of The Order for Civil Merit of The Bavarian Crown. Persons in office, who have rendered themselves worthy of THIS ORDER but do not possess the rank of real Collegial Counsellors, receive the GOLD MEDAL for CIVIL MERIT.

The *gold*, as well as the *silver* MEDAL for Civil Merit, contains on the obverse, the bust of The Founder, with the inscription, "*Max. Jos., King of Bavaria,*"— and, on the reverse, the inscription, surrounded with a laurel wreath "*To Merit for Prince and Country.*"

It is worn at the button-hole of the left side of the breast by a riband an inch wide, consisting of two white and three light blue narrow stripes.

THE GOLD MEDAL of HONOUR of THE ORDER of LEWIS,—This forms an Under-division of THE ORDER of LEWIS, founded by His Majesty the reigning Monarch, King LEWIS, on the 25th of August, 1827, —which Distinction can be attained by every Servant who has passed *Fifty years* in the service of The Court, State, Army, or Church, with complete satisfaction,—but who, at the same time, cannot lay claim to the rank of Officer in the *Military* or Counsellor in The *Civil* service.

The Medal contains on the obverse, the bust of The Founder, with the inscription, "*Lewis, King of Bavaria,*"—and, on the reverse, encircled by a laurel wreath, the words, "*For* 50 *years of Honourable Service.*"

It is worn at the button-hole, on the left side, by a crimson and light blue riband.

THE GOLD and SILVER MILITARY MEDALS for NON-COMMISSIONED OFFICERS and PRIVATES,—These Medals contain on one side, the bust of King Maximilian Joseph the First, of Bavaria,—and, on the other, the Royal Arms, supported by an upright Lion, armed with a Sword, and the Motto, "*To Courage.*"

This Medal is worn at the button-hole, on the left breast, by a black, white, and blue riband.

The Medals for The Medical Staff of The Army,—These gold and silver Medals were founded, on the 12th of November, 1812, by King Maximilian, in reward for distinguished services by The Medical Staff of The Army.

They bear on the obverse, the bust of the King, with the Motto, "*Maximilianus Josephus Rex Bojoariæ,*"—and, on the reverse, between a wreath of laurel and oak leaves, the inscription "*Ob milites inter prœlia et arte et virtute servatos.*"

They are worn at the button-hole, on the left side of the breast.

THE KINGDOM OF BELGIUM.

The Order of Leopold.

By an Ordinance, of the 11th of July, 1832, it is decreed,—

1. That a National Order is created, which is intended to recompense services rendered to The Country. It is called, "The Order of Leopold."

2. The King is Grand Master of The Order.

3. The Order is divided into Four Classes, *viz.*, Grand Crosses, Commanders, Officers, and Knights.

4. Nominations to The Order belong to The King.

5. Every Member of the Chambers who accepts The Order for other than Military motives, shall be subjected to a re-election.

6. The Motto of The Order is "*Union constitutes Power.*"

7. Every Military person who is a Member of This Order, and of a lower Rank than that of an Officer, enjoys a pension of 100 francs, which ceases on his being promoted to the Rank of an Officer in the Army.

8. The quality of Member of The Order, and the Pension attached to it, is lost or suspended in the same manner as, by the penal laws, the rights of Belgic Citizenship are lost or suspended.

9. By an Ordinance, of the 8th of November, 1832, the administration of THIS ORDER is united with the Department for Foreign Affairs.

By an Ordinance, of the 3rd of August, 1832, it is decreed,—

1. That THE DECORATION of THE ORDER shall consist of a white enamelled Cross, having a garland of Laurel and Oak between each of its four points,—on one side, in the centre, a black shield enamelled, surrounded by a red circle between two small circles of gold, with the cypher of The King composed of two L L's and two R R's,—on the reverse, the Arms of the Kingdom, with the MOTTO prescribed by the Law, "*Union constitutes Power,*" in the Exergue, in letters of gold, the whole surmounted by a Royal crown.

2. The Riband shall be scarlet, watered.

3. The marks of Distinction shall be, for GRAND CROSSES,—the Decoration of The Order, of that side bearing The Motto, "*Union constitutes Power,*" upon a Silver STAR, embroidered upon the left breast of the coat. The Star shall be of the diameter of 10 centimètres 4 millimètres. Grand Crosses at the same time wear THE JEWEL of THE ORDER, suspended from a riband 7 centimètres and a half wide, passed as a scarf over the right shoulder towards the left side.

For COMMANDERS,—The Decoration of The Order, as above,—embroidered upon the coat,—but the Star of the diameter of 7 centimètres 2 millimètres, and the Jewel attached to a riband of the width of 5 centimètres, and worn round the neck.

For OFFICERS,—The Decoration of The Order, suspended from a Riband 4 centimètres wide, surmounted by a Rosette and passed through the button-hole.

For KNIGHTS,—The Decoration of The Order, suspended as above, but without any Rosette to the riband, which shall be of the width of 3 centimètres.

THE DECORATION is of *Gold,* for the three first Classes,—and, of *Silver,* for the Knights.

4. Grand Crosses, moreover, wear on days of Ceremony, THE GRAND COLLAR of THE ORDER, which is of gold, and divided into three parts, alternating,—i. e., the Crown, the Lion, and the two letters L. and R. doubled.

5. The mark of Distinction of The Order worn by the Military, consists of two Swords placed in support of the Crown in the Jewel of The Order.

Grand Crosses and Commanders shall bear on the Star, the Swords in gold, crossed, under the shield.

7. Members of THIS ORDER to take rank after the constituted Authorities on all Publick occasions.

8. Commanders, Officers, and Knights to be saluted with the *port d'armes*,—and arms to be presented for the Grand Crosses.

GRAND CROSSES.

The Right Honble. Sir ROBERT ADAIR, G. C. B.

The Right Honble. Lord PONSONBY, Ambassador Extraordinary and Plenipotentiary at Constantinople.

COMMANDERS.

Lord WILLIAM RUSSELL.

OFFICERS.

Colonel CARADOC.

Captain Sir JOHN ROSS.

Sir GEORGE BAILLIE HAMILTON.

Colonel MAXWELL WALLACE.

KNIGHTS.

JOHN MARTIN, Esq., the distinguished Painter.

JAMES CLARK, M. D.

Sir EDWARD THOMASON.

GEORGE STEPHENSON, Esq., C. E.

ROBERT VICTOR SWAINE, Esq., " in reward of his Services, as CONSUL for the Kingdom of BELGIUM, at *Hamburg*,"—the Diploma being dated, the 26th of June, 1838.

The Iron Cross.

This mark of Distinction was instituted by Law, dated the 8th of October, 1833, for those Citizens, who, from the 25th of April, 1830, to the 4th of February, 1831, were wounded or gave any striking mark of courage in the various combats for National Independence, or had rendered any great service to their Country. By the same Law, this Cross is declared to be conferred in the name of the People of Belgium on the Members of The Provisional Government.

The Ordinance, of the 25th of October, 1833, names certain persons as a Commission to examine and present claims to this Distinction, *&c.*

By an Ordinance, of the 30th of December, 1833, it is declared, that this Mark of Distinction shall consist in,—

1. A Cross of Iron, of four points,—the shield to bear the Belgic Lion, in gold, surrounded by a circle,—and, on the reverse, "1830."
2. A Medal of Iron, bearing on one side the Belgic Lion,—in the Exergue, "*To the Defenders of their Country*,"—and, on the other, nine shields of the arms of each of the nine Provinces of the Kingdom, —in the centre of these shields, a Sun, and the date 1830, with the words, "*Independence of Belgium*," in the Exergue.

The Cross and Medal to be worn upon the left breast,—the Cross to be suspended from a watered riband of the width of 3 centimètres 2 millimètres of a red ground, having on each side a yellow and black border, each of the width of 3 millimètres.

The Medal shall be suspended from a riband of the same width as that of the Cross, of a black ground, with yellow and red borders, each of the width of 3 millimètres.

The Military honour of *port d' armes* shall be rendered to persons decorated with the Cross or Medal.

By an Ordinance of the 21st of February, 1835, The Iron Cross *of the Second Class* was suppressed, and persons then possessing The Cross of the Second Class, were authorised to wear The Order *of the First Class.*

The Commission, mentioned above, was dissolved subsequently to the month of April, 1835.

The following are the Names of the Members of The Provisional Government, on whom this Decoration was conferred,—

Le Comte Félix de Merode,
Le Baron d'Hooghoorst,
Messieurs Gendebien, Sylvain Van de Weyer, de Gotter, Jolly, de Cossin, Van der Linden.

This Cross has not been granted to any Englishman.

THE DUCHY OF BRUNSWICK.

The Ducal Order of Brunswick of Henry the Lion.

This Order was founded by William, Duke of Brunswick and Lunebourg, as a reward for those who had distinguished themselves in his service,—who had acquired a reputation for high Civic or Military talents, —or, who had shone in the Arts and Sciences.

On the 25th of April, 1834, The Duke published the following Statutes,—

1. The Order to be considered as founded, from the 25th of April, 1834.
2. To be named "The Ducal Order of Brunswick of Henry the Lion."
3. The reigning Duke to be always Grand Master.
4. The oldest Member of The Ministry to be Chancellor,—And the affairs of The Order to be transacted by The Ministry.
5. The Order to comprise four Classes,—*viz.*, Grand Crosses, Commanders of the first and second Classes, and Knights.

6. Every Subject of Brunswick to be eligible,—but no one to be admitted to a higher class, until he shall have been a Knight.

7. The Order gives precedence before persons of the same Rank, not decorated.

8. The Decoration of The Order shall consist of an octagonal Cross, of clear blue enamel, with a red escutcheon in the centre. The points of the Cross surmounted by small knobs of gold. The obverse of the Cross represents the Crest of the escutcheon of the Arms of Brunswick with it's Insignia, so that the Casque appears upon the lower point, the column with the Leaping-Horse and the two Sickles upon the centre escutcheon, the Peacock's feathers on the points of the right and left, and the tail of the Peacock with the Star upon the upper point of the Cross. A golden Lion walking upon two branches of laurel, masked, by the Ducal crown surmounts the whole. In the space between the branches the Ducal cypher shall be placed, in gold, likewise surmounted by a crown. The red escutcheon on the reverse of the Cross, bears the Motto of The Order, "*Immota fides,*" in letters of gold, and the cypher MDCCCXXXIV., the date of the foundation, is contained in the circle of gold surrounding the above mentioned Motto.

This Decoration shall be of three sizes,—Grand Crosses shall wear it attached to a red riband of a hand's breadth, with yellow borders, passed as a scarf from left to right.—Commanders of the first and second class, shall wear a similar riband of 2½ inches wide, passed round the neck,—And Knights, a like riband of two fingers width, passed through the button-hole.

Grand Crosses shall also wear on the left breast, an octagonal Cross, the silver rays of which shall contain the above named Cross of gold, in clear blue enamel. In the centre, on a silver ground, shall be the Ducal cypher in gold, surmounted by a crown, and surrounded by a red riband bearing the Motto of The Order in letters of gold.

Commanders of the first class shall also wear on the left breast a Silver Cross, of eight points,—in the centre of which shall be the Motto of The Order in letters of gold, on a red ground, surrounded by a riband of gold bearing the date MDCCCXXXIV. Between the branches of the Cross, the Ducal cypher, in gold, shall be placed, surmounted by the Ducal crown.

On days of great Solemnity the Grand Crosses shall wear the *Insignia* of The Order suspended from the neck by a Chain of gold. The links of this Chain to be composed alternately of three different pieces,—1. An Escutcheon of the Arms of Brunswick and Lunebourg surrounded by banners,—2. Two Lions, of gold, facing the Escutcheon, —3. The Escutcheon of the Cross of The Grand Crosses,

9. Nominations to all the Classes of The Order shall be by *Patent,* signed by The Duke. The Names of the Members and their Arms shall be registered in a book appropriated to this object.

10. Relates to the dismissal of Members for improper conduct.

11. Relates to the return of *The Insignia* to the Chancery of The Order, on the death of the Possessors.

12. Members may add *The Insignia* to their Armorial Bearings.

13. By this Article it is declared. that a *Cross of Merit* shall be founded, in connection with this Order.

14. This Cross shall comprise two Classes, to be called "*Possessors of The Cross of Merit of the first and second Class.*"

15. The Decoration shall be worn, attached to the riband of a Knight of The Order of Henry The Lion. For the first class it shall consist of a Cross, of *gold,* the angles ornamented with a crown of green oak. On the Escutcheon, the Ducal cypher and crown, and, on the four sides, the words, "*Immota fides.*"

The same Decoration, in *silver,* but without the oaken crown, is appropriated to the second class.

16. The Possessors of The Cross of Merit shall cease to wear it, when named Knights of The Order.

17. Relates to the mode of distribution of The Cross.

18. Article 7. to be applied to this Cross.

19. The riband of The Cross shall not be worn *without* The Cross.

Crosses of Distinction for Military Service.

William, Duke of Brunswick and Luneborg being desirous of conferring a Mark of his satisfaction upon his Officers, issued the following Regulations, on the 1st of April, 1833,—

1. We confer on all Officers, who have served honourably for 25 years and upwards, a Mark of Honour, as a Publick recompense for their services,—

2. His Decoration shall be worn at the left breast, attached to a riband of Royal blue, with a yellow border,—and shall consist of a golden Cross, with purple points,—the Escutcheon bearing on the obverse, Our cypher surmounted by a Crown,—and, on the reverse, the number 25,—

3. The calculation of the years of service shall be made by the rules that govern the regulation of the pensions of Officers, but years of war shall not be reckoned as double,—

4. This Decoration shall only be conferred upon Officers and Functionaries of the Military Administration of the same Rank, who, at the annual distribution, shall be in active service, either in the Corps of the Line, the Fortresses, or the Military Administration,—

5. The Possessor of this Decoration shall lose the right to wear it, if dismissed from the service by a Council of War, or otherwise more severely punished,—or if, after quitting the service, he shall be punished for any disgraceful offence,—

6. Decorations thus forfeited, and those whose possessors shall have died, shall be sent to the Minister at War.

7. The distribution of these Decorations shall take place annually on the 25th of *April*. For this purpose a Commission of Officers shall be

appointed every year, at the beginning of the month of *March*, who shall examine the claims of persons to this CROSS, and submit the definitive List for Our decision on the first week of *April*.

THE DUKE, in order to confer a similar Mark of Distinction on his Under-Officers and Soldiers, issued the following Regulations, on the 1st of April, 1833,—

1. That a Mark of Distinction be conferred on the Under-Officers and Soldiers, who have served with honour, during a certain number of years without interruption,—

2. That this Mark of Distinction comprise three Classes,—and be worn at the left breast, attached to a riband of Royal blue, bordered with yellow,—

That THE DECORATION be,—

For the first Class, a CROSS of *Silver*, with a round Escutcheon, surrounded by rays, bearing on the obverse, the cypher of The Duke, crowned,—and, on the reverse, the number 25.,—

For the second Class, a CROSS of *Silver*, bearing on a round Escutcheon, on the obverse, The Duke's cypher,—and, on the reverse, the number 20.,—

For the third Class, a *silver* BUCKLE, bearing an oval Escutcheon, with The Duke's cypher on the obverse,—and, on the reverse, the number 15.

3. That the first Class be conferred on Under-officers and Soldiers, who have completed 25 years of faithful service,—the second class, on those of 20,—and the third, on those of 15 years of similar service.

4. That the principles adopted in fixing the *Pensions* of Under-officers and Soldiers shall be followed, in calculating the length of service,—but years of War shall be reckoned only as simple years of service.

5. That, after 20 years of service, the possessors of this Decoration shall enjoy a monthly increase of pay,—those of the first Class receiving one crown *per* month after 25 years' service,—and those of the second, half that sum after 20 years' service.

6. That the possessor of this Mark of Honour be at liberty to wear

the Decoration, even when he shall have quitted the Military Service,—but that he shall cease to receive the pay attached to it.

7. That the person who obtains the superior Class shall resign the inferior, and also the superior, if he obtain the Cross of Officers.

8. Persons sentenced to Military punishment, shall lose their rights to claim or wear this mark of Distinction.

9. Decorations so forfeited, and those of deceased persons, to be placed in the hands of The Minister of War.

10. Provides for the distribution of THE ORDER, on the 25th of *April*, as in the preceding Regulations for the Officers.

CROSS of HONOUR FOR THE CAMPAIGN of 1809.—This Cross, intended for the Troops of the late Duke FREDERICK WILLIAM, was founded on the 30th of October, 1824, by Duke CHARLES, and conferred on Officers, Under-Officers, and Soldiers, who had followed the late Duke FREDERICK WILLIAM in *England*, and who, at the above date, were in The Ducal Military or Civil Service of Brunswick. The Decoration consists of a CROSS, of four points,—that, for the Officers, of *gold*,—and, in *bronze*, for the Under-Officers and Soldiers. A crown of laurel traverses the angles of this Cross, which on the obverse, displays the date 1809, with the inscription, "*For Fidelity and Valour*," —on the reverse, the Leaping-horse of Brunswick, with the inscription "*Charles Frederick Augustus William*,"—and is worn on the left breast, attached to a sky blue riband.

MEDAL of HONOUR for THE CAMPAIGNS of PORTUGAL and SPAIN,—in which the Regiment of Hussars, and that of Infantry of the late Duke FREDERICK WILLIAM, have participated, (being in the service of *Great Britain*), from 1810 to 1814.

This Medal was also instituted, on the 30th of October, 1824, by Duke CHARLES, and distributed amongst the Officers, Under-Officers, and Soldiers, who had fought in *The Peninsula* against the French,—and who, at the commencement, were in the Ducal Military or Civil service of Brunswick.

THE MEDAL is of *silver*, for the Officers,—and, of *bronze*, for the Under-Officers and Soldiers. On one side is the inscription "*Penin-*

sula, surrounded by a crown of laurel,—on the other, are trophies, having in the centre the cypher of The Duke CHARLES. It is worn, attached to the left breast by a crimson coloured riband.

MEDAL of WATERLOO,—instituted in the month of June, 1818, by THE PRINCE REGENT of GREAT BRITAIN and HANOVER, in memory of the Campaign of 1815, by Decree, to the following effect,—

Founded in perpetual remembrance of the Campaign of 1815, and of the glorious conduct of The Ducal troops of BRUNSWICK. THE MEDAL *to be cast from the Cannon taken from the enemy,* adorned with the effigy of The Duke FREDERICK WILLIAM, who perished in the field,—to be conferred on all the persons of this Corps, who took part in the battle.—

1. The Medal to be the same for all ranks.
2. To be worn at the third button-hole or close to it, on the left breast, attached to a riband of yellow and blue stripes of a quarter of an ell long,—the riband not to be worn *without* the Medal.
3. and 4. The Christian and Surname of the Possessor to be engraven on the Exergue, and also his rank during the Campaign, particularly, on the 16th and 18th of *June,*—and no other person to wear the Medal.
5. and 6. The Medal to be conferred on all persons, who were in the ranks of The Ducal troops, during the period from the 15th of *June* to the 7th of *July,* 1815.
7. Describes those who shall be considered as belonging to the Corps.
8. The Medal, without the riband, shall be conferred on the heirs of those who perished in the field, or died subsequently.
9. The following persons to be excluded,—
   a. Those who were dismissed the service on account of their conduct, on the 16th and 18th of June, and those who retired to avoid similar punishment.
   b. All who were absent from the ranks, on the 16th 17th and 18th of *June,* without lawful cause,—and

those whose general conduct at this time should be blameable by a Council of War.

c. All those who had deserted subsequently to the 15th of *June*, 1815., and who had not returned under the general amnesty of the 28th of *May*, 1816.

d. All those who had been found guilty of any dishonourable action, since the 15th of *June*, 1815.

10. If any person possessed of this Medal, shall be tried for any offence, the tribunal pronouncing judgment shall likewise declare, whether he be worthy still to bear this mark of Honour.

11. Persons to whom the Medal may not have been sent, from ignorance of their residence, may make their claim through The Commander, who was at the head of their Division in 1815.

THE KINGDOM OF DENMARK.

The Order of the Elephant.

The Order of The Elephant is one of the most Antient, and, after The English Order of The Garter, and that of The Golden Fleece, the most distinguished among The Orders of Europe. It's continuity having been uninterrupted, and it's distribution sparing, and awarded with judgement.

The Foundation of This Order dates from those times, when all orders of men, deeming *The Crusades* the only road to Heaven, enlisted themselves under these sacred Banners, and were impatient to open the way with their sword to *The Holy City*,—at once deserting their Native country, and waging in remote parts of the World a warfare, which, it was conceived, must of necessity be called "*The Holy War*," since it was designed not for *Conquest*, but for the conversion of *Unbelievers*. This devotion, this self-offering in distant regions of the Earth, for a Cause universally considered a just one, formed so essential a part of the necessary qualifications of a brave man, that every one who aspired to Glory and to Honour, found it indispensable to make a Campaign against The Saracens, and to moisten with his blood the arid sands of *Palestine*.

In order to perpetuate the remembrance of these Expeditions, and to form a still stronger union against the common Foe, Societies and Fraternities were established, Emblems for distinction were selected and adopted, as the decorations of Clothes and of Weapons, —and hence arose, in the eleventh, twelfth, and thirteenth Centuries, many a Chivalric Order, that still flourishes with undiminished splendour, and among which must be numbered The Illustrious Order of The Elephant.

The precise time of it's Origin has not been exactly ascertained, — and, even the Historians of Denmark differ in opinion on this point,—some of them maintaining, that Canute the Sixth, King of the Danes, established it at the close of the Twelfth Century, on his return from a successful expedition against *The Saracens*,—others, on the contrary, naming as it's Founder, King Christian the First, who lived about the middle of the fifteenth Century.

The Danish Government in it's Official announcements assumes the beginning of the fifteenth Century, as the date of the Origin of The Order. It was renewed by King Christian the First, in the year 1458, —and, as the duties of The Knights approached those of the Church, and the defence of the Christian Religion was one of their most important obligations, His Majesty caused the new Regulations of THE ORDER to be confirmed by various Bulls of Popes PIUS and SIXTUS the Fourth, in 1462 *and* 1464.

The Knights were then termed, in the Danish language, "*The Society and Brotherhood of The Holy Virgin Mary*,—their Number was fixed at *Fifty*,—and to the COLLAR of THE ORDER was attached the image of *The Virgin Mary*, with the infant Saviour upon her arm.

These Ordinances, however, were entirely changed by THE STATUTES, which King CHRISTIAN the Fifth gave to THE ORDER, on the 1st of December, 1693, the import of which is to the following effect,—

The number of The Knights is *Thirty*, exclusive of the Princes Royal, who are Knights by birth, but who cannot enter into The Order, before they have attained their *Twentieth* year. The same Regulation applies also to the connexions of The Royal Family,—but other individuals must have attained their *Thirtieth* year.

Every Knight who is a Native, must declare himself of *The Evangelical Religion*, and must previously possess the Second Danish ORDER of THE DANNEBROG. If, however, he does not possess the latter ORDER, he receives it eight days previously, but resigns it again, on receiving THE ORDER of THE ELEPHANT,—as no other ORDER can be borne at the same time with it.

The King gives to every Knight the Title of "*Sir*."

THESE STATUTES are considered to be valid at present,—but deviations have been made from them, both in the number of Knights, and the liberty of wearing the two Danish Orders together.

THE BADGE of The Order has undergone various alterations. At present it consists of a white Elephant, on his back a castle, and a conductor upon his neck, all enamelled proper, and on the side of the Elephant, a Cross of *The Dannebrog*, in *diamonds*. It is usually worn, pendent to a rich, broad, light blue, watered riband, which passes scarfwise over the left shoulder to the right hip.

But, on State occasions, it is worn upon the breast, pendent to a Collar of gold, composed of Elephants and Towers alternately, enamelled proper.

The Knights also wear a Star, of eight points, embroidered in *silver*, on the left side of the outer garment.

The Motto of The Order is, "*Magnanimi Pretium.*"

Knight.

Arthur, Duke of Wellington, K. G., 4th of July, 1815.

The Order of the Dannebrog.

Waldemar the Second, King of the Danes, in the commencement of the Thirteenth Century, having waged war against the Pagans of The Baltic, his army in one of the battles was compelled to give way to the enemy, by whom they were attacked, and were on the point of flight,—when, at the prayer of their Bishop for Heavenly aid, a *Red Banner* with a *White Cross* descended from the Clouds, and a voice from above proclaimed to the Christian combatants a glorious victory, provided they adhered to that sacred sign. Animated by this Heavenly token of regard, the fugitives turned again, pressed in full confidence of Victory into the ranks of the foe, and completely overthrew them. In remembrance and celebration of this important event, as well as in recompense of those who had distinguished themselves in the battle, Waldemar founded The Order of The Dannebrog, and bestowed it immediately upon Thirty-five of his bravest Warriors.

Such is the Origin of this Order according to Tradition, concealed as it is in the romantick garb of the miraculous,—for it's establishment in fact is concealed in the darkness of antiquity, and there is an utter want of authentick historical information upon the subject. That WALDEMAR was THE FOUNDER appears, however, highly probable, and hence, THE DANISH GOVERNMENT assumes the year 1219, as the period of it's Institution. And a *Banner* regarded as that sent from Heaven, long floated victoriously at the head of the Danish hosts, like *The Oriflamme* of France.

In the Fifteenth Century THE ORDER fell into decay, and at length became wholly obsolete,—when, after a lapse of two hundred years, an idea of it's Restoration was conceived by Baron von GRIFFENFELD, who having arranged the Ceremony of the Coronation of CHRISTIAN the Fifth, formed the plan of *reviving* THE ANCIENT ORDER of WALDEMAR,—by means of which those persons might be rewarded and distinguished, who were not in a situation to receive THE ORDER of THE ELEPHANT.

On the 12th of October, 1671, after the celebration on the previous day of the birth of the Crown Prince FREDERICK, the ceremony of renewing THE ORDER of THE DANNEBROG took place, when it was conferred upon nineteen persons, of whom GRIFFENFELD was one.

THE STATUTES, however, did not appear until Twenty-one years after, being dated the 1st of December, 1693. Up to the year 1808, these Statutes remained without increase,—at that period, the Constitution of THE ORDER was extended by the reigning King, FREDERICK the Sixth, and Royal Letters Patent relating to the same, were issued on the 28th of June, of that year,—according to which, it's present arrangement in general is, as follows,—

"When in the midst of a burdensome war," HIS MAJESTY observes, "the Sceptre of our forefathers was placed in our hands, We felt ourselves strengthened by reposing confidence in Providence, by reflecting on the just Cause we had to defend, and by contemplating the People whom We had been called upon to govern. We bore in mind what that People had always been, and what We ourselves had known them to be,—true to their God and King, prudent and industrious in Peace, brave in Danger, ready to sacrifice all for the good of their Native Country,—and We saw with Paternal joy that, whether in the turmoils of War or in the quiet occupations of Peace, the coming days not less than the past would be rich in Honourable Deed, in splendid Action, and that the Subjects of The Danish Monarchy would continue to be conspicuous among the Nations of Europe for unshaken Attachment to their King and Country.

"All which We have further seen confirmed by many indubitable Proofs.

"From the time, therefore, that We entered on the Government, it became one of Our most earnest wishes to establish a pledge of mutual attachment between Us and our People, as a testimony to all how highly We esteem, and how anxious We are to reward, genuine Merit in every class of the Citizens of the State.

"Being convinced that, next to the incitements of Religion and Morality, the upright man regards HONOUR as the purest and noblest incentive to good actions, and that next to the pleasing consciousness of having done good, he finds his noblest reward in the Esteem of his Prince and Fellow-citizens, We felt that Our Paternal object would best be obtained by an external Badge of acknowledged Civic worth.

"We have for this purpose made choice of *The Cross of The Dannebrog*, which, ever since the days of WALDEMAR *the Victor*, has been honoured in Our Flag and Banner, and, as a Badge of Knighthood, has graced THE TRUE MEN of the Danish Monarchs.

"Accordingly in now giving an increased extension to THE ORDER of THE DANNEBROG, which was revived by the Great Grandfather of Our Father's Sire, it is Our design, in conformity with his intention, to increase still more the Lustre of THE ORDER, by honouring with it every one who, in his situation and sphere, may be found deserving of honourable distinction.

"The special duties which are incumbent on the Members of THE ORDER, are prescribed to them in THE STATUTES of The Order,—Our most gracious Will, in respect to its present arrangement in general, is hereby made known to all Our Subjects.

"Access to THE BADGE of THE DANNEBROG is thrown open to every one of Our beloved and true Subjects, without regard to Rank or Age. It is to be acquired by distinguished services and actions,—as, for example, by superior proofs of Bravery, Judgement, and Hardihood in the Warrior,—Fidelity, Judgement, and Zeal in the man of Office,—by sacrifices made for King, Country, and Fellow-citizens,—by successful execution of difficult Undertakings, the result of strenuous efforts and beneficial to the State,—by such attainments in the Arts and Sciences, as confer Honour on the Nation,—by ingenious Inventions, whereby new sources of Wealth are thrown open to the State,—also by new, successful and generally useful establishments in the Agriculture, Industry, and Trade of the Country.

"The affairs of THIS ORDER, and also of that of THE ELEPHANT, shall be managed by an Assembly of Knights, to be called THE CHAPTER of THE ORDER, where, on solemn occasions, We ourself, as Chief of The Order, will preside.

"THE ORDER of THE DANNEBROG is to be divided into Four Classes,—all whose Members are KNIGHTS of THE DANNEBROG, yet with different Titles, Prerogatives, and Decorations.

"Every Knight must have passed through the lower Classes of THE ORDER, before he can be admitted into the higher. It shall, however, be in the power of The Chief of The Order, in the case of very signal exploits, to honour with the *Insignia* of the higher Classes, even those who have not passed through the lower Classes of The Order.

"Knights of the *First* Class of THE ORDER of THE DANNEBROG are called, 'Grand Commanders,'—and are Members of The Chapter. They have the Title of "*Excellency*," rank immediately after Field Marshals and Lord High Admirals, and receive Military honours as Lieutenant Generals.

"Whenever We may find good to confer this degree of The Order of The Dannebrog on a Prince of the Blood Royal, or on a Knight of The Elephant, who either has been previously honoured with some other degree of The Order of the Dannebrog, or has not yet been received into this Order, it is Our most gracious pleasure, that the Badge of Knighthood so granted to them, shall be considered as a special proof of Our Royal favour and distinction.

"The Insignia of The Grand Commanders are, as follow,—They wear the Cross of The Dannebrog richly set with Diamonds in a white watered silk riband with red edges, pendent from the neck,—and the Star of The Dannebrog on the left breast.

"Knights of the *Second* Class are called, 'Grand Crosses.' To this Class belong all those who are at present honoured with The Order of The Dannebrog. Their Rank is highest in Class II, of the Table of Precedence. They receive Military honours equal to Major Generals.

"The Insignia of The Grand Crosses are the following,—They wear the Cross of The Dannebrog in the white riband now in use, crossing the body from the right shoulder,—as also the Star on the left breast. But, if they are Knights of The Elephant or Clergymen, they are then to wear the riband with it's Cross, pendent from the neck.

"Knights of the *Third* Class are called, 'Commanders of The Order.' They receive Military honours equal to Staff Officers.

"The Commanders of The Order wear the Dannebrog Cross in a white watered silk riband with red edges, round the neck,—also an embroidered Dannebrog Cross, without beams, on the left breast of the Çoat.

"Those who belong to the *Fourth* Class of The Order are called, 'Knights of The Dannebrog,' and receive Military honours equal to Subaltern Officers.

"The Knights bear The Dannebrog Cross in a narrow white

watered silk riband, with red edges, on their left side in one of the button-holes of the coat.

"Knights of the Third and Fourth Classes of The Order of The Dannebrog have no special Rank, as such,—but, as being men who have been honoured by the King, they shall on all occasions,—when not actually engaged in the duty of their respective services,—be looked upon and treated, as the first among their equals.

"Every one who is graced with a higher degree of The Order, must lay down the *Insignia* of the lower degree. But Knights of The Dannebrog who are graced with The Order of The Elephant, shall continue to wear the *Insignia* of that Class of The Order of the Dannebrog, of which they were Members.

"Besides the *Four* Classes of The Knights, it is further Our pleasure to honour with The Silver Cross of The Dannebrog, all such persons as We may deem worthy of such Distinction. This Badge of Distinction is to be worn in a narrow white watered silk riband, with red edges, on the right side in one of the button-holes of the coat.

"Those who are thus honoured, are to be called. 'Dannebrogs Men,' and are to receive Military honours from single Centinels by shouldering arms,—and, as being men who have been honoured by The King, they shall on all occasions,—when not actually engaged in the duty of their respective services,—be looked upon and treated, as the first among their equals.

"For the purpose in other respects of bringing all the Classes and Divisions of The Order into a reciprocally honouring connection, We will, as Chief of The Order, alternately wear each of the *Badges* of Honour fixed for the several classes of The Order, as well as *The Silver Cross* of The Dannebrogs Men,—and We likewise most graciously command, that all Members of The Chapter of The Order, in addition to their other Insignia, shall always wear the Badge of The Dannebrogs Men.

"The 28th of June in each year, as being the Birth-day of The Danish King Waldemar the Second, and as the day on which these Our Letters Patent are issued, shall be appointed as a day of Solemnity common to The Orders of The Elephant and of The Dannebrog.

"On that day The Chapter shall be holden in the Church of the

Palace of *Fredericksborg*, where all it's Members, as also the resident Knights and Dannebrogs Men, shall attend,—The Knights of The Elephant, as well as The Grand Commanders and Grand Crosses of The Dannebrog, in the Robes fixed for them by the Statutes.

" The reigning King's birth-day is to be an Anniversary day for the Knights of both Orders. At this Solemnity the Knights of The Orders of The Elephant, and of the two first Classes of The Order of The Dannebrog, shall attend in their Robes in The Chapter in *Rosenborg* Palace, —and proper places there shall likewise be assigned to The Knights of the other Classes, and to The Dannebrogs Men.

" The 15th of April, as being the birth-day of King CHRISTIAN the Fifth of blessed memory, the second Royal Benefactor of THE ORDER of THE DANNEBROG, is to be the Special Order-day for Knights of THE DANNEBROG,—on which day The Knights, as well as The Dannebrogs Men, in the same way as above-mentioned shall meet at *Rosenborg* Palace.

" THE CHAPTER shall keep a watchful eye on the behaviour of all the Members of The Order, and shall give information to Us, if any one should offend against the duties of his Order. The Chapter is empowered to reconcile such disputes as may arise mutually between The Knights of The Order. It shall receive through it's Chancellor, or Vice-chancellor, all proposals relating to THE ORDER.

" If any Dannebrogs Man or Knight should be arraigned or prosecuted before the Tribunals for any action, which might involve Corporal punishment, or the loss of Honour or Publick respect, it shall be the duty of The Tribunals without delay to report such to Our Chancery, in order that The Chancery may lay the representation of The Chapter of The Order before Us for our decision, as to whether the person so arraigned or prosecuted, shall, until farther notice, be prohibited from wearing The Badge of The Order.

" And no sentence that inflicts corporal or degrading punishment, shall be carried into execution on a Member of The Order, until, in pursuance of a Resolution adopted by The Chapter and most graciously sanctioned by Us, the Badge of his Order has been taken from him.

" Given in our Royal residence City *Copenhagen*, the 28th of June, 1808.

FREDERIK R."

The Decoration of the first Class is a Cross, richly set with *brilliants.*

The Star is of silver, surmounted with the Ensign of The Order.

The Collar, which is of gold, is composed of the letters W. C. alternately, each crowned with a Royal crown of Denmark,—between the letters, a Cross enamelled white, and in the C. a figure 5.,—the W. alluding to Waldemar *the Institutor*, and the C. and 5. to Christian the 5th., *the Reviver* of The Order.

The Decoration of the second Class is a Cross, adorned with *fourteen brilliants.* The Star is the same, as that of the first Class.

The Decoration of the Third and Fourth Classes consists of a smaller Cross, of gold, enamelled white, and bordered with red,—on the obverse, in the centre, are the initials of Charles the 5th, crowned, with the words, " *Gud og Kongen*," *(God and the King)*, distributed on the points of the cross, and between the angles are crowns, of gold,—in the centre of the reverse, are the initials of Waldemar, crowned, and on the points are inscribed the dates of the foundation and restoration of The Order,—the whole surmounted with the cypher F. R. VI., and the Royal crown, of gold.

The Motto of The Order is, " *Pietati et Justitiæ.*"

The celebrated Danish Poet, Ingemann, has a fine Poem upon this miraculous Standard, which is thus spiritedly translated by Dr. Bowring,—

Wave proud o'er Codan's surges
  Thou Dannebrog blood-red!
Thy glance through night emerges,
  No lightening scathes thy head.
Thou, over heroes rearing
  A pyramid of fame,—
Thy splendid Cross upbearing
  To heaven our Denmark's name.

From heaven thou didst come hither,
Thou Denmark's hallow'd gem,
And call'dst her warriors thither,—
Earth cannot equal them.
Yet,—long as song rejoices
The listening land and sea,
Or Northern bards have voices,
Shall glory wait on thee.

Call out, in battle's thunder
On Zuûl, thine hero bold,—
Storm forth in strains of wonder
The deeds of Tordenskiold!
Fly, fly on heavenly pinions
Their glory to proclaim,
And teach the stars' dominions
Thy noble Hoidfeld's name!

Each star, when thou alightest,
Name some heroic man,
And tell the best and brightest
Of Denmark's CHRISTIAN!
He stands in bright apparel
On triumph's splendent car,
Where crown'd with deathless laurel
Our Rud and Absalon are.

With palms of victory waving,
The Dannebrog! he sees,
Thy white Cross battle-braving,
For battles victories.
High o'er the loud winds bear thee
Thy sons around thee call,
Let Ocean's verges hear thee,
And glance and gleam on all!

So, proud through Denmark flying,
  So, proud to India fly,
Where barbarous lands are lying
  O'er mountains far and high,—
There tell thy deeds of glory
  Thy heroes' bold emprise,
And gladden with the story
  Thine own blest Paradise!

See! see! thou art succeeding!
  They kindle at the word!
See! for thy glory bleeding
  They rush upon the sword!
O'er every Ocean scatter'd
  Thy name of power shall spread,
Till Northern helms are shatter'd,
  And Danish hearts are dead.

KNIGHTS.

Sir RICHARD DOBSON, Knt., M.D., conferred upon him by His Majesty The King of Denmark, on the 27th of November, 1832, and graciously permitted by His Sovereign, WILLIAM the Fourth, to wear the Decoration.—*See*, RUSSIA, ORDER of St. WLADIMIR, of the fourth class.

NICHOLAS CARLISLE, Esq., K. H., K. I. C., D. C. L., &c., SECRETARY of THE SOCIETY of ANTIQUARIES of LONDON.

THE ORDER OF FIDELITY.

THIS ORDER was founded by SOPHIA MAGDALENA, Queen Consort of King CHRISTIAN the Sixth, on the 7th of August, 1732, the Anniversary of her Marriage, in commemoration of the Connubial Happiness and virtuous Love which she had experienced in the Nuptial State, and which lasted 38 years. But, at her death, on the 27th of May, 1770, it became *extinct*.

THE ROYAL HOUSE OF DENMARK is, therefore, one of the few Antient and Illustrious Families, which confer but a limited number of these HONORARY DISTINCTIONS.

Military Medals.

Medal of Honour for the Battle of Copenhagen,—After the Battle, so honourable to The Danes, which was fought on the 2d of April, 1801, *off Copenhagen*, against an enemy of vast strength, The King caused a Medal of Honour to be struck for those, who had displayed proofs of attachment to the defence of their King and Country,—those for the Officers, in gold,—and the others, in silver. This Medal is attached to a red riband, in which is woven the White Cross of *The Dannebrog*, as an emblem of The Danish Standard,—On one side is a Lion, on the prow of a Vessel of War in the antique style, holding in one claw the Arms of Denmark, and in the other, a Club which it raises in a posture of defence, with the inscription, "*The Second of April*,"—On the other side are The Royal Insignia, the crown, the sceptre, and the glaive, with the words, "*The King honours and the Country is grateful*,"—On the Exergue, are the name and rank of the person, upon whom the Medal has been conferred. No more have been struck off than the number destined for distribution. This mark of Distinction being temporary, it may naturally be supposed that, after the expiration of 37 years, they are rarely seen.

Medals intended for the Workmen of The Royal Docks,—By a decree of the 29th of January, 1801, issued by His Majesty Christian the Seventh, a Medal was instituted to reward and excite the emulation of the Royal Workmen in watching over the preservation of The Royal Marine,—to the Workmen of The Royal Docks were added the Under-Officers of the corps of Artillery, and Seamen when they had been Cannoneers or simple Sailors, the Artificers in the construction of Models, Inspectors of Fire Engines, the Docks, &c.

Artificers, Mariners, Cannoneers of the first Class, Carpenters, Master-workmen, and Under-Officers of workmen, cannot obtain this

Decoration until after 25 years of service,—Ropemakers, Joiners, &c., after 30 years of service. Irreproachable conduct, during this period, is strictly insisted upon. The possessors of the Decoration are entitled to wear it at all times, even when superannuated for age or infirmity.

The Master-workmen who are decorated with this Medal, enjoy the Rank of Lieutenant, with the permission to wear an Uniform similar to that of such officer.

Marks of Distinction for Service,—These Medals are of *two kinds*, and were instituted by an order of The King, on the 23rd of April, 1817,—one of them, for 8 years, and the other, for 16 years of faithful service.

The *first* which is made of *bronze*, is worn suspended from a gilt chain, and is distributed to each Under-Officer, or Military Musician of equal rank, who, after 8 years of service and good conduct, enters into a new engagement for 8 years.

The *second* is also of *bronze*, and is given to all those who engage themselves for a second time, after 16 years of service. It is worn at the same time, with and above that obtained for 8 years service.

Civic Medals.

The Medal, *Pro Meritis*,—This Medal which was struck in 1771, by order of King Christian the Seventh, for distribution, as a mark of Royal good will, amongst persons distinguished by their Merit, is cast in gold and silver, but is not worn ostensibly. On the Obverse is the Portrait of the King,—and on the Reverse, two Cornucopiæ, surrounded by a crown of Laurel and Oak, with the Inscription, "*Pro Meritis.*"

The Medal, with the Inscription, *Fortjenst*,—This Medal, which was also instituted by King Christian the Seventh, in 1792, as a recompense for Services rendered to the Industry or those Civic Virtues, which exercise an influence over the general good, is not worn. On one side is the Portrait of The King,—and on the other, the inscription, "*Fortjenst*" *(Merited)*, with a crown of Oak.

The Medal for Noble Actions,—There being no Medal proper to reward Noble Actions, prior to the year 1793, The College of Finance proposed to strike one,—the Obverse of which should be the same, as that *Pro Meritis*,—and the Reverse should display a Crown of Oak leaves, with the words, "*For aedel Daad*" *(For a Noble Action)*.

This Medal, the obverse of which was changed in 1811, for the likeness of the reigning Sovereign, is furnished with a ring, and may be worn.

Medal for assistance afforded to Drowning Persons.—Assistance afforded to Individuals in danger of being drowned, was formerly rewarded by the Medal, "*For aedel Daad.*" It was, however, considered more advisable to recompense this species of Merit with a particular Medal,—and, on the proposition of The College of Fi-

NANCE, the King caused one to be struck in 1812,—the Obverse of which was the same, as that "*For aedel Daad,*"—and the Reverse displayed a Crown of aquatic Flowers, with the Inscription, "*Recompense and Praise spring flourishing from the Gulf of Danger.*"

The Decree founding this Medal, declares, moreover, that the name of the Individuals upon whom it is conferred, shall be engraved upon it,—and that the same shall be done upon the Medals, "*Pro Meritis, Fortjenst,* and *For aedel Daad.*"

THE KINGDOM OF FRANCE.

THE ORDER OF THE HOLY GHOST.

THE ORDER of THE HOLY GHOST, the most Illustrious Order of Knighthood in France, was instituted by HENRY the Third, King of France, on the 30th of December, 1578, in commemoration of *Whitsunday,* that day having been particularly auspicious to him,—for, on that Festival, in the year 1551, he was born,—on the same day, in 1573, he was elected King of Poland,—and on Whitsunday, 1574, he was called to the Throne of France, on the death of his Brother, CHARLES the Ninth, —In honour, therefore, of that day, and in remembrance of this Union of two Crowns on one head, an event so remarkable for himself and for the honour of his Dynasty, he founded THIS ORDER, which, in relation to the Festival of Whitsuntide, bore the name of THE HOLY GHOST.

With these intentions he further united the designs of

ingratiating to himself the minds of the Nobility, who, by means of the League and the faction of The House of GUISE, had been alienated from The Sovereign,—as well as of founding a NEW ORDER, in competition with that of SAINT MICHAEL, which had then lost all its Dignity,—and which New Order should acquire greater respect, in consequence of being very sparingly bestowed.

Some Historians likewise maintain, that the *Secret Love* which then prevailed to the greatest extent in The Court of France, had a still larger share than even Religion or Politicks in the foundation of THE ORDER. The proofs of this assertion they derive from the different divisions and colours of the COLLAR of The Order, such as it originally was, — indeed, a knowledge of HENRY'S character renders this supposition highly probable, and the idea is confirmed by the fact, that his Successor, HENRY the Fourth, at once abolished these secret allusions, and ordained THE COLLAR of THE ORDER to be, what it now is.

THE STATUTES of THE ORDER accord in general with those of other ORDERS,—but they contain this peculiarity, that The King of France, as Chief and Grand Master of The Order, is bound on the day of his Anointing and Coronation, in presence of the Archbishop of Rheims, or of him by whom the Ceremony of Anointing is performed, to swear to the observance of the Ordinances of The Order, according to the form prescribed in the Ordinances themselves, without being able to evade this obligation by any cause whatever. This Oath is inserted in the Coronation Service, and is inscribed with the other Oaths, which the Kings are compelled to take before they are crowned. The form of the

Oath contains the most essential parts of the Statutes of The Order, as declared by HENRY the Third, in the year 1578.

HENRY the Fourth, however, altered several of these Statutes, twelve years afterwards. He ordained, that Foreigners, both Princes and other Persons, might be received into The Order. If it were bestowed upon a reigning Prince, he should have the rank of a *Commander*,—and a Knight of The Order should be sent to him with the Insignia, to invest him therewith and to clothe him with the Mantle,—On the other hand, he was bound within the term of a year, to thank the King through an Ambassador for such Distinction. Other Foreigners, who were not reigning Princes, were compelled the very first year of their appointment, to appear in person before The King, to receive from him the Insignia, and to take the Oath prescribed, unless they were expressly dispensed from so doing.

The attainment of THE ORDER of THE HOLY GHOST supposes the possession of the Second French Order, that of SAINT MICHAEL. Hence the King, either the day before the Investiture or the same morning, in his Closet, bestows on those who are appointed to The Order of THE HOLY GHOST, that of SAINT MICHAEL, if they do not already possess it.

The number of Knights empowered to bear the Title of "*Knights of The Royal Orders*," is limited to One Hundred, and is never exceeded. The Statutes of Foundation bestowed upon them many Privileges and Rights, of which, however, in modern times many have become obsolete. Among other honours they enjoy that of dining on Festivals of The Order at the same table with the King. From the Endowments of The Order, every Knight at first received 4000 livres annually. Since 1764, however, the Twenty oldest receive annually 8000 livres,—and, at present, the Thirty oldest receive 6000, and the others, 3000 livres. There are various obligations, such as their attending Mass every day, and their partaking of the Divine Ordinances on Festivals, the repetition daily of a number of Prayers and Hymns or the seven Penitential Psalms,—but, if they find these observances too severe, they are bound to bestow daily at least one alms upon the Poor.

In the number of One Hundred Knights are comprised the Princes of The Blood, the other Princes of the Royal Family, the Ecclesiastics, and the Grand Officers, such as Commanders, and in general only the French, and not the *Foreign* Knights, who also receive no Revenue.

This Order is simply called in France, " *Le Cordon bleu.*"

The Knights are styled, " *Chevaliers des Ordres du Roy,*"—and the Commanders, if Ecclesiastics, " *Commandeur de l'Ordre du St. Esprit,*" —if Laymen, " *Commandeur des Ordres du Roy.*"

No one can be invested with The Order of The Holy Ghost, unless he professes the faith of the Church of Rome,—and proves his Nobility of three descents by the father's side.

The Device, or Motto, is, " *Benoist St. Esprit.*"

The Badge of The Order is a Cross, of eight points, enamelled green, with a border of white,—and, in the centre, enamelled crimson, is a Dove, the emblem of The Holy Ghost, *argent,* having a red bill and feet,—in the four angles, are four fleurs de lis, of gold, and the eight points are terminated with as many brilliants. The Reverse of the Badge bears the image, in silver, of The Holy Archangel *St. Michael,* treading the Dragon under foot,—probably to indicate the intimate connection of this Order with that of St. Michael.

The Badge is usually worn, pendent to a broad sky blue, watered riband, scarfwise over the right shoulder,—and the Star, embroidered in silver, on the left breast.

The Collar consists of three alternate links,—the first is composed of a golden helmet, with a plume of white feathers,—it is surrounded with arms and military trophies, in red, white, and blue,—The second link is a flame of gold, terminating in eight points, and in the middle is a red fleur de lis,—The third link is formed by the letter H., signifying the name of The Founder, Henricus,—it is surmounted with a royal crown, and the whole is surrounded with a celestial glory.

On days of Ceremony the Knights are obliged to appear in the Habit of The Order, which is extremely splendid. It consists of a doublet and breeches of white satin, a long Mantle of black velvet, lined with orange-coloured crape, embroidered all over with golden flames, the seams worked with gold lilies, love-knots, and the letter H. (*Henry*), and which is so drawn up on the left side, that a part hangs before in a point down to the shoes, on which the Badge of The Order is embroidered in a large size. Over this Mantle, a smaller

Mantle of cloth of silver hangs down to the breast, on which lies the Collar of The Order, with the Badge appended to it. On the head is worn a black cap with white feathers, of which colour also are the stockings and shoes.

Ecclesiastics of this Order wear the Badge on a smaller riband, round the neck,—and, on the reverse, not the figure of *St. Michael,* but, as on the obverse, a *Dove.*

The Usher and Herald also wear the Badge, round the neck,—but without the Star on the breast,—which, however, the Ecclesiastics retain.

The Festival of The Order is celebrated on the 1st of January,—and the previous day is Vespers, which the Knights are compelled to attend. The usual place of assembly is the Church of *The Augustines* in Paris.

The Order of The Holy Ghost, the Motto of which is "*Duce et Auspice,*" flourished uninterruptedly to the time of the French Revolution, when it was abolished by *The National Convention* by a special decree,—as the then prevailing *mania* of the French people for Liberty and Equality, allowed of no such external distinctions. It continued to be suppressed under the Government of the Emperor Napoleon,—and it was only in the year 1814, when King Louis the Eighteenth entered Paris, that The Order was revived in it's ancient Rights and with it's ancient Royal Splendour.

In one of the Letters of Lord Chesterfield to his Son (vol. 2, pp. 210-11), he says, "I am very glad you went to *Versailles,* to see the Ceremony of creating The Prince *de* Condé, '*Chevalier de l'Ordre,*'—and I do not doubt but that, upon this occasion, you informed yourself thoroughly of the Institution and Rules of that Order. If you did, you were certainly told, it was instituted by Henry the Third, immediately after his return, or rather his flight from Poland,—he took the hint of it at Venice, where he had seen the original Manuscript of an Order of the *St. Esprit, ou Droit desir,* which had been instituted in 1352, by LOUIS *d'Anjou,* King of Jerusalem and Sicily, and husband to JANE, Queen of Naples, Countess of Provence.

"THIS ORDER was under the Protection of *St. Nicholas de Bari,* whose image hung to the Collar. HENRY the Third found THE ORDER of ST. MICHAEL prostituted and degraded, during the Civil Wars,—he, therefore, joined it to his NEW ORDER of ST. ESPRIT, and gave them both together,—for which reason, every Knight of *The St. Esprit* is now called, "*Chevalier des Ordres du Roi.*"

"The number of The Knights hath been different, but is now (in 1752) fixed to One hundred, exclusive of The Sovereign. There are many Officers, who wear the riband of this Order, like the other Knights,—and what is very singular is, *that these Officers frequently sell their employments, but obtain leave to wear the Blue Riband still, though the Purchasers of those Orders wear it also.*"

KNIGHT.

ARTHUR, Duke of WELLINGTON, K. G.

The Order of Saint Michael.

Although of more ancient origin than The Order of The Holy Ghost, that of Saint Michael is, however, inferior in Rank, and takes the *second* place only among the French Orders.

It was founded, on the 1st of August, 1469, by Louis the Eleventh, at the Castle of *Amboise*, in honour of *St. Michael*, as Protector of the Kingdom. The cause which induced it's establishment, is not clearly known. Mathieu in his Life of Louis states, that His Majesty wished by it's means to appease the great Nobles of the State who were then excited against him, to attach them to himself, and to put an end to the many divisions among them.

To the Knights he gave a gold Chain, weighing 200 crowns of gold, which consisted of alternate Pearls and Gordian Knots, and to which a Medal was appended, representing The Archangel *Michael*, piercing the Dragon. The Motto was, "*Immersi tremor Oceani,*"—which had reference to the tradition, that as often as the Enemies of France approached by water to *St. Michael's Rock* in the Ocean, The Archangel Michael appeared, raised a powerful storm, and scattered the foes.

The Knights were obliged to wear this Chain constantly, unless, upon a journey, or hunting, or at home, when they were allowed to wear the Medal, pendent to a black riband. In general the Rules of the Order were of singular strictness and severity.

The number of Knights was fixed at Thirty-six,—who were required to be of the best Families. This provision, as well as the rest of the Ordinances, remained a long time in force, and was observed with due formality,—by which means The Order, under the three immediate Successors of Louis, preserved it's respect. Subsequently, however, especially under FRANCIS the Second and CHARLES the Ninth, the Ordinances were wholly disregarded. It was forgotten, that it is only by a sparing distribution, an ORDER acquires and maintains Dignity and Honour,—and so prodigally and indiscriminately was the number of Knights increased, that THE ORDER entirely lost it's value, and, in allusion to it's particular Ensign *(The Collar)* was contumeliously denominated, "*The Order of beasts of burden of all kinds.*" Under HENRY the Second *it might be purchased,*—and Queen CATHARINE of MEDICIS bestowed it profusely among the adherents of her Party.

To raise it again to dignity and respect, King HENRY the Third, when he founded, THE ORDER of THE HOLY GHOST, in 1578, enacted, that this *New Order* should not be received without that of *St. Michael,* and that every one destined for The Order of The Holy Ghost should, if he did not already possess it, receive it the day previous to his receiving The Order of The Holy Ghost, and might then assume the denomination of "*Knight of The Royal Orders.*"

By this intimate connection with THE ORDER of THE HOLY GHOST, that of ST. MICHAEL certainly acquired fresh respect, and flourished without interruption,—but the old evil was again introduced,—it was again bestowed most profusely, the established Rules were neglected, and a crowd of persons who could make no legitimate pretensions, were enrolled in THE ORDER.

With a view to check anew this irregularity, LOUIS the Fourteenth, in the year 1664, named a Special Commission, before which all the Knights of *St. Michael* were required to prove their Nobility, and other indispensable Qualifications,—in default of which, they were compelled to lay aside their Badge. By means of His

Ambassadors, the King also caused Foreign Courts to prohibit all persons from wearing The Order, who could not undergo such a strict Ordeal.

This purification and expulsion His Majesty followed up in the year 1665, by a formal *Renewal* of The Order,—and it was then decreed, that the number of native Knights, with the exception of those of The Holy Ghost, and comprising six Ecclesiastics, should never exceed *One Hundred*,—That the Catholick Religion,—thirty years of age,—true noble descent,—and service, either Civil or Military, of ten years, should be indispensable qualifications and conditions,—and also, that the change of Religion should occasion the loss of The Order.

It was also commanded, that the Badge should always be worn, pendent to a black riband, scarfwise,—and the Chain formerly worn, was no more mentioned.

As early as the reign of Louis the Fourteenth, the strictness of the proof of noble descent was relaxed,—for, two Artists received the honour,—and, afterwards, men of Literature were admitted,—but, if these parties were not of noble rank prior to their admission, they were ennobled at the time. This regulation, however, was not observed in future years,—and the Order acquired by degrees, the destination of serving only as a distinction and reward of Literature and the Arts, which it now possesses.

The Order of Saint Michael shared the fate of all the other Orders at the French Revolution,—Louis the Eighteenth, during his Exile, however, bestowed it on several Individuals,—and when he returned to the Throne of France, after the fall of Buonaparte, on the 16th of November, 1816, he issued a Royal Ordinance, by which The Order of Saint Michael, after laying dormant for more than Twenty years, was again formally revived,—And it was at the same time decreed, that it should serve especially as a recompense and encouragement to those of His Majesty's Subjects, who shall be distinguished in Literature, Science, and the Arts, or by discoveries, works, and enterprises useful to the State,—

The ancient number of 100 Knights was again confirmed,—but, in this number, *Foreigners* are not included,—

All requests of Admission into The Order of St. Michael are to be addressed to The Minister of The Household, who shall submit a report to The King, and propose such as may be qualified for Admission.

The Badge is a Cross of gold, of eight points, enamelled white, and between the four angles, golden lilies are introduced. In the centre of the Obverse, upon a circular shield, in bright colours, is *The Archangel Michael* slaying the prostrate Dragon with a spear. It is worn, pendent to a broad black, watered riband, sashwise over the right shoulder.

Several distinguished persons of the Protestant Faith, —several men of great Learning,—and many celebrated and ingenious Artists have been honoured with this Order.

Michael *de* Montaigne, the ornament of The French Nation,—And Robert Dudley, Earl of Leicester, the favourite of Queen Elizabeth, were Members of it.

When treating of The Orders of Chivalry in France, Mr. Evelyn observes, "I shall not much amuse you with those Orders which are so far antiquated, that even the Heralds themselves can scarcely render us any certain account,—Such is that, which is named, *De la Genette*, instituted by Charles Martel,—or, *The Order de l' Estoile*, by King John,—The Order of *The Croisant*,—*Porc Espic*,—nor much concerning *The Order of St. Michael* itself, although not many ages since first instituted (in 1469), and for a long while the principal Order in the Kingdom,—composed but of Thirty-six, because *(as the manner of this*

Nation is, to be as soon weary of their inventions, as children are of rattles) they begin to have this Order already in contempt,—Albeit the Chain and pendent Badge be commonly reserved in their Coat Armour, together with that which is now in vogue (in 1562)."

He then describes the Institution of the "*Ordre du S. Esprit,*" and the ceremonies of the Inauguration, as "it remains yet the Order of greatest esteem."—EVELYN'S Miscellaneous Writings, published by UPCOTT, in 1825, 4to, *p.* 77.

The Royal Military Order of Saint Louis.

King Henry the Third, of France, early conceived the idea of rewarding Military persons of Merit of all ranks, and for this object not only founded an Order, which he named "The Order of Christian Love,"—but also built an Hospital, for the maintenance of old and invalid Soldiers. The Badge of The Order was an embroidered Cross, of an anchor-like form, of white silk, with a blue border,—in the midst of which is a lozenge of blue taffety, with a gold Lily, surmounted by the words, "*Pour avoir bien servi.*" But neither under this Monarch, nor under Henry the Fourth, or Louis the Thirteenth, did this foundation and establishment acquire that perfection and extension, which were requisite to render them of real utility,—although the latter Sovereign, in 1633, founded a Society of Knights, under the title of "A Commandery of St. Louis."

It seemed to have been reserved for Louis the Fourteenth to bring to perfection this useful Undertaking, so benevolently begun by his Predecessors. He it was, who not only built the splendid *Hotel of the Invalids* in Paris for this object,—but who also, in 1693, founded a new external Badge for The Military, to which in honour of his Ancestor Louis the Ninth (who died in 1270, and was placed by Pope Boniface the Eighth among the Saints of the Romish Church, in 1297), he gave the name of "THE ORDER of SAINT LOUIS,"— and which subsequently in 1719 was expressly confirmed by LOUIS the Fifteenth, on account of it's essential utility.

THE STATUTES of THIS ORDER are a profession of the Roman Catholick religion, defence of the honour and rights of the Sovereign, fidelity in Service, virtuous and blameless Conduct, and twenty-five years service as an Officer, of which one year in War is reckoned as two,— this rule, however, is departed from in some cases, such as in that of extraordinary distinction.

The Members form three Classes,—*viz.*,

GRAND CROSSES, limited to *Forty*,—

COMMANDERS, limited to *Eighty*,—

KNIGHTS, as many as The Grand Master may please to name.

No one, however, can attain to a higher class, without having previously holden a lower.

During the reign of LOUIS the Sixteenth, in 1779, it was arranged, that of the 40 Grand Crosses, 30 should be destined for the Land Service,—4 for the Household Service,—and 6 for the Sea Service,— Of the Eighty Commanders, 57 should be for the Land,—8 for the Household,—and 15 for the Sea service,—At the same time, the Officers of The Order, of whom there were previously several, were limited to a Secretary General, who should be both Secretary and Keeper of The Archives, a Treasurer, and an Usher.

THE ORDER of SAINT LOUIS, like all THE FRENCH ORDERS, was regarded as suppressed and extinguished upwards of Twenty years, from the commencement of The Revolution to the Accession of LOUIS the Eighteenth. This Monarch, however, reinstated it in all it's ancient Rights, and commanded, on the 30th of May, 1816, that, beside the above mentioned Officers of The Order,—whose appointments are now vacant,—The

Keeper of the Seals of France should discharge the duty of Keeper of the Seals, and that the Secretary of War should take charge of the Income and Expenditure of The Order.

Like most Military Orders, that of St. Louis is not confined to mere *Insignia* of honour, but has *Revenues* also annexed to it. Particularly, since the year 1759, it has possessed considerable Funds, from which every Grand Cross receives annually 6000,—every Commander, from three to 4000,—and every Knight according to his priority in The Order, or according to the decision of The King, from 200 to 1000 francs.

The Badge of The Order is a gold Cross, of eight points, enamelled white, in the four angles of which are golden lilies,—in the middle, on the obverse, is an oval enamelled shield, with the image of *St. Louis*, who, in full armour, clad with a Royal mantle, holds in his right hand, a wreath of laurel, and in his left, a crown of thorns and the nails of the Passion,—around the Saint, runs a blue circle, on which are these words, in gold, *Lud.(ovicus) Magn.(us) Instit.(uit)* 1693. On the Reverse, is a gold flaming Sword, erect, the point through a chaplet of laurel, bound with a white riband, and surrounded with this Motto, on a blue enamelled circle, "*Bellicæ virtutis Præmium.*" Whether the inscription on the obverse was the same originally, and Louis bestowed upon himself the flattering title of "*Great,*"—or, whether this adulation was first added by one of his Successors, is not known. It is, however, probable that his insatiable Ambition and boundless love of Fame, induced him to assume this arrogant epithet,—but his last moments display a greatness of mind worthy of his exalted situation,—his advice to the Infant that was to succeed him, was to avoid that Glory which he had hoped to have attained by War, and to consider the Happiness of the People as the principal object of his Government.

The Grand Crosses, and Commanders, wear The Badge on a broad flame-coloured, watered riband, sashwise over the right shoulder,—besides which the Grand Crosses wear, embroidered on the left side of their outer garment, a gold Star, of eight points, with *fleurs de lis* at the angles, and the figure of *St. Louis*, with the Motto, on the centre.

The Order of *St. Louis* has no Collar.

The Knights wear the Badge, pendent to a flame-coloured, watered riband, at the left button-hole.

On solemn occasions the Knights of The Order of St. Louis rank with the Members of The Legion of Honour, according to the Class of their reception,—so that The Grand Crosses of the former take place with the Grand Crosses of the other,—The Commanders, with the Grand officers,—and the Knights, after the Commanders, but with the Officers, and before the Knights of the Legion of Honour.

When the Duke of Orleans received The Grand Cross in the year 1814, the Ceremony was performed in the following manner,—The King stood before the Throne, holding his drawn Sword in his hand, The Duke knelt before him, the Minister of war read over the Oath, which the Duke took, and was then embraced by His Majesty,—The Duke then rose, but knelt again, and respectfully kissed the hand of the Sovereign.

Since the year 1816, the collective Knights of St. Louis have established an Institution for the Education of the children of poor Members of The Order, at *Senlis*, which they have confided to The Benedictines of St. Maur.

Grand Cross.

Sir Edward Codrington, Vice-Admiral of The Blue, G.C.B., Grand Cross of St. Louis, and Knight of St. George of Russia, of the second class,—Battle of *Navarino*.

KNIGHTS.

HON. WILLIAM ANSON, Captain in The Royal Navy, C.B.—Battle of *Navarino.*

Captain EDWARD CURZON, of The Royal Navy, C. B., Cross of St. Louis, and St. Wladimir, of the third Class,—Battle of *Navarino.*

Captain LEWIS DAVIES, of The Royal Navy, C.B.—Battle of *Navarino.*

Captain RICHARD DICKINSON, of the Royal Navy, C.B.—Cross of St. Louis, and of St. Anne, of the second Class.—Battle of *Navarino.*

Captain GAWEN W. HAMILTON, of The Royal Navy, C.B.—Battle of *Navarino.*

Rt. Hon. HENRY Viscount INGESTRIE, Captain in The Royal Navy, C.B. —Battle of *Navarino.*

GEORGE B. MARTIN, Commander in The Royal Navy, C.B.—Battle of *Navarino.*

Hon. J. A. MAUDE, Captain in The Royal Navy, C.B.—Battle of *Navarino.*

Captain JOHN A. OMMANEY, of The Royal Navy, C.B.—Battle of *Navarino.*

Captain The Hon. FREDERICK SPENCER, of the Royal Navy.—Battle of *Navarino.*

Captain Sir E. LYONS.

Captain S. LUSHINGTON.

The Order of Military Merit.

As, by the Statutes of The Military Order of St. Louis, all those who were not *Roman Catholics,* were excluded from possessing it,—and, as it was impossible to depart from this rule, and yet many *Protestants* had rendered themselves worthy of such a distinction, Louis the Fifteenth, of France, was induced, in the month of March, 1759, to found an Order of Military Merit, limited to such persons as were not of the Communion of Rome.

With the exception of this Regulation, regarding their similarity of Faith, this Establishment is almost entirely the same, as that of The Order of St. Louis.

The Members consist in like manner of three Classes,—*viz.,*

Grand Crosses, *four.*

Commanders, *eight.*

Knights, an *unlimited* number.

The Order consisted at first of mere external distinctions.—But Louis the Sixteenth, in 1785, bestowed *Revenues* upon it, and consequently enhanced it's value. From that period, all the Members have enjoyed *Pensions,* like those of The Order of St. Louis,—in which, however, *Foreigners* cannot participate,—neither are they comprised in the fixed number of Members of the two first Classes.

The Badge of The Order is a Cross, of eight points, enamelled white, having on one side a drawn sword, placed erect, and surrounded with this Motto, "*Pro virtute bellica,*"—and on the Reverse, a chaplet of Laurel, with this inscription, "*Lud*(*ovicus*) XV. *instit*(*uit*) 1759."

As regards other particulars, it is worn by all Classes, in the same way as The Order of St. Louis, and was originally attached to a sky blue riband,—but, since the 25th of November, 1814, when this, like all the other French Orders was restored, it has been worn, pendent to a riband of a flame colour.

THIS ORDER has no COLLAR.

GRAND CROSS.

Lieut. General Sir JAMES LEITH, Commander of His Majesty's Forces in the West Indies, G.C.B., and Honorary Knight Commander of The Royal Portuguese Military Order of The Tower and Sword, to accept and wear the Grand Cordon of the Order of Military Merit of France, with which His Most Christian Majesty hath been pleased to honour him, "as a mark of the high sense entertained by that Sovereign of the great zeal, ability, and energy manifested by him in the month of August 1815, throughout the arduous operations whereby, with the Forces under his command, he effected the surrender of the important Island of *Guadaloupe*, and secured the French Colonies in *The West Indies* to the Crown of France.—Royal Licence, 20th of November, 1816.

COMMANDER.

Sir PHILIP CHARLES HENDERSON DURHAM, K.C.B., Rear Admiral of the Red, "in testimony of the high sense entertained by His Catholick Majesty of the distinguished services rendered to his cause by him, while commanding The British Naval Forces acting in co-operation with The British army in the arduous operations, which led to the capture of the important Island of *Guadaloupe* in August, 1815, by which *The French Colonies* in *The West Indies* were secured to The Crown of France.—Royal Licence, 28th of June, 1817.

KNIGHT.

Major THOMAS MOODY, of The Corps of Royal Engineers, "as a mark of the high sense entertained by His Most Christian Majesty of the Military services rendered by him at the Capture of *Guadaloupe*, while serving as Aid-de-Camp to the late Lieut. General Sir JAMES LEITH."—Royal Licence, 26th of May, 1820.

The United Orders of Saint Lazarus, and Our Lady of Mount Carmel.

The Order of Saint Lazarus arose in The Holy Land, contemporaneously with that of St. John of Jerusalem, and is consequently of very ancient origin. In what year, however, and by whom it was founded, is not known.

The Members at first differed in no respect from *The Knights Hospitallers*, and, like them, they made it their duty to receive and attend in their Houses, sick and poor Pilgrims, especially such as were afflicted with *Leprosy*. They even received *Lepers* into their Order, and The Grand Master of The Hospital at Jerusalem was required to be a *leprous Knight*. The Knights who were not Lepers, and could bear arms, rendered much service to the Christian Princes in the conquest of The Holy Land, on which account those chivalrous Princes took The Order under their protection, and conferred many benefactions upon the Houses, which The Order possessed in *Syria*.

King Louis the Seventh, the Younger, of France, on his return from The Holy Land, in 1154, brought back with him to France a number of Knights of Saint Lazarus, assigned to them the Castle of *Boigny*, near Orleans, bestowed upon them many immunities and possessions, among which was a Castle and a Church in the present Suburb of *St. Lazare* at Paris, which doubtless received it's name from this circumstance. His Majesty also conferred upon them the superintendence over all the Hospitals and Houses for the Sick in the whole Country. In the middle of the thirteenth century, however, the Knights were entirely expelled from Syria, came in great numbers to Europe, spread themselves over all Countries, and erected Hospitals every where,— over which The Head of The Order at *Boigny* sought to maintain the Supremacy, and for this purpose styled himself, GRAND MASTER of THE ORDER of ST. LAZARUS, on both sides of the sea. It was about this time, that they entreated Pope INNOCENT the fourth, to release them from the obligation of electing only a *Leper*, as their *Grand Master*, as the Infidels had slain all the leprous Knights. INNOCENT granted this Petition, after it had been ascertained by previous investigation, that it could be done without fear of offending God!

At the end of the Fifteenth century the original benevolent Institution declined. The Knights entirely forgot their duties, and instead of attending the Sick, squandered the very considerable Revenues, which the benevolence of many pious and charitable Persons had allotted to them. This was particularly the case in Italy. Pope INNOCENT the Eighth, by a special Bull, in the year 1490, entirely abolished THE ORDER of ST. LAZARUS, as an Establishment which no longer answered it's original intention, took away all it's possessions, and united them with those of THE ORDER of MALTA. This abolition of The Order was not, however, recognised in France. Here it existed as before, and the Successor of INNOCENT, LEO the tenth, restored it also soon after in Italy. Up to the year 1572, it flourished without interruption,—when Pope GREGORY the thirteenth, after the death of the then Grand Master, united it with THE ORDER of St. MAURICE, which Duke EMMANUEL PHILIBERT, of Savoy, had founded in the same year, and bestowed on this Prince the perpetual Grand Mastership of this double Order, which will be described among the Orders of SARDINIA.

Against this proposed alteration of THE ORDER of ST. LAZARUS the Members of it in France protested in the most vehement manner,— and the then Grand Master, FRANCIS SALVIATI, held at *Boigny*, in

1578, a general Chapter, in which THE ORDER maintained the possession of it's estates in France, and separated itself entirely from THE ORDER of ST. LAZARUS in Italy. This separation of the two Orders continued, however, only 30 years. It ultimately experienced the same fate as many ancient foundations, which, being exhausted by time, either fall altogether, or appear again in new forms.

In the year 1607, King HENRY the Fourth of France founded THE ORDER of OUR LADY of MOUNT CARMEL,—by which he wished to give a Publick Memorial of his Piety and devotion to The Holy Virgin. Pope PAUL the fifth confirmed this Institution by a special Bull,—and, by the same document, granted The King the power of naming The Grand Master of The New Order,—permitted the Knights to marry,—and even to take a Widow to wife,—imposed on them the vow of obedience and conjugal chastity, as well as of bearing arms against the enemies of the Church,—and some other Ordinances, respecting Prayers and Penances.

With a view to provide the New Order with *Revenues*, HENRY abolished the post of Grand Master of the original ORDER of ST. LAZARUS, in April 1608, and bestowed it's whole Possessions and Revenues on the New Order. As he named the previous Grand Master of The Order of St. Lazarus, PHILIBERT of *Nerestan*, Grand Master of the New Order, gave it the name of "THE ORDER of OUR LADY of MOUNT CARMEL and of ST. LAZARUS of JERUSALEM,"—bestowed upon it the perpetual care and management of all Hospitals, Lazarettoes, and similar Establishments, and particularly enjoined the care and attendance of *Lepers*,—this establishment might, therefore, be said to be rather a revival of THE ORDER of ST. LAZARUS, under an altered form, than a complete abolition of that Order. The reason, however, of this revival still remains unexplained.

Under LOUIS the Fourteenth, the Order attained to yet greater dignity and wealth, than at any former period. In 1672, it acquired the perpetual management of the Infirmaries, Hospitals, Lazarettoes, and other places in the Kingdom, where Hospitality was not observed. It united also with the two Orders, the possessions of other Orders of Knights, Hospitals which were on this account reckoned extinct and abolished in France. These were in particular, — The Order of *The Holy Ghost* of *Montpellier*, of *St. Jacob* with the Sword, of *The Holy Sepulchre*, of *St. Christina* of *Somport*, of *Our Lady*, surnamed of *Germany*, of *St. Jacob* of *Haut Pas* or of *Lucca*, and of *St. Louis* of *Boucheraumont*.

The Estates and Incomes of these Orders, Infirmaries, Hospitals, Lazarettoes, and other places, were formed by the King into COMMANDERIES, and he and his Successors, as uncontrolled Superiors of both Orders, were to apply the same to the advantage of those Military Officers, who should be received into THE ORDER, and should bestow on them such annual Revenues from the Commanderies, as they were able to support.

The King also wished, that these Commanderies should contribute something in the way of taxes and imposts to the maintenance of the Hospitals in his Armies, and in the Frontier Towns,—as he considered, that such application of the Funds was conformable to the intentions of Founders of pious Houses, now that there were scarcely any *leprous* persons in the Kingdom. At the same time, it was His Majesty's wish, that those who were afflicted with this loathsome malady, should, in conformity with it's Institution, be maintained together in one place at the expense of The Order. This accession of dignity and extension lasted, however, only Twenty years. In 1693, all that had been added to it, in 1672, was again taken away,—and The Order retained only, what it had previously possessed. It was, indeed, confirmed by Louis the Sixteenth,—but for forty years no Chapter has been holden. And as the posts of Grand Master and of the Officers are vacant, and there remains only a Chaplain of The Order, and that the Order itself has not been conferred on any one since the year 1788, it appears to be hastening to extinction, and to belong to the Class of expiring Orders.

The Badge is a gold Cross, of eight points, alternately purple and green, with fleur de lis of gold in the four angles. On the obverse, in the centre, is the image of The Virgin Mary, surrounded with rays of gold, on a ground of pale violet,—and on the reverse, on a green ground, is Lazarus rising from the grave. It is worn round the neck, pendent to a broad green riband, with a Star, of eight points, similar to the obverse of the Cross, embroidered in green and gold, on the outer garment on the left side,—or, on days of Ceremony, on a purple Mantle.

The Order of The Legion of Honour.

The Order of The Legion of Honour was instituted by The Emperor Napoleon. It survived his downfall, being preserved and confirmed by the existing French Government with some alterations,—as the youngest of all the French Orders, it takes the lowest rank among them.

When Buonaparte, while yet First Consul of The French Republick, called forth by degrees various establishments, which, in opposition to the Republican spirit, aimed at external splendour, ostentation, and brilliancy of their associations, The decorative Orders were not forgotten. To re-establish the old Royal French Orders, which were suppressed at the beginning of the Revolution, appeared to be unadvisable, on account of the hatred then cherished against every thing which recalled the days of Royalty. Of the utility of an Institution, by which he endeavoured to unite reward with personal attachment, Napoleon was well aware,—and, accordingly, on the 15th of May, 1802, he proposed to The Legislative Body the establishment of such an Order, under the Name of "The Legion of Honour." The project occasioned several sharp debates, but the friends of Buonaparte prevailed, owing in a considerable degree to the Eloquence of the then acknowledged

Orator of the Government, RÖDERER, and LUCIAN BUONAPARTE. The project having passed, became a Law, and by the 21st of February, 1803, The Institution was fully established.

All those Warriors, who had hitherto received *Sabres of Honour*, were at once declared Members, and as Merit alone was to confer a claim to the distinction, it's bestowal was extended to every deserving Citizen of the State, without respect to birth, class, rank, or religion. A numerous Council of management exercised a superintendence over the important Revenues which The Order received,—And the Members, who were called Grand Officers, Commanders, Officers, and Legionaries, were divided into Sixteen Cohorts, each of which numbered 407 Members, so that the sum total of the Members was 6512, exclusive of Foreigners. The Cohorts were distributed throughout the whole of France: each had a chief place, a Superior, a Chancellor, and a Treasurer. NAPOLEON himself was Grand Master. When he soon after declared himself EMPEROR, The Order underwent some alterations with regard to it's management, and the .Motto on the Decoration,—but it preserved in other respects, and in it's essential points, it's original purpose.

THE THREE ORDERS which were instituted by NAPOLEON, were on the principle of abjuring Feudal Customs,—but, when he had got his Court established in a splendid style, he made his Knights *hereditary*, allowing the Dignity to descend to Heirs, when the holder could form a *Majorat*, or entail on his heirs a Revenue specified by Law. He made his Dukes, Counts, and Barons, hereditary in the same way,—the Majorat or Revenue to be entailed, being different for each Dignity.

After the fall of BUONAPARTE, policy demanded the preservation of THE LEGION of HONOUR. King LOUIS the Eighteenth, soon after his Accession in 1814, formally received it into the number of Royal French Orders, and made at the same time several alterations in it's establishment, and it's decorations,—and it's Constitution, as it now exists, was more clearly defined by later Ordinances of the 17th of February, 1815, and the 27th of March, 1816,—those of the latter are, as follow,—

As soon as Providence had replaced Us on the Throne of Our Fathers, amid the calls of a People who were ever dear to Our heart, We made it our duty to preserve that Legion of Honour, which rewarded in a way congenial with French manners, all kinds of Services rendered to the Country. How could we regard with indifference an Institution which gives to the highest Power the noblest motives that influence the National character, which multiplied among our Warriors those prodigies of valour that lent such splendour to the French arms, and which excites among all classes of Citizens an emulation that cannot but heighten the fame of the Monarchy? In accepting this Institution for Ourselves and Our Successors, We make it our own work, and are convinced that the Name of HENRY the Fourth, which no Frenchman can pronounce without emotion, will only make it still dearer to the Nation over which this Prince reigned so gloriously. In thus confirming The Institution of THE LEGION of HONOUR, it has pleased Us to give to Our Subjects a new proof of Our Royal affection,—

This Preamble is followed by the new and present establishment of THE ORDER, in these words,—

THE KING of FRANCE is Chief and Grand Master of The Order.

All the Honorary distinctions, already granted to Members, are preserved,—and the present Members farther receive their incomes, according to their respective ranks. New Members, however, at first and until farther arrangements are made, receive no revenue, as the prescribed number of Members has been so far exceeded, that the possessions of The Order do not suffice to allow all to participate in them, conformably to the Regulations. While the Supernumerary Members survive, the new Members receive nothing,—but those who ascend in the degrees of The Order, receive the incomes hitherto paid to them.

The Members no longer consist of *Four*, but of *Five* Classes,—*viz.*,

GRAND CROSSES, 30.

GRAND OFFICERS, 160.

COMMANDERS, 400.

OFFICERS, 2000.

KNIGHTS, unlimited in number.

Among these prescribed Members, however, the members of The Royal Family, the Princes of The Blood, and all Foreigners, are not included. No Native can be admitted to The Legion of Honour, but by the lowest rank, that of Knight. To become an Officer, he must have been a Knight at least four years,—To be a Commander, he must have been two years an Officer,—and to become a Grand Officer, he must have been three years a Commander,—While he can only aspire to the First Class, after having been five years a Grand Officer. In cases of extraordinary distinction, these rules are, however, departed from.

Every Member who is in the Service of the State, either Naval or Military, must take the following Oath, kneeling,—

I swear to be true to my King, my Country, and to Honour,—instantly to reveal whatever may be known to me, that is contrary to the service of His Majesty or the welfare of the State,—to receive from no Foreign Prince any appointment, maintenance, or pension, without the express consent of His Majesty,—strictly to observe the French laws, orders, and rules,—and, in general, to do all that Duty enjoins to a true and honourable Knight of THE LEGION of HONOUR.

In time of Peace no one can be received into THE ORDER, unless he has particularly distinguished himself during 25 years in the Civil or Military service,—he then can make pretensions to it.

Promotions take place twice a year,—on the 1st of January, and the Festival of *St. Henry*, the 15th of June, that renowned Prince being the Protector of THE ORDER.

In time of War, additional Promotions are made. The appointment of Military Officers always takes place *on Parade*,—that of Civil persons, in *The Publick Sittings* of The Tribunal of the first instance. *Foreigners* receive only Letters of advice, and Decorations without brevets. In the Military service, every Campaign is reckoned as two years,—but a single Campaign is only reckoned as one year,—exceptions to this rule are only made by special Ordinances.

When Promotions and Appointments are to be made, the King determines beforehand the number of Members to be named for every Class. Then follows the Partition of the Appointments by The Grand Chancellor, among the different Ministers in these proportions,—*viz.*, $\frac{1}{40}$ to the Ministry of The King's Household,—$\frac{2}{40}$ to that of Justice,—$\frac{1}{40}$ to that of Foreign Affairs,—$\frac{6}{40}$ to that of Domestick Affairs,—$\frac{2}{40}$ to that of Finance,—$\frac{20}{40}$ to that of War,—$\frac{5}{40}$ to that of the Marine,—$\frac{\frac{1}{2}}{40}$ to that of Police,—and $\frac{2\frac{1}{2}}{40}$ to that of The Grand Chancellory of The Order.

The administration of the Revenues of The Order, together with all matters relating to The Order, is in the hands of The Grand Chancellor, who holds the rank of a Knight of the first class, and is always chosen from among them. With him is appointed a Secretary General, as an Assistant, who transacts all business in case of his illness or absence. On this report the King regulates the finances of The Order.

On the new arrangement of The Order, the right of Members to belong to The Electoral Colleges was abolished, as contrary to the Constitution of The Order. At the same time, The Grand Council, The Division into Cohorts, The Grand Secretary, and The Establishments for the education of Orphans of Members of The Order, were also abolished.

All other Ordinances, regarding The Order, not previously revoked, were suffered to remain in force.

THE DECORATION of THE ORDER is, at present, a Gold Cross, enamelled white, with five wings, each with two points, on which are small gold balls. Behind these five wings, runs a green wreath of laurel and oak. The round gold shield in the middle, bears on the Obverse, the head of King Henry the Fourth of France, surrounded with the words in golden letters, on a dark blue ground, "*Henri 4. Roi de France et de Navarre.*" On the Reverse, are three Lilies, encircled with the words "*Honneur et Patrie.*" Above the cross is the Royal Crown.

Under the reign of Buonaparte, the likeness was that of himself, encircled with his name,—and, on the Reverse, was the French Eagle, with outstretched wings.

The First Class, or The Grand Crosses, wear this beautifully formed Badge on a deep red, watered riband, sashwise over the right shoulder,—and on the left side of the Coat, a similarly formed Star, in the middle of which is the likeness of Henry the Fourth, encircled by the Motto, "*Honneur et Patrie.*"

The Grand Officers wear the cross smaller, in the left button-hole,—and on the right side of the Coat, the same Star as the first class, but of smaller dimensions.

The Commanders wear it round the neck,—

The Officers wear it in the left button-hole, with a loop,—and

The Knights wear it in the same manner, but without the loop, and of silver.

The first two Classes have at Court and on occasions of Ceremony, the same Rank and honourable Distinctions, as The Grand Crosses of The Order of St. Louis. On Publick Ceremonies, the first Class ranks with the first class of The Order of St. Louis and of Military Merit,—The second Class, with the Commanders of the same Order, —then follows the third class, and then the fourth, with The Order of St. Louis, all according to the order of their reception.

At the Interment of Members of the two first Classes, the same Ceremonies are observed, as at that of a Lieutenant General, unless they have filled a higher Military Rank.

At the Interment of a Commander, the Ceremony is the same, as at that of a Colonel,—at that of an Officer, as at that of a Captain,—at that of a Legionary or Knight, as at that of a Lieutenant.

To the first Class, Sentinels present arms,—to the rest, they only shoulder arms.

The Establishment for the Education of daughters of The Order, which was first placed at *Ecouen,* and is now at *St. Denis,* is under the Superintendence and Direction of The Grand Chancellor. It is arranged to receive 500 Pupils,—of whom, 400 are taught and supported *gratuitously,* and 100 are maintained by their friends.

The Duchess of Angoulesme is Protectress of the Establishment. Under her directions are placed, one Superintendent, 8 Dignitaries, 10 Ladies of the first class, 36 of the second class, and 20 Novices. With some slight difference, they all wear by way of honourable distinction a peculiar decoration on a red and white riband, consisting of a white Cross, of gold, displaying The Virgin Mary on one side, and the three French Lilies on the other. These, however, can only be worn out of the House, when their possessor has discharged the duties of her post with faithfulness and assiduity for Twenty years.

Grand Cross.

Sir Astley Paston Cooper, Bart., G. C. H.

Commander.

Thomas Fellowes, Captain in The Royal Navy.—Battle of *Navarino.*

Knights.

Sir Thomas Lawrence, Knt., Principal Painter in Ordinary to His Majesty, and President of The Royal Academy, " in consideration of his having recently painted for His Majesty the Portraits of His Most Christian Majesty CHARLES the Tenth, King of France and Navarre, and of His Royal Highness THE DAUPHIN,"—Royal Licence, 14th of July, 1826.

GEORGE MANSEL, Esq., Captain in The Royal Navy, "for his Services during the operations of the French forces at *Algiers*."—Royal Licence, 14th of November, 1831.

Lieutenant BALDWIN WAKE WALKER, of The Royal Navy, "in approbation of his Services at the taking of the Fort of *Patras*."—Royal Licence, 1st of December, 1831.

Sir ARCHIBALD WILLIAM CRICHTON,—*See*, RUSSIA, Order of ST. ANNE, of the Second Class.

Sir CHARLES WILLIAM DOYLE, Knt.,—*See*, HANOVER.

Captain H. I. CODRINGTON.

Captain ALFRED LUCKRAFT.

Commander SIDNEY C. DACRES.

Lieut. GEORGE LOGAN, of the Royal Marines.

Crosses and Medals.

Cross of July,—Louis Philippe, King of The French, by an Ordonnance, dated the 30th of April, 1831, ordered,—

1st. That the Decoration instituted by the Law of the 13th of December, 1830, in order to perpetuate the memory of the glorious days of The Revolution of 1830, should be called "*The Cross of July.*"

2d. That it should consist of a Star of three points, in white enamel, mounted in Silver. The centre of the Star divided into three glories, enamelled in the National Colours, surrounded by an oaken crown, to bear on the face the 27th, 28th, 29th July, 1830,—and for the legend, "*Given by the King of the French.*" The reverse divided like the centre of the face, to bear the Gallick Cock in gold, with the legend, "*Country and Liberty.*"

3d. That the Cross shall be suspended from a watered riband, of an azure blue colour, of 37 millimètres wide, with a red border of two millimètres in width, on each side of the riband, at a distance of two millimètres from the edge.

4 & 5. Provide, that the Citizens decorated with this Order, shall take the Oaths of Fidelity and Obedience, — and that Military honours shall be paid to The Cross.

Medal of July,—Louis Philippe, by an Ordonnance, dated the 30th of August, 1830, declared, that a Medal should be struck in memory of the events of the 26th, 27th, 28th and 29th of July, 1830, as a reward for those who were wounded in defence of The National cause at *Paris* on the above days,—and that the necessary operations should be entrusted to a Commission named by The King.

Medal for Veterans.—Louis the Fifteenth, by an Ordonnance, dated the 16th of April, 1771, conferred marks of Distinction on the

Soldiers of his army. To those who had served 24 years he gave a Medallion, upon which are two swords crossed and bound together by a riband, on a red ground, and surrounded by an ornamented border. Many Knights of *St. Louis*, on the suppression of ORDERS replaced their decorations by this Medal. Those Military men who had served 48 years, wore two Medallions.

There are still a few old Soldiers, who are decorated with this Medal.

DECORATION of THE LILY.—The Count d'ARTOIS, on his re-entry into France in the year 1814, conferred upon The National Guards, on the 2d April, 1814, the permission to wear a white, watered riband, with a *Silver fleur de lis* suspended from it, and surmounted by a crown, —as a distinguishing mark of the Services, which they had rendered.

To each individual was delivered a Brevet conferring the right to wear this decoration at the button-hole of the coat,—subsequently it was successively granted to the Army, and a part of the Nation.

By an Ordonnance of the King, of the 5th of August, 1814, a blue border of two millimètres was added to the riband of The National Guard of Paris,—and, by another Ordonnance, of the 15th of April, 1816, The King supplied the place of the *fleur de lis* by a STAR, enamelled white and blue, bearing on one side the portrait of LOUIS the Eighteenth, with the MOTTO, "*Fidelité dévouement,*" (*Fidelity, devoted Attachment*),—and, on the reverse, a Lily, with the date of the following remarkable days, 12th of April, 3d of May, 1814, 17th of March, 3d of July, 1815. It is suspended from a riband which is divided into three equal parts, the centre white, and the two others royal blue.

Subsequently to this period, different Ordonnances have successively granted particular ribands to The National Guards of each Department of the Kingdom.

DECORATION of ROUEN.—This Medallion, conferred on the Royal Volunteers formed in the Department of the Lower Seine, bears on the obverse, the head of a Warrior helmeted, with the legend, "*Virtue, Courage, Heroism,*"—and, on the reverse, a *fleur de lis*, and the legend,

"*Pledge of Union,*"—it is surrounded by a wreath of Laurel, and surmounted by a Crown.

Decoration of Bayonne.—This Decoration, conferred on the National Guard of this Town, bears on the obverse, a *fleur de lis*, on an oval shield, surmounted by a Crown, with the legend, "*National Guard of Bayonne,*"—and, on the reverse, an escutcheon, surrounded by the words, "*Never polluted.*"

Decoration of Bordeaux.—In 1814, the Guards of Honour who received His Royal Highness The Duke d'Angoulesme, were allowed to exchange the green Scarf which they wore on the arm, for a Decoration, under the name of a Bracelet, consisting of a Medallion surmounted by a Crown,—in the centre, a Lily, and the legend, "*Bordeaux,* 12th *of March,* 1814."

Decoration of Lyons.—Some persons who were present at the siege of Lyons in 1793, wear a Star, of four points, enamelled white, with a *fleur de lis* in each angle,—in the centre, the date 1793,—and around it, the legend, "*Siege of Lyons,*" on a blue ground.

Decoration for The Royal Volunteers.—This Decoration, conferred in 1815, is a Star, of four points, enamelled blue, and surmounted by a crown,—in each angle a *fleur de lis,*—with the date 1815 on a white band across the centre,— nd around it, the legend, "*Royal Volunteers,*"—on the reverse, is the Royal effigy, with the inscription, "*King, Country, God.*"

THE FREE CITY OF FRANCFORT ON THE MAYNE.

MEDALS.

THE SENATE instituted a MEDAL for The Volunteers of THE FREE CITY of FRANCFORT, who fought for the Deliverance of Germany in 1813 and 1814.—And another for The Volunteers and Troops of the Line, who took part in the Campaigns of 1814 and 1815.

GENERAL PRINCE REUSS-GREIG HENRY XIII., Governor General of The General Government of Francfort, instituted a CROSS in 1814, and conferred it on all the Officers of the Troops of The General Government of Francfort. The letters, "FR. F. W. *Al.*," upon this Cross, signify, "*Franz Friederich Wilhelm Alexander.*"

THE CIRCLE OF FRANCONIA.

THE TEUTONICK ORDER.

THE TEUTONICK ORDER, which is a Military Religious Order of Knights, was established towards the close of the Twelfth Century, and was so called, because it consisted principally of *Germans*, or *Teutones*.

This Order is said to have had its Origin in the following manner, —The Christians, under GUY of LUSIGNAN, laying siege to *Acre* or *Acon*, a City of Syria, on the borders of The Holy Land,—at which siege were present, RICHARD King of England, PHILIP AUGUSTUS of France, &c., some Germans of *Bremen* and *Lubec*, touched with compassion for the sick and wounded of the Army, who wanted common necessaries, set on foot a kind of Hospital under a Tent, which they made of a Ship's sail,—and here betook themselves to a charitable attendance on them.

This circumstance suggested a thought, about the year 1190, of establishing a *Third* Military Order, in imitation of *The Templars*, and *The Hospitalers*.

The design was approved by the Patriarch of Jerusalem, the Archbishops and Bishops of the neighbouring places, the King of Jerusalem, the Masters of The Temple and the Hospital, and the German Lords and Prelates then in The Holy Land,—And, by common consent, FREDERICK, Duke of Suabia, who was then at their head, sent Ambassadors to his brother HENRY, King of the Romans, to solicit the Pope to confirm THE NEW ORDER.

CALIXTUS the Third, who then governed the Church, was favourable to the request, which he granted by a Bull of the 23d of February, 1192, and the New Order was called "THE ORDER of TEUTONICK KNIGHTS of THE HOUSE of ST. MARY of JERUSALEM."

This, which was once one of the most powerful ORDERS in Europe, is now divided into Two BRANCHES,—

THE FIRST BRANCH is for *Roman Catholicks*, who take the Oaths of Celibacy, and have a House at *Mergentheim*, a Town of Germany, on the South side of the *Tauber*, wherein to transact their affairs. It is the residence of The Grand Master.

The ENSIGN worn by this BRANCH is, a Cross Patonce, *sable*, thereon a Cross fleury, *or*, over all on the centre an escutcheon, *or*, charged with The Imperial Eagle, *sable*. This Ensign is worn round the neck, pendent to a gold chain.

THE SECOND BRANCH is for *Protestants*. They have a House at *Utrecht*, wherein to transact their business, and in which the Secretary resides.

The method of Election into this Branch is somewhat singular. The Nobles of *Holland*, if they propose a son to be a Knight, go to Utrecht, and enter his name in the Register, and pay a large sum of money to the use of the Poor, who are maintained by THE ORDER,—and the person whose name is entered, succeeds in rotation.

When a Knight dies, he that is first on the List is summoned to attend The Chapter, and to bring with him proof of his Nobility for four generations, both on his father's and mother's side,—if not, he is struck out of the List.

The Ensign of this Branch is a Cross, *pattée*, enamelled, *argent*, surmounted with another, *sable*,—above this Cross, is a ball twisted white and black. It is worn, pendent to a broad black, watered riband, which is worn about the neck. The same Cross is embroidered on the left breast of the upper garment of each Knight.

The Emperor Joseph the Second diminished very considerably the possessions of The Order, situate in his States,—and the Treaty of *Luneville*, in 1801, deprived it of two principal Bailliwicks,—those of *Alsace* and *Lorraine*.

The Illustrious Equestrian, Secular, and Capitular Order of Saint Joachim.

This Order owes it's origin to no Crowned Head, but has been recognised, both in Great Britain and upon The Continent, as an Order of Knighthood.

It was set on foot in the year 1755, and derives it's chief claim to notice, from having our great Naval Hero, Lord Nelson, numbered among it's Members.

An objection, however, has been raised against the sanction of this Order, as being one of an inferior description, and not worthy of being ranked among the Orders of Knighthood,—as well in consideration of it's foundation, as the description of Persons by whom it is distributed.

The Badge, appertaining to the Knights, Commanders, and Grand Commanders, consists of a gold Cross with eight points, the whole enamelled in white. In the middle, and on both sides, is a Laurel Crown enamelled on a flat circle. Upon the Obverse, is the figure of *St. Joachim* in a green dress, with a white sash round the body,—from his left shoulder is suspended a white scrip, and, upon his head, he wears a green cap, according to the Oriental fashion. In his left hand is a shepherd's crook. Upon the Reverse, is a green *Cross-pattée.* This Cross hangs by a large gold ring from a

Knightly helmet of massy gold, and is worn suspended from a dark green, watered riband.

A circumstantial History of this Decoration has recently issued from the accurate and elegant pen of my friend, GEORGE FREDERICK BELTZ, Esq., K. H., LANCASTER HERALD, in his "*Review of* THE CHANDOS PEERAGE *Case*," p. 224.

When the King's permission was, in 1802, solicited for the Noble Admiral (NELSON) to avail himself of this boon, some apprehension was entertained lest future inconvenience might have arisen from giving, by such a recognition, a certain sanction to an ORDER of so inferior a description, as not to have been ever heard of at The British Court. Upon an Official inquiry into the History and Character of The Fraternity, it then appeared, that it had been set on foot, in 1755, by an Association of younger Members of some Princely and Noble Houses in Germany, who at first called themselves by the strange, almost blasphemous Title of "*Knights of The Order of Defenders of The Honour of Divine Providence*,"—that FRANCIS, one of the Princes of *Saxe-Cobourg*, was the first Grand Master elected by The Association,—that he resigned in 1773, —and that, upon the death of the second Grand Master, The Count *de* MONTFORT, in 1780, GEORGE CHARLES LEWIS, Count of *Leiningen-Westerbourg*, was elected Grand Master, under whose Presidency, in 1785, The Title of The Order was changed to that of "*St. Joachim, the Blessed Father of the Holy Virgin Mary.*" Upon the death of The Count, in 1787, the Office was conferred upon his son, CHARLES GUSTAVUS, who, having been deprived of his Estates in Germany on the French invasion, in 1793, under General CUSTINE, was conducted a Prisoner to Paris, and died there in the Abbey of *St. Germain*, in June 1798,—when he was succeeded in the Dignity of Grand Master by his son FERDINAND CHARLES, Count of *Leiningen*, who, in 1802, held the Office.

The affairs of the *soi-disant* Order were, for many years, administered by LEVETT HANSON, Esq., a Gentleman of Yorkshire, who resided entirely abroad, adopted the Style of "*Sir*," and transported the Chancery and Archives of The Order from city to city. Having succeeded in inducing Lord NELSON to enter into The Brotherhood, the Royal Licence permitting His Lordship so to do, with some Account of The Order, was printed and circulated, and subsequently embodied in a more general Work, which appeared under the title of " HANSON's *Orders of Knighthood.*" It was long understood, that *moyennant* a certain not inconsiderable deposit at a Banking-house in Pall-Mall, the Distinction was at the service of any one who might have a fancy for it,—and that Letters-missive were soon forthcoming from *Sir* LEVETT, containing due notification of Election by " *The Equestrian, Secular, and Chapteral Order,*" at it's last sitting at *Bamberg, Hamburgh, Lubeck,* or wherever that Personage happened, at the time, to be domiciliated.

KNIGHTS GRAND COMMANDERS.

HORATIO, Viscount and Baron NELSON, K. B., Vice-Admiral of The Blue, also Duke of BRONTÉ in Sicily, Grand Cross of The Order of St. Ferdinand and of Merit, and Knight of The Order of The Crescent,—by Warrant and the Royal Signet and Sign Manual, dated 23d of July, 1802, authorized to receive and wear the Ensigns of Knight Grand Commander of The Equestrian, Secular, and Capitular Order of *Saint Joachim,* on the nomination signified to him by FERDINAND CHARLES, reigning Count of *Leiningen-Westerbourg,* Grand Master of The Order, by his Letter, dated at *Westerbourg,* in General Chapter, on the 3d of April, 1802.

HORATIO NELSON, Esq., commonly called Lord Viscount MERTON, only Son and Heir Apparent of The Right Honble. and Reverend WILLIAM, EARL NELSON, then " *a Minor,*"—by Warrant under the Royal Signet and Sign Manual, bearing date the 29th of March, 1806, to receive and wear the Ensigns of this Illustrious Order, conferred upon him by The General Chapter of The Order,—Royal Licence, 25th of April, 1806.—Died, in January, 1808.

PHILIP D'AUVERGNE, Prince *de* BOÜILLON, Rear Admiral of The Blue Squadron of His Majesty's Fleet,—to receive and wear the Ensigns of Knight Grand Commander of The Equestrian, Secular, and Capitular Order of *Saint Joachim*, to which he was elected at a General Chapter holden at *Erlangen*, on the 27th of September, 1803, by the Count of LEININGEN-WESTERBOURG, Grand Master, and the Capitular Knights of the said Order.—Royal Licence, 27th of March, 1807.

CHARLES IMHOFF, Esq.,—now Sir CHARLES, and Major General,—a Lieutenant Colonel in the Army,—to receive and wear The Ensigns of The Equestrian, Secular, and Capitular Order of *Saint Joachim*, of which he had been elected a Knight Grand Commander at a General Chapter of the said Order holden at *Bamberg*, on the 20th of June, 1806.—Royal Licence, 18th of May, 1807.

KNIGHTS COMMANDERS.

Sir ROBERT KER PORTER, K. H.,—Knight Commander,—in May, 1807.—Knighted, 2d of April, 1813.

Sir SAMUEL EGERTON BRYDGES, Bart.

THE REPUBLICK OF GENOA.

The Order of Saint George.

The Republick of Genoa *formerly* bestowed The Order of Saint George upon those, who had done great and essential Services.

The Knights wore a gold Chain round the neck, from which was suspended a Saint George's Cross of gold, and red enamel.—And a Cross, of the same form and colour, was embroidered on their Mantles, and upper Apparel.

This Order has been long *dormant*.

THE KINGDOM OF GREECE.

The Order of The Redeemer.

The Order of The Redeemer was instituted by Otho the First, Prince Royal of Bavaria, King of Greece,—who, by the Statutes, dated, *Nauplia*, the 1st of June, (20th of May, O. S.), 1833, on the recommendation of His Ministers of State, determined and ordered, as follow,—

Article 1. An Order of Merit shall be established for The Kingdom of Greece, in remembrance of the deliverance of the Country,—an event which, by the aid of Divine Providence, is as miraculous, as it is fortunate. It shall bear the name of "Order of The Redeemer."

Art. 2. The Order of The Redeemer is divided into Five Classes,—*viz.*, 1st Class, Knights of The Silver Cross,—2d Class, Knights of The Gold Cross,—3d Class, Commanders,—4th Class, Grand Commanders,—and 5th Class, Grand Crosses.

Art. 3. The number of Members admissible into the first Class,—The Knights of The Silver Cross,—is unlimited.

The number of the other classes is fixed, as follows, and can never be exceeded,—KNIGHTS of THE GOLD CROSS, 120,—COMMANDERS, 30,—GRAND COMMANDERS, 20,—and GRAND CROSSES, 12.

The Princes of The Royal Family, as well as those Foreigners on whom a higher honour shall be conferred, shall not, however, be included in the above number.

Art. 4. THE BADGE of THE ORDER consists of a white, enamelled Cross, of eight points, surmounted with the Royal Crown.

The middle of the Cross, decorated with a wreath of Oak and Laurel, shows on one side, the Greek Cross, with the shield of a heart shape, as they are borne in the Royal Coat of Arms, and this inscription in the Greek language, "*Lord, thy right hand is glorified with power.*" On the Reverse, is the bust of The Founder, with the inscription, "OTHO, KING OF GREECE."

Art. 5. The Knights of THE SILVER CROSS wear on the left breast, THE BADGE of THE ORDER, enamelled in Silver, appended to a blue, watered silk riband, with a narrow white border.

The Knights of THE GOLD CROSS wear THE BADGE of THE ORDER, enamelled in gold, in the same manner, as The Knights of The Silver Cross.

The Commanders wear on the breast, THE BADGE of THE ORDER, enamelled in gold, on a similar, but rather broader riband, appended round the neck.

The Grand Commanders wear THE BADGE, enamelled in gold, after the pattern given for The Commanders,—and also, on the right breast of the coat, a STAR, of eight points, embroidered in silver, on which lies a Cross, according to the form prescribed in the fourth Article, and with the inscription there detailed, in gold letters, enamelled blue.

The diameter of the Star is fixed at seven centimetres.

The Grand Crosses wear THE BADGE of THE ORDER, enamelled in gold, on a broad silk riband of the colour above described, hanging from the left shoulder to the right side,—and also, on the left side of the coat, on the breast, a STAR, embroidered in Silver, of the pattern prescribed for The Grand Commanders, but with a diameter of ten centimetres.

Art. 6. All Members of THE ORDER of THE REDEEMER, who are

Greek Subjects, are to wear the *Insignia* appointed for their Class, whenever they appear in presence of The King or of the Royal Princes, or by Royal Invitation, or summons at a Publick Ceremony or Assembly.

Independently of which cases, Grand Commanders and Grand Crosses are bound to wear THE BADGE, prescribed for the Knights of The Gold Cross, whenever they are not adorned with THE BADGE of their own class.

Art. 7. THE KING is GRAND MASTER of THE ORDER, and wears according to His pleasure, THE BADGE of either Class.

The bestowal of THE ORDER rests wholly with His Majesty.

Art. 8. THE ORDER is conferred for life.

On the decease of a Member, the *Insignia* bestowed upon him, shall be returned to the Minister of Domestick and Foreign Affairs.

Art. 9. THE ORDER of THE REDEEMER shall only be conferred,

1st. Upon *Greek Subjects*, who, either during the Struggle for the Deliverance of Greece, have rendered important services, and contributed essentially to it's freedom and preservation,—or, who, in future, in any branch of the Publick Service, in the Army or the Navy, in Diplomacy, in Law, or in any Publick Office, or in any Situation of Publick confidence, in Art or Science, in Agriculture, Trade, or Commerce, shall have rendered service to The Throne, the honour of the Greek Name, and the welfare of their Father-land.

2d. On *Foreigners*, who, either possess the above qualifications,—or, who, by their peculiar Merit, seem calculated to promote the Dignity and Honour of THE ORDER.

Art. 10. No *Greek Subject* can be received into a higher Class of THE ORDER, without having passed through the lower. The only exception to this rule has consisted in the first establishment of the higher Classes of THE ORDER.

On the Promotion to the higher Classes, the claims of distinction, WORTH and MERIT, shall alone be considered, and all regard to Rank and Station shall be excluded.

Art. 11. The bestowal of THE ORDER on Greek Subjects, as well as their Promotion to the higher classes, can be conferred only by a Decree, signed by The King, and countersigned by The Secretary of Domestick and Foreign Affairs,—in which the Merits of the Individual are to be particularized.

The Order can be conferred upon *Foreigners*, only by Royal Letters.

Art. 12. A suitable Endowment shall be apportioned to The Order of The Redeemer, in order to provide an adequate yearly Income to a certain number of Members of each Class.

The marks of Honour to be paid to the Members of The Order, will be fully determined hereafter.

The Ministry of Domestick and Foreign Affairs is charged with the Publication and Execution of this Decree.

Grand Crosses.

Vice-Admiral Sir Pulteney Malcolm, G. C. B., and Knight Grand Cross of The Most Distinguished Order of St. Michael and St. George, " in approbation of His Majesty Otho, King of Greece, of his Services while Commander in Chief of the British Squadron in The Mediterranean, particularly during the Blockade of *The Morea*, and at the Surrender of the Fortresses of *Modon*, *Coron*, and *Patras*."—Royal Licence, 16th of Februray, 1835.

Admiral Sir Edward Codrington, G. C. B., Grand Cross of St. Louis and of The Legion of Honour, and Knight of The Second Class of St. George of Russia.

Commanders.

Leicester Stanhope, Esq., (commonly called The Honble. Leicester Stanhope), a Colonel in The Army, and C. B., " a testimony of the Royal approbation of His Majesty Otho, King of Greece, of his conduct while in the Service of *Greece*."—Royal Licence, 30th of April, 1838.

Lieut. Colonel Hon. John Hobart Caradoc, K. H.

Rear Admiral Sir F. L. Maitland.

Captain Sir Thomas Fellowes.

Captain Hon. Sir I. A. Maude.

Captain Hon. Fredk. Spencer.

Captain Edward Curzon.

Captain Lewis Davies.
Captain Viscount Ingestrie.
Captain Richard Dickinson.
Captain Sir Edward Lyons, Knt.
Captain Stephen Lushington.
Captain J. N. Campbell.
Captain R. L. Baynes.
Captain Sir John Franklin.
Captain Henry John Codrington.
Captain Alfred Luckraft.
Captain George Aug. Eliott.
Captain W. B. Walker.
Commander Sidney C. Dacres.
John Liddell, M. D.

Silver Cross.

Edgar Garston, Gent., " in testimony of the approbation of His Majesty Otho, King of Greece, of his conduct while serving in the Greek Army, during the war of *Independence*."—Royal Licence, 3d. of June, 1837.

Lieut. George Logan, of the Royal Marines.

THE KINGDOM OF HANOVER.

The Royal Hanoverian Guelphic Order.

It appears by the preamble to the Statutes of this Order, that his late Majesty George the Third having for some time entertained the intention to increase the means, by founding an Order for Hanover, to be enabled to give a publick testimony of His favour to faithful Servants of the State, and Individuals devoted to His exalted person,—and also to reward distinguished services to the Country, His Royal Highness George, then Prince Regent, had, in the Name and on the behalf of His Majesty the King his Father, deemed the present time to be particularly favourable to the accomplishment of that intention,—

The period at which the Hanoverian States had returned, after a long hostile Usurpation, under the Government of their legitimate Sovereign, gave The Prince Regent an opportunity to consider, with lively interest, the past years of their misfortune, in which all Classes of His beloved German Subjects had given the most undoubted proofs of unshaken Fidelity and Attachment, and recalled to His memory more particularly that year in which, on the occasion of the Expulsion of the Enemy from His States, so many Individuals risked their lives and property, and many had the good fortune to distinguish themselves particularly by their Actions, —if THE PRINCE REGENT was previously induced by all these considerations, and at the time of the Extension of the Boundaries of His German Dominions, and their Elevation to the Rank of a KINGDOM, to institute this already proposed ORDER, and the never to be forgotten Fame which His HANOVERIAN troops obtained at the battle of *Waterloo* in Flanders, on the 18th of June, 1815, had called upon Him more particularly and earnestly to mark this so glorious an Epoch for HANOVER, by the Foundation of this ORDER for henceforth and for ever.—

Wherefore, in consideration of so important a subject, HIS ROYAL HIGHNESS had taken upon himself, in the Name of His Majesty THE KING, the Office of GRAND MASTER of this ORDER, and as he thereby declared Himself again and publickly as it's CHIEF, SOVEREIGN, and GRAND MASTER, and were to be considered so by every person, He had commanded certain STATUTES and REGULATIONS to be drawn up, considered the same most maturely, and approved of the same, by His virtue and power as GRAND MASTER, so that they might, in all the affairs concerning THE ORDER, be taken as an unchangeable guide, and were to be preserved for ever in His archives of THE ORDER,— His gracious pleasure was, therefore, that,—

THIS NEW ORDER of KNIGHTHOOD was to be considered as instituted from the 12th of August, 1815, and to be called "The Guelphic Order," with a view to preserve on this day the remembrance of several important events.

For the more particular splendour of this Order, The Prince Regent decreed, that,—

The office of Grand Master of the same was for ever to be vested in the Crown of Hanover, and for no reason whatsoever to be separated from the same by himself or his Successors.

The Members of the Order were to consist of Three Classes, namely,—Grand Crosses,—Commanders,—and Knights,—and

A Medal was also to be instituted for Non-commissioned Officers and Soldiers, who should have distinguished themselves before the Enemy, by courage or judgement.

The Badge of The Order is a Maltese Cross, surmounted by two Swords, and the Crown of Hanover upon the upper limb of the Cross, and between each limb a Lion *passant guardant* in the centre,—in a Medallion, is *The White Horse* of Hanover on a field, *gules*, encircled by the Motto of The Order, "*Nec aspera terrent*," which is again encircled by a wreath of *Laurel*. In the Badge of the Civil Knights the Swords are omitted, and the wreath is of *Oak*.

This Badge, when not worn from The Collar of The Order, which is composed alternately of Lions *passant guardant*, and double cyphers of G. R., is worn by the Knights Grand Crosses scarfwise, from a sky blue riband. The Grand Crosses also wear on their left breast, the Star of the Order.

The Commanders wear the Badge, somewhat smaller, round the neck, pendent from a riband two and a half inches broad, and the same Badge worked in *silver*, with two Swords in *saltier*, but, without the Crown, on the left side of their coat.

The Knights wear the Badge to a Riband two fingers broad, and gold buckle, from their button-hole.

The Medal is worn to a Riband of the same breadth from the button-hole, and is accompanied with an annual *Pension* of two dollars monthly.

The Order is instituted as well for Civil as Military Service. Nevertheless, there is to be, as far as the Decoration is concerned, on the STAR and the BADGE the difference, that those Military persons who have obtained the Order for distinguished and prudent conduct before the enemy, are to wear on the Star and the Badge, instead of the Wreath of *Oak* a Wreath of *Laurel*, and two Swords under the Crown and over the Badge.

THE MOTTO of The Order is, "*Nec aspera terrent,*" in letters of gold, within a sky blue circle, surrounded by *a wreath of Laurel*, if accompanying the Arms of a *Military* Knight,—but if those of a *Civil* Knight,—then to be placed within the said Circle and Motto, surrounded by *a wreath of Oak*.

No person admitted into any of the Classes of THE ORDER is entitled to assume the appellation of "*Sir,*" unless he has received the Accollade.

In order to perpetuate the remembrance of THE INSTITUTION of THIS ORDER, HIS ROYAL HIGHNESS most graciously ordained, that THE FESTIVAL of THE ORDER shall be celebrated solemnly and yearly on the 12th of August (his birth-day), as the day of the Foundation, in His Capital of Hanover.

The Knights are authorised to surround their Arms with the *Motto* and *Collar* of THE ORDER.

But THE KINGDOM of HANOVER being separated from THE BRITISH Crown by the demise of His late Majesty King WILLIAM the Fourth, THE ROYAL HANOVERIAN GUELPHIC ORDER must now be classed with those of *Foreign States*. I, therefore, confine my enumeration of it's Members only to those, upon whom the Honour has been conferred, since the Accession of His Majesty ERNEST-AUGUSTUS to the Throne of HANOVER, on the 20th of June, 1837.

In the month of August, 1837, the following alterations respecting the wearing of the Decorations of The Royal Hanoverian Guelphic Order were determined by The King of Hanover,—

1. Knights Grand Crosses to wear the Riband over the right

shoulder towards the left side, instead of being placed over the left shoulder, as formerly.

2. The Riband from which the Cross of a Knight Commander is suspended from the neck, is to be fastened together with hooks and eyes, and not, as formerly, with a clasp,—the Decoration to be worn close under the cravat.
3. The Cross of the third class, formerly worn with a buckle and swivel, to be worn now suspended from the button-hole with the Riband drawn through the ring at the top of the Cross,—which ring is to be made large enough for that purpose,—and the buckle and swivel to be henceforward discontinued.

By an official notice in THE HANOVERIAN GAZETTE, it appears, that, on the 12th of August, 1837, being the Anniversary of the creation of THE GUELPHIC ORDER, a CHAPTER was holden in The Royal Palace, His Excellency Count KIELMANNSEGGE presiding. The new Nominations, since the 12th of August, 1836, were announced, as well those by WILLIAM the Fourth as by the present KING. It appeared, that almost all the Nominations by WILLIAM the Fourth were *English* Subjects,—those of the present King, all *Germans*,—except,—

The Right Honble. JOSEPH PLANTA, Grand Cross.

Captain GEORGE GOSLING, R. N.

Lieutenant GORDON, R. N.

Captain STEPHENS, Aid-de-Camp to The Duke of Cambridge.

Sir CHARLES WILLIAM DOYLE, Knt., a Lieutenant-General in the Army, C. B., to accept and wear the Insignia of The Grand Cross of The Royal Hanoverian Guelphic Order, the Insignia of The Royal French Order of The Legion of Honour, and the Supernumerary Cross of The Royal and Distinguished Spanish

Order of Charles the Third, "in consideration of his distinguished Services in the field of battle, and during *The Peninsular War*.—Royal Licence, 28th of November, 1838.

Captain THOMAS BARKER DEVON, of The Royal Navy, to accept and wear The Commander's Star of The Royal Hanoverian Guelphic Order (in lieu of The Cross of a Knight of the said Order formerly conferred upon him), "in approbation of active, gallant, and zealous services rendered to Hanover by him in the rivers *Elbe*, *Weser*, and *Ems*, in the year 1813."—Royal Licence, 7th of December, 1838.

Various other BRITISH SUBJECTS have been nominated by THE KING of HANOVER.

The Hanoverian Gold Medal of the Guelphic Order.

This Gold Medal, being the *only one* hitherto conferred upon an Englishman, and having been struck specially at *Hanover*, by the command of His Majesty King George the Fourth, requires to be particularly recorded,—

The Obverse presents a Bust of His Royal Highness The Prince Regent, laureated, with the legend, "Georg. Prinz. Regent 1815."

The Reverse contains this inscription, "Verdienst ums Vaterland," encircled with a richly clustered wreath of Laurel. Round the rim of the Medal is engraven, "K. G. Britan. General. Hos. Commiss. John Charles Stahlschmidt."

The Medal, which is richly chased and faced with crystal, is attached to the light blue riband, and buckle of The Guelphic Order.

The Gentleman who has had the singular good fortune to receive this Medal, is John Charles Stahlschmidt, Esq., of *St. John's*, Lambeth, and of *Albany Cottage*, Weybridge, both in the County of Surrey, to whom the agreeable intelligence was announced by Count Münster, Resident Hanoverian Minister of State at The British Court, and, in that capacity, Chancellor for the time being of The Guelphic Order,—in a letter, dated *Putney Hill*, the 7th of October, 1825, in the following obliging terms,—

" That His Majesty had most graciously been pleased to grant him THE HANOVERIAN GOLD MEDAL, as a reward for his meritorious services, rendered in the years 1813-1814, when, acting as Deputy Purveyor to the Forces, he was attached to The Hanoverian Army,—and, in the year 1816, when a Board of Officers was assembled at Hanover, for the examination of Invalid Soldiers of The King's German Legion."

The Grant of this Medal was, in the customary manner, Officially announced in The Royal Hanoverian Gazette, of the 16th of August, 1826, with the other appointments of Members of The Guelphic Order, which had taken place during the preceding year.

In the Journal *du Departement des Deux-Néthes*, dated, *Antwerp*, the 28th of July, 1815, The Civic Commissioners, appointed by The Mayor of Antwerp for the organization of Hospitals for the general reception of the Wounded, after the Battle of *Waterloo*, make, in their Official Report to that Functionary, the following observation,—" We know not how sufficiently to express our acknowledgements to General MACKENZIE, Messrs. HIGGINS, Mc DOUGLE, RAITT, SAUNDERS, STAHLSCHMIDT, MACHELL, and MITCHELL, and, in general, to all the British Authorities, who, on every occasion when we have claimed their assistance, ministered abundantly to our wants."

The Physician in Chief of The Hospitals, L. H. VRANCKEN, M. D., one of the Members composing the Medical Jury of The Department *des Deux-Néthes*, writes to Mr. STAHLSCHMIDT from *Antwerp*, on the 14th of August, 1815, in the following gratifying manner,—

" At the moment I am about to cease my Functions, as Physician in Chief of The British Hospitals of French wounded Prisoners of War, united in different Establishments under my Inspection and Superintendence, since the celebrated Battle of *Waterloo*, to the number of 2383, of which more than 2000 are convalescent and cured,—I cannot take leave of you, without committing great injustice towards your merits, unless I convey to you publickly my acknowledgement of the zeal and extraordinary activity with which you have constantly seconded me in the organization and administration of the before-mentioned Establishments, the regularity of which must, in common justice to you, be attributed to yourself alone.—The Regulations there established will always be present in my memory,—

"I am, therefore, Sir, in duty bound, to convey to you my acknowledgements for your highly meritorious conduct, assuring you at the same time of my approbation of your administration under circumstances of such peculiar difficulty."

On the occasion of Mr. STAHLSCHMIDT's repassing *Antwerp*, on his road from *Paris*, to join The Board at Hanover, CHARLES FREDERICK RAITT, Esq., the Superintendent and Agent for French Prisoners, addressed a Letter to him, dated *Antwerp*, the 28th of January, 1816, to the following effect,—"I cannot permit the opportunity of your being here, on your road to *Hanover*, to pass by, without, in the strongest manner, bearing testimony to your zeal and good conduct, during the whole period you had the charge of The French Prisoners, —and, I trust, that your Merits, wherever your services are called upon, will be as conspicuous as they have been here, and finally lead to the highest Rank in your Department."

And, further, CHARLES GROSKOPFF, M. D., the Principal Medical Officer in charge of The General Hospital Establishments of The Hanoverian Troops in the Vicinity of Paris, writes also to Mr. STAHLSCHMIDT, from *Rueil*, near *Paris*, on the 3rd of December, 1815, as follows,—"I cannot relinquish the charge of The General Hospital at *Rueil*, without testifying to you my approbation of your conduct in the Quality of Deputy Purveyor to that Establishment. Accept my sincere acknowledgements for the support which you have afforded me in the discharge of a duty, which the nature of the circumstances and the superficial knowledge of the new Hanoverian Levies of our Hospital Regulations, tended to render particularly difficult."

An extract from the Report of Sir JOHN MEADE, K. H., President of The Board of Officers assembled at *Hanover*, addressed to Sir JAMES MC GRIGOR, Director General of The Army Medical Department, and dated, *Osnabrück*, the 6th of June, 1816, expressly states, "I should consider myself as guilty of an injustice did I not *again* bring Mr. STAHLSCHMIDT to your recollection, whose merits are beyond all praise."

These honourable Testimonials are recapitulated and confirmed by Sir JAMES MC GRIGOR, Director General of the Army Medical Department, who, on the 1st of December, 1824, states officially, "that Mr. STAHLSCHMIDT was employed in The General Hospital Staff of The Army in *The Netherlands* and *France*, during which period he executed the duties entrusted to him, with great zeal and ability,—

"Mr. STAHLSCHMIDT had charge of the Purveyors Department of The Hanoverian Troops quartered in the Vicinity of Paris, for which duty his knowledge of the German language and proficiency in The British Hospital Regulations particularly qualified him.

"The same qualifications led to his appointment with The Board of Officers constituted for the purpose of invaliding The King's German Legion in the year 1816, and from the various Reports which have reached me from his Superior Officers, I consider that his merits, with particular reference to these two Services, have been peculiarly conspicuous and satisfactory."

SUCH TESTIMONIALS may dignify any Family!—and these will assuredly be cherished with the most affectionate regard and care by the latest Posterity of this excellent Gentleman.

THE FREE HANSE TOWNS.

MEDAL.

THE FREE HANSE TOWNS of HAMBURG, LUBECK, and BREMEN, determined in the year 1814, to institute a MEDAL of War for those individuals of THE HANSEATIC LEAGUE and THE NATIONAL GUARDS, who had made The Campaign.

At *Bremen* the foundation was made by a Decree, of the 3d of April, 1814.—At *Hamburg*, on the 9th of March, 1815,—and, at *Lubeck*, on the 7th of June, 1815.

The MEDAL is to be worn on the left breast, or at the button-hole, attached to a red and white riband.

THE ELECTORATE OF HESSE CASSEL.

THE ORDER OF THE GOLDEN LION.

THIS ORDER was founded, on the 14th of August, 1770, by FREDERICK the Second, Landgrave of HESSE-CASSEL, who married The Princess MARY, daughter of His Britannick Majesty, GEORGE the Second, and was a Knight of The Most Noble Order of THE GARTER.

HIS HIGHNESS, actuated by those generous inclinations which ever disposed him to recompense Virtue and Merit, and to increase the Splendour of his House, named it "THE DOMESTICK ORDER of THE GOLDEN LION," and chose his Ancestress, *Saint Elisabeth*, as the Patroness of it.

Up to the conclusion of the year 1816, it continued unaltered,—but it's original Constitution was about that time modified by The Elector WILLIAM the first, who gave it NEW STATUTES, which appeared in print on the 1st of January, 1818.

In pursuance of these Regulations, the Sovereign of the Country is the Chief and Grand Master of THE ORDER,—and the Princes of the reigning House, by virtue of their birth, are Grand Crosses, and, on arriving at years of maturity, are decorated with the *Insignia.*

The Members of THE ORDER who hitherto formed but one Class, now consist of four Classes,—*viz.*, GRAND CROSSES,—COMMANDERS, of the first Class,—COMMANDERS, of the second Class,—KNIGHTS, whose number is *unlimited.*

It is bestowed without respect to religious belief, birth, or ancestry, to *Natives* and *Foreigners,* both *Civil* and *Military,* partly as a reward and honour for distinguished merits, partly as a token of friendship and esteem.

Such as are Subjects of The Landgrave must have served their Sovereign, during a certain number of years, with zeal and fidelity,—And with regard to the Military, they must be at least Lieutenant Generals, and have previously received The Order "*Pour la Vertu Militaire.*"

With respect to such as have Offices at Courts, or Civil Employments, they must be persons of a Rank equally honourable and distinguished.

The reigning Landgrave, as Grand Master, invests all newly created Knights of The Order, according to accustomed Ceremonies. In case the newly chosen Knight cannot attend in person, the Insignia are sent to him, accompanied by a Princely rescript, in the hand writing of the Sovereign,—or, they are, in his Name, dispatched to the Knight Companion elect, together with the Statutes, and are presented to him by a Knight of THE ORDER, whom the Sovereign has commissioned to that effect.

The Knights Companions are individually bound, not only to attach themselves to the Person of the Sovereign, but likewise to promote and advance all such matters, as may tend to His honour and advantage. They are under the strongest obligation to succour each other, and to protect, assist, and defend the poor, the helpless, the widow, and the orphan, so far as their means will allow them to do so.

THE BADGE for The Grand Crosses, is an octagonal Medal, enamelled red, in the centre, a Lion rampant, of gold, ducally crowned,—on one side is the MOTTO, "*Virtuti et Fidelitati*,"—and, on the other, "FREDERICUS II., D. G., *Hassiæ Landgravius inst.* 1770." It is worn, pendent to a deep crimson coloured, watered riband, of the breadth of the hand, scarfwise over the right shoulder.

They likewise wear upon the left side of their Mantles, and usual upper apparel, a STAR, of eight points, embroidered in silver,—in the middle whereof, is the Hessian Lion, striped across, with red, in a blue field,—and around the same, the MOTTO is embroidered in letters of silver, upon a circle of deep coloured crimson velvet.

THE BADGE of the other Classes is totally different from the preceding. For the COMMANDERS of the first and second class, it is a gold Cross, enamelled crimson, with a white margin, and surmounted by a Royal crown, of gold. The middle space which lies on a Silver Star, of eight points, displays on the obverse, the golden crowned Lion on a blue ground, around which, in a crimson circle with an orange border, is the MOTTO of THE ORDER, in letters of gold,—on the reverse, is the intertwined cypher W. K. (*William, Elector*), of gold, in a blue field and bordered by a gold ring. This Cross is worn by the second and third Classes, round the neck on the breast. By way of distinction, however, the second Class wear on the left side on the coat, a SILVER CROSS, of eight points, in the midst of which, surrounded by a gold border, is the gold erect, crowned Lion, in a blue field.

THE BADGE of the Fourth Class or Knights has, indeed, no relation to the form of the Cross of the second and third Classes,—the middle space is, however, oblong, and the SILVER STAR is wanting, as well as the ROYAL CROWN surmounting the Cross. It is worn by this Class on a narrow riband in the left button-hole.

THIS ORDER differs from all others in the manner by which the several Classes are distinguished from each other, for in no other is there any actual difference in the Decoration,—it is always of the same form, and differs in size only with each Class,—but this Order presents the singular instance of a complete departure from the ancient hereditary and regular method. That the Decoration of The Commanders is also of a far more pleasing and tasteful form than that of The Grand Crosses, admits also of no doubt.

The Officers of The Order are, a Chancellor, Master of the Ceremonies, Counsellor, Treasurer, Keeper of the Wardrobe, and Herald.

Each Knight is entitled to combine the *Insignia* of The Order, with the usual Armorial Bearings of his Family.

Within the space of three months, after the death of each Knight, his Family is bound to transmit, as well the Statutes, as the Badge of The Order, to the Directorial Counsellor.

The Order of Military Merit.

This Order was founded, on the 5th of March, 1769, by The Landgrave Frederick the Second, of Hesse Cassel, for the reward of Military Merit among his Officers.

The number of the Members who consist of one Class only, is unlimited,—and the Order is conferred upon Officers of *Foreign* Troops. The bestowal of it rests entirely with The Grand Master, who is always The Sovereign of The House of Hesse Cassel.

No Revenues are associated with it.

The Badge is a Gold Cross, of eight points, enamelled pink, and covered with a Royal crown, and on the four points the cypher originally of the Founder F. L. *(Frederick, Landgrave)*, but now of the present Grand Master W. K. *(William, Elector)*, with the Motto, "*Virtuti*," is displayed in letters of gold. Between each of the angles of the Cross, is the erect and crowned gold Lion of Hesse.

This Cross is worn round the neck, pendent to a plain sky blue riband, with a rich border of silver embroidery. It must be worn at all times, and upon all occasions.

In time of Peace, this Order is conferred only upon Field Officers.

The Order of The Iron Helmet.

This Order, like it's model that of *The Iron Cross*, arose at the time of the great struggle of the People for Freedom and Independence.

It was founded, on the 18th of March, 1814, by The Elector William the First, of Hesse, for the reward and distinction of Military services rendered to the Country by his Subjects,—but Military men of *Foreign* armies are not excluded from receiving it.

It consists of three Classes,—*viz.*,—Grand Crosses,—Knights, of the First Class,—Knights, of the Second Class.

The Badge is the same for all Classes, but is double the size for the Grand Crosses. It consists of a black *Brabant cross of cast iron,* on the obverse of which, in the centre, is an open Helmet, by the side whereof on the cross is the cypher of the Founder W. K. *(William, Elector),* and beneath, the date of the year 1814. The reverse is smooth.

This Cross is worn round the neck by The Grand Crosses, on a red riband, with a white border.

The Knights of both Classes wear it in the button-hole,—and the first Class has by way of distinction, a Cross of the riband of the Order on the left breast.

The Grand Cross is only given to him who, as Commander of Hessian troops, has essentially contributed to gaining a Battle, or who has successfully defended or captured a Fortress.

As the Statutes expressly enjoin, that The Order of the Iron Helmet shall not be given after the conclusion of the War, it has consequently not been bestowed since the Peace.

MEDALS, AND MARKS OF DISTINCTION.

THE MEDAL of MERIT,—in silver, was instituted towards the end of the year 1820 by The Elector, WILLIAM the first, for the purpose of creating a mark of honourable Distinction for Military men, Servants of the State, and the Prince of inferior rank and Municipal officers, who, by long and faithful service, useful inventions, courage in time of danger, or other actions of a similar kind, may have deserved well of their Prince, the Publick, or their Fellow Citizens.

THIS MEDAL was renewed and altered in 1821, by WILLIAM the Second,—and, at the close of the year 1831, was changed by The Co-Regent and Electoral Prince into a CROSS, named "THE CROSS of MERIT." The conditions upon which it was to be obtained, however, remained the same as those attached to The Medal. THIS CROSS, which is of gold or silver, and of four corners rounded, bears on one side of the Medallion, the cypher of The Elector, WILLIAM the second, —on the other, that of The Co-Regent,—and on the two sides of the four points, the inscription, "*For Merit and Fidelity.*" It is worn at the left button-hole, attached to the riband of THE CROSS of a Knight of THE ORDER OF THE GOLDEN LION.

THE MEDALS of COMMEMORATION and HONOUR,—These were instituted by The Elector, WILLIAM the second, on the 14th of March, 1821, for The Hessian Soldiers who had fought in the two German wars of Deliverance of 1814 and 1815. They were conferred upon Military persons of every rank, with the indication of the post which they then filled,—whether in fighting in the ranks, connected with the Staff, in the Military administration, or having entered the field under the Banners of their Country, passed the *Rhine*, and not been guilty of any dishonourable action.

THE MEDAL FOR THE COMBATANTS,—is *cast from the Metal of the Cannon taken from the Enemy,* and is so specified in an inscription on the Exergue,—That of persons who had *not* been engaged, is cast in *iron*,—both of them are worn at the left button-hole.

THE GRAND DUCHY OF HESSE DARMSTADT.

The Order of Louis.

This Order was founded, on the 25th of August, 1807, by The Grand Duke Louis the First, to honour and reward Merit, both *Civil* and *Military*, from the highest to the lowest degree.

On the 14th of December, 1831, Louis the Second issued special Regulations for it's management,—and changed the Name from " The Grand Ducal Hessian Order of Merit," to that of " The Order of Louis."

It consists of five Classes,—*viz.*, Grand Crosses,—Commanders, of the first Class,—Commanders, of the second Class,—Knights, of the first Class,—and Knights, of the second Class.

The Badge which is a Cross, of eight points, is enamelled black, with a red border, set in gold,—in the centre of the obverse, is the gold letter L., on a red ground, encircled by a white margin, on which are the words, " *For Merit*,"—on the reverse, on a black ground, is a green wreath, interwined with laurel and oak, for Soldiers and Citizens, and encircling the words, " *God, Honour, and Fatherland.*"

The Riband on which it is worn, is black, with a red border.

The Cross and Riband are of the same form for all classes, but diminish in size with every descending class.

The first class wear the Cross on a riband of a hand's breadth, over the right shoulder,—the second, round the neck. Both Classes are adorned on the breast with a SILVER STAR, of eight points, the centre of which contains the reverse of the Cross.

The third class wear it also round the neck, but without a Star,—and the fourth and fifth, in the button-hole.

THE DUCHY OF LUCCA.

The Cross of Saint George.

Charles Louis *de* Bourbon, Infant of Spain and Duke of Lucca, by a Decree, of the 1st of June, 1833, founded at *Vienna* an Order, named, " The Cross of St. George for Military Merit,"—which comprises two Classes, and is worn at the left breast.

The Decoration consists of a Cross, of Silver, with the cypher C. L. in the centre, surmounted by a crown. It is attached to a white watered riband, edged with red.

The Cross of the first Class is conferred only upon Superior General Officers, and other Officers of great merit. That of the second Class is conferred upon Officers, and also upon Under-Officers and Privates.

This Decoration is sometimes conferred, ornamented with *diamonds*, —but this is very seldom done, and only for extraordinary Merit.

Medal.

The same Duke also founded at *Vienna*, on the 1st of June, 1833, a mark of Distinction, called " *The Medal for Seniority*," as a reward for *thirty years* of faithful service. It is worn on the left breast, pendent to a riband of blue and yellow stripes.

THE DUCHY OF NASSAU.

Decorations.

There are three Ducal Decorations of Nassau,—

A Medal *for the battle of Waterloo.*
A Medal *of Military Merit,*—and
An Honorary Military Distinction *for Service.*

Duke Frederick Augustus, on the Institution of the Medal of Military Merit, published an Edict to the following effect,—

That *Foreigners,* as well as *Natives,* may obtain this mark of Distinction, which should consist of a Medal of gold and silver.

That the Medal to be conferred on Under Officers from the rank of Serjeant Major, and Soldiers, as a reward for individual brave actions in time of war.

That the *golden* Medals be given for the most daring exploits,—and the *silver*, for those of less brilliancy.

That the Medal shall be conferred personally by the Chief Officer, and to be worn at the button-hole, attached to a deep blue and yellow riband.

That the possessor of the *silver* Medal shall enjoy an increase to the amount of half his usual pay,—and the possessor of the *gold* Medal double pay,—and which increase shall be paid, so long as the party shall continue in the service as on the peace establishment.

That the Medal and additional Pay shall be enjoyed, even when the possessor shall have been promoted to the rank of an Officer.

Persons convicted of an offence by a Council of War, and punished, shall lose their Medal, and all concomitant advantages,—as also, *such as sell, or lose the Decoration at play.*

The Medal of deceased possessors descends to their wives and children only.

Duke WILLIAM, by an Edict, dated, at *Biebrich,* on the 25th of February, 1834, published on the foundation of THE MARK of HONOUR *for Military Service,* ordered,—

That this Distinction should consist,—For OFFICERS, of a CROSS, of *gold,* bearing on the obverse, in the medallion a W., — and, on the points, the Words, "XXV. *years of faithful service.*" The reverse to bear in the medallion the words, "25. *February,* 1834," the date of the first distribution. This Cross to be worn, attached to the left breast, by a blue riband,—

For UNDER-OFFICERS, MUSICIANS, and PRIVATES, of a CROSS, of *silver,* with the same inscription as above, with the exception of the Roman cypher on the upper point, which should be according to the Classes, XXII, XVI, and X.,—The riband of the first class, to be the same as that for the officers,—that for the second, to have on the same ground, *a yellow ray,*—and that of the third, *two rays* of this colour.

This Decoration may be worn, after the possessor has obtained his discharge.

Under-Officers, Musicians, and Privates, who, after the termination of their period of service, shall voluntarily continue in the Army, shall receive for the first Class, an increase of pay of *one-third,*—for the second, a *sixth,*—and for the third, a *twelfth.*

THE KINGDOM OF THE UNITED NETHERLANDS.

The Military Order of Wilhelm.

This Order, created by the Law of the 30th of April, 1815, was instituted as a reward for Military Services rendered to the King and the Country. The Grand-Mastership is inseparable from the Crown of The Netherlands.

The Order consists of four Classes,—*viz.*, Grand Crosses, Commanders, and Knights, of the third and fourth Class.

The Decoration consists of a Cross, of eight points, enamelled white, with *golden* balls, those for the fourth class of *silver*, covered and angled by two branches of laurel, disposed like a Cross of Burgundy, and connected by a golden *briquet* (steel for striking light). On the reverse, a Medallion, enamelled blue, in the centre of which is a W., in a wreath of laurel.

On the rays of white enamel is the inscription, "*Voor moed, beleid, trouw,*" *(for Courage, Prudence, and Fidelity)*. The whole surmounted by a crown of *gold*,—that for the fourth class being of *silver*.

Grand Crosses wear upon the left breast a Star of silver, and the

Badge of The Order attached to a wide orange coloured riband, bordered with Nassau blue, passing from the right shoulder to the left side.

Commanders wear the Cross embroidered on the left breast of the Coat, and the Badge suspended from the riband of The Order passed round the neck.

Knights of the *third* Class wear the Badge suspended from the button-hole by a riband two fingers wide.

Knights of the *fourth* Class wear the Badge from a riband of the width of one finger.

Military men of the Army and Navy, not being Officers, who obtain the Decoration of Knights of the *fourth* Class, receive an increase of pay to the amount of one-half, and double pay, if they enter into the *third* Class.

The number of The Knights is unlimited.

Grand Cross.

The Right Honble. Edward, Viscount and Baron Exmouth,—*See,* Spain, Order of Charles the Third.

Commander.

Vice-Admiral Sir David Milne, K. C. B.

Third Class.

Charles Count Alten,—*See,* Russia, Order of St. Anne.

Colonel Sir A. Anderson, C. B.

General Lord Hill.

Thomas Baker, Esq., Post Captain in The Royal Navy, and C. B., "in consideration of the signal services rendered by him, in the month of November, 1813, to His Majesty The King of The Netherlands, then Prince of Orange, by effecting, with great promptitude and ability, at that critical period, the disembarkation of a Body of Seamen on the Coast of Holland, thereby eminently contributing to the success of the efforts then making by The Dutch Nation to recover their Independence."— Royal Licence, 7th of March, 1818. Now Rear Admiral.

Lieut. Colonel COLIN DUNDAS GRAHAM, late of The Scotch Brigade, and Deputy Governor of St. Maw's Castle, " as a mark of the approbation of His Majesty The King of THE NETHERLANDS of his services, while in The Scotch Brigade, formerly in the service of Holland, and subsequently at *The Helder* in 1799, and since The Revolution in 1813."—Royal Licence, 30th of May, 1818.

Sir GEORGE ADAM WOOD,—*See*, Austria, Order of MARIA THERESA.

EDWARD BRACE, Esq.,—*See*, Spain, Order of CHARLES the THIRD.

FOURTH CLASS.

HENRY ALLEN JOHNSON, Esq., " in testimony of the approbation of His Majesty The King of THE NETHERLANDS, of the signal intrepidity displayed by him in the field against the Enemy, while acting as Aid-de-Camp to His Royal Highness The Hereditary Prince of ORANGE, during the Campaigns in *The Peninsula*."—Royal Licence, 12th of August, 1817.

Major General Sir C. B. VERE.

CHARLES HAMILTON SMITH, Esq., Major in The Army, " in testimony of the high sense which His Majesty The King of THE NETHERLANDS entertains of the services rendered by him at *Zierikzee*, and at the taking of The Island and Fortress of *Tholen*."—Royal Licence, 10th of June, 1823.

Lieut. Colonel GEORGE MUTTLEBURY, late of the 69th (*or*, South Lincolnshire) Regt. of Foot, and C. B., " in consideration of his distinguished Services and conduct in the Battles fought in *The Netherlands* in the the year 1815."—Royal Licence, 10th of November, 1831.

Sir HENRY HOLLIS BRADFORD, K. C. B.—Died near Brussels, 17th of December, 1816.

Hon. WILLIAM STUART, Major General, and C. B.—*Waterloo.*

A Dispatch from Field-Marshal The Duke of WELLINGTON, dated, *Head-Quarters, Paris,* the 8th of October, 1815, contains the following List of Officers, upon whom His Majesty THE KING of THE LOW COUNTRIES had conferred Decorations of different Classes of *The Wilhelm's Order,*" "in testimony of His Majesty's approbation of their services and conduct, particularly in the late Battles fought in *The Netherlands.*"

Third Class WILHELM'S ORDER.

Lieut. General Sir HENRY CLINTON, G.C.B.
Major General Sir COLQ. GRANT, K.C.B.
Major General Sir COLIN HALKETT, K.C.B.
Major General Sir GEORGE COOKE, K.C.B.
Major General Sir JAMES KEMPT, K.C.B.
Major General Sir WILLIAM DORNBERG, K.C.B.
Major General Sir PER. MAITLAND, K.C.B.
Lieut. General CHARLES COUNT ALTEN, K.C.B.

Fourth Class WILHELM'S ORDER.

Honourable Colonel STEWART, 1st Guards.
Colonel FRA. HEPBURN, 3d Guards.
Colonel FRED. ARENTSCHEILDT, 3d Hussars.
Colonel A. B. CLIFTON, 1st Dragoons.
Honourable Lieut. Colonel W. ELPHINSTONE, 33d. Foot.
Lieut. Colonel E. O. TRIPP.
Lieut. Colonel Sir CHARLES BROKE, K.C.B.
Lieut. Colonel Sir HENRY BRADFORD, K.C.B.
Lieut. Colonel Sir GEORGE BERKELEY, K.C.B.
Lieut. Colonel LORD GREENOCK.
Lieut. Colonel R. NIXON, 1st Foot.
Lieut. Colonel G. MUTTLEBURY 69th Foot.
Lieut. Colonel HARRIS.
Lieut. Colonel J. ROSS, 95th Foot.
Lieut. Colonel BUSCHE, 1st Light Battalion, K.G.L.
Lieut. Colonel GEORGE BARING, 2d Light Battalion, K.G.L.

The Order of The Belgic Lion.

This Order, instituted by the Law of the 29th of September, 1815, is particularly destined as a recompence for those, who shall have given proofs of Attachment to their Country, of Zeal and Fidelity in the performance of their duties as Citizens, or of extraordinary success in the Sciences and Arts.

The Grand Mastership is inseparable from the Crown of The Netherlands.

The Order comprises three Classes,—*viz.*, Grand Crosses, Commanders, and Knights.

Those who deserve Distinction for any meritorious actions, may be associated with The Order, under the denomination of *Brothers*.

The Decoration of The Order consists of a Cross, of eight points, enamelled white, with golden balls, and a golden W. in the angles,—in the centre, an escutcheon, enamelled blue, with the legend, " *Virtus nobilitat,*"— and, on the reverse, the Lion, placed as in the Arms of the Kingdom, the whole surmounted by a Royal crown of gold.

The Riband of The Order is Nassau blue, bordered with orange.

The Grand Crosses wear that side of the Decoration of The Order on which is the device, without the Crown, upon a Star of gold, embroidered on the left side of the coat, and the Badge of The Order suspended from a riband four fingers wide, passed from the right shoulder to the left side.

Commanders wear the Cross of The Order embroidered upon the coat, without the Star, but surmounted by the Crown, and The Badge suspended from the neck by a riband of three fingers wide.

KNIGHTS wear the Decoration of THE ORDER suspended from a riband, passed through the button-hole.

THE BROTHERS, instead of the *Cross*, wear a *silver Medal*, having on one side the emblem of THE ORDER,—and, on the reverse, the device, suspended from a riband of Nassau blue of the width of one finger and a half, and striped with orange in the centre. *The Brothers* enjoy an annual *Pension* of 200 florins, the moiety of which is continued to their Widows.

THE STATUTES of THIS ORDER have not yet been promulgated.

Medals, and Crosses of Honour.

Medal of Doggersbank.—William the Fifth, Prince of Orange, in the month of October, 1781, in memory of the Naval Combat which took place, on the 5th of August of the same year, at *The Doggersbank*, between the Dutch fleet commanded by Admiral Zoutman, and the English fleet under the command of Admiral Parker, distributed a Medal, of an oval shape, having on one side a Marine victory holding in the right hand a crown of laurel, and in the left a Palm branch, standing upon the prow of a Vessel, bearing the inscription, "*Doggersbank,*"—in the field of the Medal, Vauc MDCCLXXXI., and as a legend, "Pax quæritur bello."—On the reverse, the inscription, "Eximiæ Virtutis Præmium," in the centre of a Crown of laurel, and as a legend, "Munificentia Principis Auriaci."

To Commanders of Vessels who were present at the Battle, it was given in *gold*, to be worn at the button-hole, attached to a riband of red, white, and blue,—in *silver*, to a riband of the same colours, to Officers,—and, in *silver*, to an orange coloured riband, to Cadets and Under-officers.

Medal of The Hague.—Conferred upon all those who first took up arms at *The Hague*, for the Expulsion of the French, and the Restoration of the legitimate Government, on the 17th of November, 1813.

The head of this Medal bears two Swords in the form of a *St. Andrew's Cross*, with the legend, "*For our Country and Orange,*"—on the reverse, the date XVII. *November* MDCCCXIII, in the centre of a wreath of Laurel and Oak, in allusion to the mixed character, *Military* and *Civil*, of this enterprise. It is worn, attached to a black and yellow riband, the colours of the Arms of *The Ville.*

Medal of Dordrecht,—Conferred upon those who, at *Dordrecht*, devoted themselves to the Expulsion of the French, and the Defence of this Town against their troops. The head of the MEDAL represents the prow of a Ship and a Cannon placed like a Cross of *St. Andrew*, and surmounted by a Mural Crown, and the legend, "*For our Walls and Hearths*,"—on the reverse, "*Dordrecht* XXIV. *Nov.* MDCCCXIII," in the centre of a wreath of Laurel and Oak.

MEDAL of NAARDEN,—Conferred upon the armed Citizens of *Amsterdam*, in memory of the Siege of *Naarden* occupied by the French, in 1813 and 1814. The head of the MEDAL contains, in the centre of a wreath of laurel and orange, the inscription, "*Siege of Naarden*, 1814,"– and, on the reverse, within a wreath of oak the inscription, "*Homage of The Central Commission at Amsterdam* to . . . . ,"— to which is added, the name of the person upon whom the Decoration is conferred. The Riband is white, black, and red.

MEDAL FOR FAITHFUL SERVICE,—King WILLIAM of The Low Countries, by an Ordinance, of the 19th of February, 1825, founded this MEDAL, which is conferred on Military men, in *bronze*, for *twelve* years, —and, in *silver*, for *twenty-four* years of faithful Service. It is worn at the button-hole, attached to an orange coloured riband, and has on one side, the Royal mantle charged with a W. crowned,—and, on the reverse, the Arms of the Kingdom, resting upon Military attributes, with the legend, "*For faithful Service.*"

MEDAL of JAVA,—This Medal was instituted by King WILLIAM of The Low Countries, by an Ordinance, of the 27th of June, 1831, as a reward for those Military men, who took part in the Campaigns of *Java* of 1825-30. It is worn at the button-hole, attached to an orange coloured riband, and bears on one side the bust of The King, with the legend, "*William* I., *King of The Netherlands*,"—and, on the reverse, the inscription, "*War against Java* 1825-30," in the centre of two branches of oak and palm.

CROSS of HASSELT,—This Cross, *cast in the metal of the Cannons taken*, on the 8th of August, 1831, at the battle of *Hasselt*, was instituted, on the 12th of September, 1831, by King WILLIAM of The Low Countries, and was distributed among the Military of every rank without exception, who took part in this event, occasioned by the revolt of *Belgium*. It is no longer granted.

MEDAL of ANTWERP,—This Medal was instituted by a Decree, of the 31st of May, 1833, by King WILLIAM, and conferred on Military men, who were distinguished by their conduct in the defence of The Citadel of *Antwerp*, in December, 1832.

THE BRONZE BUCKLE,—Instituted, by a Decree, of the 31st of May, 1833, by King WILLIAM, and intended as a Distinction, for Honourable actions. On one side is The Citadel of *Antwerp* under the form of a Pentagon,—and, on the reverse, a Crown of laurel, surmounted by a W. crowned, with the inscription, "*December*, 1832." It has been conferred upon those Military men, who formed part of the Garrison of The Citadel and the Forts dependent upon it,—as also, upon the Naval men belonging to *The Flotilla* before the City. It is worn, attached to a riband of Nassau blue.

THE KING, by a Decree of the 24th of January, 1839, founded a MEDAL, "*for Courage and Fidelity*."

It will be conferred upon Military men in the army of THE DUTCH INDIES, who do not belong to an European Corps, or who are not of European Origin.

THIS MEDAL, within the legend, "*In the Name of the King*," will bear these words, "*for Courage and Fidelity in The Dutch Indies*." The Medal will be worn upon the left breast, suspended from a riband of Nassau blue. The Medals will be, of *silver* and *bronze*. The *silver* Medal will be conferred, as a reward for very distinguished actions,—or, when the person decorated with the *bronze* Medal, shall continue to give proofs of Courage and Fidelity.

An increase of pay amounting respectively to *one half*, or *one third*, will be attached to the possession of the *silver* or *bronze* Medal.

THE GRAND DUCHY OF OLDENBURG.

THE FAMILY ORDER OF MERIT OF DUKE PETER FREDERICK LEWIS.

THIS ORDER was instituted, on the celebration of the 25th Anniversary of the return of THE GRAND DUCAL FAMILY to *Oldenburg*, in November, 1838.

The number of Members is at present fixed at 48,—*viz.*, 6 Grand Crosses, *two* of whom are Capitulars, and has each a Prebend of 500 rix dollars *per annum*,—6 Grand Commanders, of whom *two* have Prebends of 400 rix dollars,—12 Commanders, of whom *four* have Prebends of 300 rix dollars,—and 24 Knights, of whom the *four* Senior have Prebends of 250 rix dollars.

The reigning Grand Duke is Grand Master,—and the Hereditary Prince, when 18 years of age, is Grand Prior.

A general BADGE of Honour in these Classes, is connected with THE ORDER.

THE CHAPTER of THE ORDER is to be holden on the 17th of January, being the Birthday of the late Grand Duke PETER FREDERICK LEWIS.

MEDAL.

The Ducal Decree, concerning THE MEDAL of MILITARY MERIT for THE OLDENBURG REGIMENT, founded by Duke PETER, on the 30th of April, 1816, declares,—

That a printed Certificate shall be delivered by The Colonel to every individual who had faced the enemy,—which Certificates shall give the possessors the right to wear the MEDAL of MERIT, *date outwards,* on the left breast, attached to a blue riband. This right to belong not only to those at present in the regiment, but also to those who may have quitted it, whether Volunteers or otherwise.

THE IMPERIAL OTTOMAN EMPIRE.

THE IMPERIAL OTTOMAN ORDER OF THE CRESCENT.

It is an event, which forms a memorable Æra, in the Annals of the Eighteenth Century, that *The Turks*, against whom the first Order of Knighthood was established, should have instituted a Military one, to recompense the bravery of a CHRISTIAN, and a HERO, and expressly to commemorate a Victory, which was gained upon their own Coasts, and upon which depended their existence as a Nation. But what could otherwise be expected from the unrivalled courage, and transcendent genius of BRITANNIA'S "darling," NELSON!—not more renowned as an HERO, than consummate as a STATESMAN,—and by whom this glorious and ever memorable Victory was obtained over *The French Fleet*, in THE BATTLE OF THE NILE, on the first of August, 1798.

Having received an official confirmation of this decisive Victory, on the 29th of August, THE GRAND SEIGNIOR hastened to express his satisfaction at this joyful event to Mr. Smith, The British Minister at The Sublime Porte, on the 8th of September following, in these words,—

"It is but lately, that, by a written communication, it has been made known, how much The Sublime Porte rejoiced at the first advice received of The English Squadron in *The White Sea* having defeated *The French* off *Alexandria*, in Egypt. By recent accounts comprehending a specifick detail of the Action, it appears now more positive, that His Britannick Majesty's Fleet has actually destroyed by that Action the best Ships which the French had in their possession. This joyful event, therefore, laying this Empire under an Obligation, and the Service rendered by Our much esteemed friend, Admiral NELSON, on this occasion, being of a nature to call for Publick Acknowledgement, His Imperial Majesty, the Powerful, Formidable, and Most Magnificent Grand Seignior has destined as a Present in his Imperial Name to the said Admiral, *a Diamond Aigrette*, and *a Sable Fur*, with broad Sleeves,— besides *Two Thousand Zequins*, to be distributed amongst the wounded of his Crew. And, as the English Minister is constantly zealous to contribute by his endeavours, to the increase of Friendship between the two Courts, it is hoped he will not fail to make known this circumstance to his Court, and to solicit the permission of The Powerful and Most August King of England, for the said Admiral to put on and wear the said *Aigrette* and *Pelisse*."

To immortalize an Event so glorious to The British Arms, Mr. Smith suggested to the Minister of The Grand Seignior, Selim the Third, the propriety of establishing an Order, to be styled, "The Imperial Ottoman Order of the Crescent." The Sultan hastened to adopt the recommendation,—and This Order was instituted upon the purest principles of ancient Chivalry, being purposely founded to remunerate Warlike achievements, and Knightly prowess.

This enlightened Monarch, whose clear intellect sought by the introduction of useful institutions from Foreign States, to establish and promote the prosperity and improvement of his Subjects,—alas! fell a sacrifice to his praiseworthy endeavours.

THIS ORDER is formed after the manner of European Orders, being divided into three Classes,—but it is destined only for the reward of Services, which *Foreigners* may have rendered to THE PORTE, and not for his own Servants and Subjects. Owing to this peculiarity it has remained single of its kind,—though this circumstance may have partially contributed to its preservation,—for it is little seen in the Ottoman Empire itself, and the Turk, attached as he is to ancient manners, seldom comes in contact with an Institution adapted to European customs.

It is usually conferred upon Military men, Ambassadors and their Suite. Our lamented Hero, NELSON, was *the first Knight Companion of it.*

The distinguished successes of THE ARMY, under the command of Generals ABERCROMBIE and HUTCHINSON, on the arid Plains of EGYPT, and the important Services of Lord KEITH and THE ROYAL NAVY, induced THE SUBLIME PORTE to extend THE ORDER much beyond the bounds of it's original Institution. It has been successively conferred upon all the Commanders in Chief, and Superior Officers, *Naval* as well as *Military*.—And the *Insignia* of the second Class have been presented to numerous British Officers, who served in that arduous Expedition.

THE BADGE bears the characteristic Turkish emblem of the Half Moon, which MAHOMED the Second assumed, after the Conquest of *Constantinople*. On a round gold shield or medallion, enamelled red, is seen on the Obverse in the centre, a STAR, encircled with rays of brilliants, and on the margin, the CRESCENT, also formed of brilliants. On the Reverse, in Turkish characters, is the name of SELIM the Third, surrounded by a wreath of ornaments. The Knights of the first Class wear it, pendent to a broad red riband, from the right shoulder to the left side,—and, on the left breast, a STAR embroidered in silver, of the form of a radiant Sun, bearing in the oval centre which is red, the Silver Star and the Crescent.

The Knights of the second Class wear the same Decoration, but of smaller dimensions, round the neck, and without the STAR on the breast.

THE MEDAL of the *third* Class, which is worn on the left button-hole, is still smaller,—it is not enamelled with red, but is of *gold*, and instead of the brilliants, are the Star and the Crescent of silver,—and the date, in *Roman* numerals, 1801.

Whether there exists a prescribed CEREMONIAL, to be invariably observed, on the INVESTITURE of a Knight, is not accurately known.—But the following Ceremonial was observed by His Highness THE CAPTAIN PACHA, when, by command of THE SUBLIME PORTE, he conferred The Imperial Order of Knighthood of The Turkish Crescent upon Sir RICHARD BICKERTON, Bart., and several other Naval Officers in the Service of HIS BRITANNICK MAJESTY,—

On the morning of the 8th of October, 1801, The Admiral, accompanied by The Turkish Admiral of The Gallies, and Suite, and those Officers of The Navy, who had been particularly selected,—proceeded on horseback from General HUTCHINSON'S Tent, to the Tent of His Highness THE CAPTAIN PACHA, and were received by the whole Turkish line, under arms, with Musick playing, and Colours flying. When they alighted and approached the Tent, which was open in the front, they observed THE CAPTAIN PACHA, seated upon a magnificent Sopha,—attended by The Pacha of Egypt, the Chief General of His Highness's Army, and The Reis Effendi. The three latter were seated upon the same Sopha, and rose at their approach,—but His Highness received them sitting. They were placed in chairs on each side of the Sopha,—the Admiral on the right of The Captain Pacha. The General officers of The Turkish Army and Navy stood at the back of the chairs,—behind them were ranged His Highness's retinue, arrayed in their different Badges of Distinction, and round the Tent, in front, were drawn up his Body Guards. The rich Dress, and venerable figure of His Highness, made a most interesting appearance.

Having been served with Coffee and Sweet-meats, the Ceremony commenced by His Highness investing The Admiral with a *Pelisse*, the *Star*, and *red Riband*, and *Medal*, of THE ORDER of THE CRESCENT,—all of which being properly arranged, he was desired to kneel, at which time The Grand Seignior's Firman was read, empowering His Highness to confer the Honour of Knighthood, which was immediately performed on the Admiral,—upon whose rising, a Royal Salute was fired, and other demonstrations of satisfaction, agreeably to The Turkish custom.

The Admiral having retired to his seat, the Senior Post Captain was invested in the same form with the *Pelisse* and gold *Medal*, and was afterwards Knighted,—and then the other Captains in succession.

Four Masters and Commanders, and Lieutenant WITHERS, were then Knighted in the same manner,—but only received a gold MEDAL of THE ORDER, without the Pelisse.

The same Ceremony had been performed on General HUTCHINSON and the General Officers of the Army, the day before.

During the whole of the Ceremony, Musick was playing. After it was finished, a long history was read, stating the Power and Magnificence of The Grand Seignior, and consequently the value which was to be set upon the different Honours then conferred. This being finished, the newly created Knights were regaled with Sherbet,—they then arose, wearing the *Insignia* with which they were invested, and departed, mounted on their horses, in the same form which they had observed upon their coming thither,—at which time, another Salute was fired.

This Ceremony was performed on the spot where the Battle of the 21st of March, 1801, was fought, which decided the fate of *Egypt*.

SUPERIOR CLASS.

The Right Honble. HORATIO, Viscount and Baron NELSON, *of The Nile*, K. B., and Vice-Admiral of The Blue,—Royal Licence, 20th March, 1802.

Lieut. General JOHN HELY HUTCHINSON, afterwards Viscount HUTCHINSON, G. C. B.

The Right Honble. GEORGE, Lord KEITH, K. B., Commander in Chief of His Majesty's Ships and Vessels in the Mediterranean.—

Royal Licence, 20th of March, 1802. *See*, SARDINIA, ORDER of ST. MAURICE and ST. LAZARUS.

His Excellency THOMAS, Earl of ELGIN, His Majesty's Ambassador Extraordinary and Plenipotentiary at The Sublime Porte.—Royal Licence, 20th of March, 1802.

Major General Sir EYRE COOTE, K.B.—Royal Licence, 5th June, 1802. Died, 10th December, 1823.

Admiral Sir WILLIAM SIDNEY SMITH,—*Egypt and Syria*, 1799.

His Excellency Sir JOHN BORLASE WARREN, K. B., Rear-Admiral of The White Squadron of His Majesty's Fleet, and His Majesty's Ambassador Extraordinary and Plenipotentiary to the Court of St. Petersburgh.—Royal Licence, 21st of August, 1802.—Died, in 1822.

Major General JOHN DOYLE, G. C. B., Lieutenant Governor of The Island of *Guernsey*.—Royal Licence, 29th of July, 1803.

General Sir HILGROVE TURNER, G. C. H., Knight of The Imperial Russian Order of ST. ANNE, of the first Class, Knight of The Imperial Ottoman Order of THE CRESCENT, and late Governor of *Bermuda*, "by verbal permission of His Royal Highness The Prince Regent," afterwards GEORGE the Fourth.

It is to the firmness and decision of this gallant Officer, that ENGLAND is indebted for the celebrated *Trilinguar* Inscription, discovered at *Rosetta*,—a Fac-simile and copies of which were extensively circulated by THE SOCIETY of ANTIQUARIES of LONDON. This famous stone now forms a prominent feature in the Hall of Egyptian Antiquities at THE BRITISH MUSEUM.

It was styled "*The Gem of Antiquity*" by the French, as being the Key to the Hieroglyphick language.

His Excellency WILLIAM DRUMMOND, His Majesty's Ambassador Extraordinary and Plenipotentiary at *The Sublime Port*.—Royal Licence, 8th September, 1803.

JOHN STUART, Esq., Major General of the Forces, K. B., "to receive and wear the Badge of The Order of The Crescent, which The Grand Seignior hath conferred on him at *Constantinople*."—Royal Licence, 15th of November, 1803. Knighted, 20th of June, 1804. *See*, THE TWO SICILIES, ORDER of ST. JANUARIUS.

Sir Richard Hussey Bickerton, Bart., Rear-Admiral of The Red Squadron of His Majesty's Fleet.—And The Order of The Crescent having been instituted by his Majesty's Ally The Grand Seignior, in commemoration of the signal and important Services rendered to The Ottoman Empire, by the glorious Achievements of the British Arms in *Egypt*,—The King was further pleased by Warrant under His Royal Signet and Sign Manual, to grant to Sir Richard, His Royal Licence and Permission to bear, in allusion thereto, a *Crescent*, and certain other appropriate honourable Augmentations to his Family Arms, together with *Supporters*.—Royal Licence, 29th of December, 1804.

The Most Honble. Richard, Marquess Wellesley, K. P.—Royal Licence, 19th of November, 1805.

Harford Jones, Esq.—*now* Sir Harford Jones Brydges,—Resident of The East India Company at *Bagdad*, and Member of The Ottoman Order of The Crescent, of the *Second Class*, to receive and wear the Ensigns of *The Superior Class* of the said Order, as a mark of The Grand Seignior's further favour.—Royal Licence, 5th of March, 1807.—Created a Baronet, 27th of August, 1807.

His Excellency The Right Honble. Charles Arbuthnot, Ambassador Extraordinary and Plenipotentiary at *The Sublime Porte*.—Royal Licence, 7th of March, 1807.

Second Class.

The Right Honble. Richard, Earl of Cavan, Major-General of His Majesty's Forces,—Royal Licence, 19th of April, 1803.

Sir John Francis Cradock, afterwards Lord Howden, K. C. B., Major General of the Forces, Quarter-Master-General of the Army in Ireland, and late Commandant of the 54th Regt. of Foot, "to receive and wear The Badge of The Order of The Crescent, which The Grand Seignior hath transmitted to him."—Royal Licence, 15th of December, 1803.

David Baird, Esq., Major General of His Majesty's Forces, "to receive and wear The Badge of The Order of The Crescent, which The

Grand Seignior hath conferred upon him."—Royal Licence, 31st of December, 1803.—Created a Baronet, on the 13th of April, 1809, and G. C. B.

HARFORD JONES, Esq., Resident of The East India Company at *Bagdad.*—Royal Licence, 24th of May, 1804. —*See, The Superior Class.*

Lieut. Colonel GIDEON GORREQUER,—*See,* SARDINIA, ORDER of ST. MAURICE and ST. LAZARUS.

General RICHARD, Earl DONOUGHMORE,—*Egypt.*

Lieut. General LEWIS LINDENTHAL,—*Egypt.*

GEORGE ALAN MADDEN, Major General, C. B.—*Egypt.*

Sir GEORGE MURRAY, Lieut. General, G. C. B., and G. C. H.—*Egypt.*

Sir HILDEBRAND OAKES, Bart.—*Egypt.*
Died, on the 9th of September, 1822.

General BROWN CLAYTON,—*Egypt.*

Captain, now Colonel, ARCHIBALD MONEY, C. B., 60th foot.—*Egypt.*

Colonel B. ANSLEY.

Colonel J. R. ARNOLD.

Major General L. BRADSHAW.

Lieut. General J. G. CLAY.

Lieut. General Sir C. DOYLE.

Colonel G. A. HENDERSON.

General Lord HILL.

General Sir W. HOUSTON.

Sir R. H. KENNEDY.

Colonel F. MAULE.

Lieut. General Sir T. G. MONTRESOR.

General Sir W. PEACOCK.

Major-General G. J. REEVES.

Lieut. Colonel GEO. THORNHILL.

The following Officers of THE EIGHTY-EIGHTH Regiment of Foot, *or,* CONNAUGHT RANGERS, each received a MEDAL from THE GRAND SEIGNIOR, for their gallant Services in *The Indian Army,* which crossed *The Desert to Egypt,* in 1801,—

Lieut. Colonel BERESFORD.
Lieut. Colonel DUFF.
Major HOUGHTON.
Captain VANDELEUR.
— SILVER.
— BERESFORD.
— TROTTER.
— Mc GREGOR.
Lieutenant BLAKE.
— MURPHY.
— BROWNE.
— CANILAC.
— OATES.
— ATKINSON.
— IRWIN.
— ADAIR.
— Mc DOUGALL.
Ensign ROGERS.
Adjutant THOMPSON.
Surgeon Mc GREGOR.
Assistant Surgeon TAUNRY.
Colonel GEORGE O'MALLEY, for services as a Subaltern in the 13th Regiment with the army in *Egypt*, in 1801.

PALESTINE.

THE SOVEREIGN ORDER OF ST. JOHN OF JERUSALEM, *or*, OF MALTA.

The general History of this celebrated ORDER,—the venerable Patriarch of European Chivalry,—is so widely extended as to need no recapitulation in this place. It may be sufficient, therefore, only to observe, that, after flourishing for many Centuries in *The Holy Land*, at *Cyprus*, at *Rhodes*, and finally, in *The Island* from which it's more modern designation is taken, it's existence as a Sovereign State was put an end to by BUONAPARTE, as is thus related by Sir WALTER SCOTT,—

"The 10th of June, 1798, brought the magnificent armament of The French before Malta, once the Citadel of Christendom, and garrisoned by those intrepid Knights, who, half Warriors and half Priests, opposed the Infidels with the enthusiasm at once of Religion and of Chivalry. But those, by whom THE ORDER was now maintained, were disunited among themselves, lazy and debauched Voluptuaries, who consumed the Revenues destined to fit out expeditions against the Turks in cruizes for pleasure, not war, and giving balls and entertainments in the Seaports of Italy. BUONAPARTE treated these degenerate Knights with a want of Ceremony, which, however little it accorded with the extreme strength of their Island, and with the glorious defence which it had formerly made against the Infidels, was perfectly suited to their present condition. Secure of a party among the French Knights, with whom he had been tampering, he landed troops, and took possession of these almost impregnable Fortresses with so little opposition, that CAFFARELLI said to NAPOLEON, as they passed through the most formidable defences,—" It is well, General, that there was some one within to open the gates to us. We should have had more trouble in entering, if the place had been altogether empty."

" A sufficient Garrison was established in *Malta*, destined by BUONAPARTE to be an intermediate Station between *France* and *Egypt*,—and, on the 19th, the daring General resumed his Expedition."—SCOTT'S Life of NAPOLEON BUONAPARTE, vol. 4. p. 61.

A majority of The Knights then retired to *Trieste*, and afterwards to *Messina* and *Catana*.

At present THE ORDER is under the protection of THE EMPEROR of RUSSIA.

The Knights had a SEMINARY in England, until the Suppression of it by King HENRY the Eighth,—they, however, continued to appoint one, to whom they gave the Title of GRAND PRIOR of ENGLAND.

Sir RICHARD SHELLEY was the last Grand Prior in England, in the reign of Queen MARY.

Their SUPERIOR here in England was the *first Lay Baron*, and had a Seat among the Lords in Parliament, —and some of their Privileges were extended even to their Tenants.

THE BADGE of THE ORDER is a gold CROSS, of eight points, enamelled white, pendent to a black, watered Riband, worn at the breast.

But this ORDER being composed of persons of different Countries, the BADGE is decorated so as to distinguish the Country of the Bearer, *viz.*, GERMANY, by an Imperial Crown and Eagle,—FRANCE, the Crown and Fleurs-de-Lis, &c.

LADIES can be admitted into THE ORDER of MALTA. According to old customs, they are styled, " *Sisters Hospitalers.*"

The presence of His Imperial Majesty, FERDINAND the First, Emperor of AUSTRIA, in *Milan*, on the gorgeous Solemnity of His Majesty's Coronation, in September, 1838, has been followed among many other acts of Clemency and Liberality, by the restoration to THE KNIGHTS of MALTA of the Estates which had been added to the Dominions of the Crown, with permission to re-establish new *Commanderies*, by voluntary contributions. The *Military* object of THE ORDER has long ceased to have any influence,—the care only of the *Sick*, as the earliest office of the first Knights at *Jerusalem*, has lately been revived, and they have founded *Hospitals* for the purpose both at *Milan*, and in some other Cities. The restoration of the Estates in *Lombardy* is of the more importance, as the other Italian States, may be thereby induced to give back the property in their possession. FRANCE possesses the richest *Commanderies*,—which it may be difficult for THE ORDER to recover.

COMMANDER and KNIGHT.

Captain HOMR RIGGS POPHAM, of The Royal Navy,—Royal Licence,

28th September, 1799.—Knighted.—Died, 11th September, 1820.

KNIGHTS.

Admiral Sir WILLIAM SIDNEY SMITH.

The Right Honble. Sir WILLIAM JOHNSTONE HOPE, Vice-Admiral of the White, and G. C. B.

Major Sir W. H. TONKIN, Russian Vice-Consul at Teignmouth and Exeter.

Sir JAMES LAWRENCE, Author of a work, "*On the Nobility of the British Gentry*," 8vo.

THE GRAND DUCHY OF PARMA.

THE ORDER OF CONSTANTINE.

THE ORDER of CONSTANTINE is one of those ORDERS, which like that of THE GOLDEN FLEECE, of ST. LAZARUS, and a few others, are equally conferred by Two Sovereigns. It is bestowed by THE GRAND DUCHESS of PARMA, and by THE KING of THE TWO SICILIES. Each may be considered to have equal rights to confer it.

The origin of THIS ORDER is extremely remote, and is involved in fable. The old Historians make CONSTANTINE *the Great*, its Founder,—and they relate, that, in the year 313, previous to a battle in which he was about to engage against the rebel MAXENTIUS, he saw in the air a CROSS of the same form, as the Cross of THE ORDER, on which these words were resplendent, "*In hoc signo vinces.*" He gained the battle, and in remembrance of his success, and of this miraculous appearance, he founded THE ORDER. This wild assertion contradicts itself,—for, in the year 313, no one had conceived the idea of founding ORDERS, and where are the historical proofs of such a circumstance to be procured?—but, although not the Founder, his name has nevertheless been conferred upon THE ORDER.

In the year 1190, The Eastern Emperor Isaac Angelicus Comnenus founded an Order, with the consent of The Pope, for the reward and distinction of those who devoted themselves to the defence of The Empire, which he named The Order of Constantine. He adopted this designation, both because the Family of Comnenus derived it's descent from the Constantine, and also to honour the memory of that great Potentate, who founded the Roman Empire on the banks of *The Bosphorus.* He also termed it, "*The Order of Angelicus,*" because this was his own name, placed it under the protection of *St. George,*" whence it is occasionally termed "*The Order of St. George,*" and submitted it to the rule of *St. Basilius.*

That much benefit was expected from it, is evident from the circumstance, that The Grand Master was to have Vicars General over the whole world. The Knights also, who consisted of Grand Priors and Knights, were compelled to prove their *Nobility* through four generations, and, at their decease, to bequeath to The Order, out of their possessions, a sum of 100 crowns.

The dignity of Grand Master of this Order was from it's commencement an Hereditary property of the Family of The Comneni, who for a long period maintained it with becoming splendour. Overwhelmed, however, amidst the ruins of The Eastern Empire, The Comneni were compelled to seek protection and support among other Princes. This occasioned the conferring of Knighthood *to become a mere means of emolument,*—whoever paid the appointed fee, received The Order, by which means it became degraded and contemptible. The last descendant of the Family, Andreas Angelicus Flavius, Prince of Macedonia,—who, deprived of his Dominions, sought the help of other Sovereigns, came in 1699 to Parma. In the prospect of dying without issue, he thought it more advisable to dispose of The Grand Mastership to the best advantage, or the right of bestowing The Order during his lifetime, than to allow it to pass after his death into the hands of Strangers, without deriving any advantage from it himself,—and, as the then Duke of Parma, Francis the First, of The House of Farnese, determined on it's purchase, the right was transferred into the hands of the latter. This disposal was formally approved by Pope Innocent the Twelfth, and afterwards by Clement the Eleventh,—and The Dukes of

PARMA thus acquired in the most complete and satisfactory manner THE GRAND MASTERSHIP of THE ORDER of CONSTANTINE.

With a view to raise the degraded ORDER, FRANCIS bestowed upon it many possessions, made *Parma* the Seat of THE ORDER, and richly endowed the Church of *Madonna della Steccata*, in that City, in which Priests and Chaplains of THE ORDER were appointed to officiate.

In the year 1731, the reigning Family of FARNESE became *extinct*, on the death of Duke ANTHONY,—when, by virtue of earlier arrangements, The Infant DON CARLOS, son of King PHILIP the Fifth, of Spain, succeeded to The Duchy of *Parma*. He immediately declared himself GRAND MASTER of THE ORDER, and arranged it's Decorations. Three years afterwards, he exchanged the Title and the Possession of *Parma* for that of *Naples*, which the Spaniards had conquered, and which he now ruled as King. He, therefore, took with him the archives of THE ORDER, renewed THE ORDER in a formal manner, and transferred it from the soil of *Parma* to that of *Naples*.

After the death of FERDINAND the Sixth, in 1759, CHARLES succeeded to the Spanish throne,—his younger son, FERDINAND the Seventh, receiving with the Crown of *Naples*, The Grand Mastership of THE ORDER, and, while in *Parma*, succeeded the brother of CHARLES, the Infant DON PHILIP. He demanded of his Nephew FERDINAND the restitution of THE ORDER of CONSTANTINE, as belonging to The Duchy of *Parma*, but the demand was constantly made in vain. As fruitless were similar claims, subsequently made by his son. THE ORDER remained with the Crown of *Naples*, was transplanted to *Sicily*, in 1806, on the occupation of *Naples* by The French, and was only restored, in 1814, on the expulsion of The French from that Kingdom.

The vicissitude of States in modern times restored *Parma* and *Placentia* to THE IMPERIAL HOUSE of AUSTRIA. The Peace of *Paris* determined both of them to be the property of The Austrian Archduchess MARIA LOUISA, Ex-Empress of France, and she declared herself formally and solemnly, on the 23d of April, 1816, GRAND MISTRESS of THE ORDER,—founding her right thereto, not merely on the circumstance of it's having for more than a Hundred years been the Hereditary property of the successive possessors of The Dukedom, and of their claims having been confirmed by The Popes,—but also, on her own descent from THE HOUSE of FARNESE by her own mother, MARIA THERESA, a Princess of *Sicily*.

Whether the King of THE TWO SICILIES offered any declarations against this resumption of THE ORDER by MARIA LOUISA, is not known. It has, however, been conferred by *both* the parties.

The four Classes, of which it consists, are—GRAND DIGNITARIES, GRAND CROSSES, COMMANDERS, and CROSSES.

The Grand Master wears a golden CHAIN, composed of 15 links,—from the centre link is suspended the Decoration of The Order, and a *St. George* on horseback, fighting the dragon. He also wears, like the other Grand Crosses, THE STAR, on the left breast.

The Grand Crosses wear the Cross of The Order, with a *St. George*, attached to a wide sky blue riband, passed as a scarf from right to left.

Commanders wear on the left breast a Star, and the Cross of The Order, suspended from the neck.

Knights wear on the left breast a Star, and the Cross at the button-hole.

The *Serving Brothers* wear the Star on the left breast, without the upper branch and without the letters I. H. S. V.,—and the Decoration of THE ORDER at the button-hole.

These letters I. H. S. V. signify, "*In hoc signo vinces.*" The Monogram between the two Greek letters A and Ω., indicates tha the Cross is the Beginning and the End of all things.

THE EMPIRE OF PERSIA.

The Royal Persian Order of The Lion and The Sun.

The Persian Kings have always attached great importance to the Royal Privilege of having a *Band of Musicians*, and of displaying at their Festivals, and when encamped with their Army, particular *Banners*.

One of the *Standards* of the Antient Monarchs of this Kingdom was the *Apron* of Kâwâh, the celebrated Blacksmith, who rebelled against Zohauk, and placed the virtuous Feridoon upon the Throne.

To understand this history of *The Apron* more clearly, we are informed by Sir John Malcolm, that there are various accounts of the Descent of Zohauk,—according to some historians, he was an Arabian, but descended from Kaiomurs,—others trace his descent to Shedad, and term him a Syrian,—and it has even been conjectured, that he was the Nimrod of the Hebrews. All agree in one fact, that he was of a cruel and sanguinary temper. He is described as having had two dreadful *Cancers* on his Shoulders, which the Persian Fabulists have changed into *Snakes*, whose hunger nothing could appease but the brains of Human beings. Two of Zohauk's subjects were slain daily to furnish the horrid meal,—until the manly indignation of KÂWÂH, a Blacksmith of Isfahan, whose two sons were on the point of being sacrificed, relieved the Empire from this Tyrant, and raised FERIDOON, a Prince of the *Puishdadian* Dynasty, to the Throne.

We are, however, told, that the whole of ZOHAUK's history is a Fable, and that the story of the *Snakes* is evidently allegorical, and was probably designed to represent the fatal consequences of yielding to evil temptation.

FERIDOON's first act was to convert the celebrated *Apron* into The Royal Standard of Persia. As such, it was richly ornamented with Jewels,—to which every King, from FERIDOON to the last of The *Pehlivi* Monarchs, successively contributed. It was called, "*The Derufsh-e-Kawanee*," *i. e.* "*The Standard of Kawa*," and continued to be The Royal Standard of Persia until the Mahomedan Conquest, when it was taken in battle by SAAD-E-WUKASS, and sent to The Caliph OMAR.

Subsequent to the introduction of the Religion of MAHOMED a variety of Colours or Flags have been adopted, which have in general been ornamented with Symbols allusive to the Deeds of that Prophet or his Descendants. Among these, a Representation of the *Zulfekâr*, or *Two-edged Sword of* ALY, has been the most common,—but, notwithstanding the attachment of the Persians to this sacred Banner, the Sovereigns of that Kingdom have, for many Centuries, preserved, as the peculiar Arms of their Country, the Sign or Figure of *Sol* in the Constellation *Leo*,—and this Device, which exhibits a *Lion* couchant and the *Sun* rising at his back, has not only been sculptured upon their *Palaces* and embroidered upon their *Banners*, but has been converted into an ORDER, —which, in the form of GOLD and SILVER MEDALS, has been given to those who have distinguished themselves against the Enemies of their Country,—Those, on whom these MEDALS have been conferred, being very proud of the Distinction, and all are extremely anxious to obtain them.

THIS ORDER, which was instituted, in 1808, by FUTTEH ALY SHAH, consists of TWO CLASSES. Additional Decorations have been created,—and THE ORDER has been conferred upon several Ministers and Representatives of European Governments in Alliance with Persia.

First Class.

Richard, Marquess Wellesley, K. G., "as a testimony of the high regard and respect which His Majesty The King of Persia feels for His Lordship's character."—Royal Licence, 29th of July, 1812.

John Malcolm, Esq., a Lieut. Colonel in the Service of The East India Company, and late Envoy and Plenipotentiary from the Supreme Government in India to the Court of Persia, "as a distinguished testimony of the Royal regard and esteem of the King of Persia,"—He had the honour of Knighthood also conferred upon him.—Royal Licence, 15th of December, 1812. Afterwards Major-General, and G. C. B.

Major Thomas Lindsay, of The East India Company's Service, "in testimony of the approbation of The King of Persia of his distinguished conduct, while employed in his actual Service."—Royal Licence, 2d of June, 1820.

Henry Willock, Esq., Major in The Service of The East India Company, and late Chargé d'Affaires to The Court of Persia, to accept and wear the Insignia of The Persian Order of The Lion and The Sun, of the First Class, with which His Majesty The Shah of Persia was pleased to honour him, "for his services in the field."—Royal Licence, 19th of July, 1826.—Knighted, 30th of June, 1827.

John Cormick, M. D., Member of The Royal Persian Order of The Lion and The Sun, of The *Second* Class, to accept and wear The Insignia of the said Order of The Lion and The Sun, of The *First* Class, with which His Majesty The Shah of Persia has been pleased to honour him, "in further testimony of His Royal approbation of his distinguished conduct, while in the actual service of that Sovereign." — Royal Licence, 14th of May, 1827.

Isaac Hart, Esq., Captain in The Army on Half-pay of the 65th (*or,*

2d Yorkshire North Riding) Regt. of Foot, "in testimony of His Majesty THE SHAH of PERSIA's approbation of his distinguished conduct, while in the actual Service of that Sovereign."—Royal Licence, 2d of April, 1828.

JOHN MACDONALD, Esq., Lieut. Colonel in The Service of The East India Company on the Madras Establishment, and Envoy Extraordinary from The Supreme Government of India to THE SHAH of PERSIA, "for his distinguished conduct before the Enemy in the Field."—Royal Licence, 10th of September, 1829.

Lieut. General Sir WILLIAM KEIR GRANT, (*late* KEIR), K. C. B., K. C. H., and Knight of the Imperial Austrian Military Order of *Maria Theresa,* "for his distinguished Services in *The Persian Gulf.*"—Royal Licence, 24th of October, 1831.

In the month of April, 1832, The King was pleased to grant to Lieut. General Sir W. KEIR GRANT, his Royal Licence and permission, that he and the Heirs Male of his body respectively upon whom the Dignity of a *Baron of Austria* should devolve, pursuant to the limitations of certain Imperial Letters Patent or Diploma, bearing date at Vienna the 22d day of January, 1831, might fully avail himself and themselves of the said Honour, that he and they might assume and use the said Title in this Country, and bear the Arms annexed thereto.

GEORGE WILLOCK, Esq., a Major in The Army, "in approbation of his conduct, while actually in *The Persian Service.*"—Royal Licence, 16th of April, 1832.

JOHN MC NEILL, Esq., "in approbation of his services, while attached to the British Force in *Persia.*"—Royal Licence, 14th of January, 1836.

SECOND CLASS.

JOSEPH D'ARCY, Esq,, Lieut. Colonel in The Army, and Captain in The Royal Regt. of Artillery, "in testimony of the approbation of His Majesty The King of Persia, of the distinguished conduct displayed by him, while employed in his actual Service."—Royal Licence, 30th of June, 1818.

JOHN CORMICK, M. D., conferred upon him by His Persian Majesty, "in testimony of his distinguished conduct, while in the actual service of that Sovereign."—Royal Licence, 20th of September, 1825.

WILLIAM MONTEITH, Esq., Captain of The Corps of Engineers in The Service of The East India Company on the Madras Establishment, "in testimony of the Royal approbation of His Majesty THE SHAH of PERSIA of his conduct, whilst he had the honour of being employed in the service of that Sovereign."—Royal Licence, 19th of February, 1827.

GEORGE WILLOCK, Esq., Major in the Service of The East India Company on the Madras Establishment, "in testimony of the Royal approbation of His Majesty THE SHAH of PERSIA of his distinguished conduct, while in the actual service of that Sovereign." —Royal Licence, 14th of May, 1827.

The Right Honble. Sir GORE OUSELEY, Bart., G. C. H.

Sir ROBERT KER PORTER, "as a particular mark of His Majesty's personal favour and esteem."

Sir JAMES EDWARD ALEXANDER, Knt., Captain in The British Army, and Lieut. Colonel in The Portuguese Service.

HENRY LINDSAY BETHUNE, Esq., of *Kilconquhar*.

WILLIAM G. BEEK, Esq., Colonel of Engineers in the Army of ABBAS MEERZA, Prince Royal of PERSIA, as a testimony of the high sense entertained by His Majesty FUTTEH ALY SHAH of the great fortitude and distinguished talents displayed by him in the capture of the Forts of *Amerabad* and *Koochan*, in 1832.

Lieutenant FRANCIS FARRANT, of the 3d Regt. of Light Cavalry, in the Service of The East India Company on The Bombay Establishment, "in testimony of THE SHAH of PERSIA'S approbation of his Services while attached to the Persian Army, and that he may enjoy all the Rights and Privileges thereunto annexed."—Royal Licence, 7th of December, 1835.

Lieutenant JOHN LAUGHTON, of The Bengal Engineers, "in approbation of his services, while attached to the Army of THE SHAH of PERSIA."—Royal Licence, 13th of January, 1836.

Captain CHARLES KINNAIRD JOHNSTONE, of The East India Company's

Maritime Service, " in testimony of his approbation of the valuable Services rendered by him, while in the actual employ of THE SHAH."—Royal Licence, 4th of March, 1837.

Lieut. Colonel CHARLES CHRISTOPHER JOHNSON, " in testimony of the approbation of THE SHAH of PERSIA of his services, while employed with the Persian Army."—Royal Licence, 1st of May, 1837.

Captain Sir K. A. JACKSON.

THE KINGDOM OF POLAND.

Preliminary Observations.

Poland, once a considerable Kingdom of Europe, but now totally dismembered and divided among the neighbouring States, presents an interesting subject of inquiry to the Philosopher and the Politician. About a Century ago, it was one of the most respectable Powers upon The Continent, — but it has now not only lost it's Independence as a Nation, but it's very Name, even, with more than savage barbarism, is declared "*to be blotted out of the Map of Europe,*" ! !

Some commiseration may, however, be expected from England for the sufferings and wrongs of this unhappy Country, which had formerly the virtue to wish one of her sons, "whose Character displays almost unvaried excellence,"—Sir Philip Sidney,—to have become their Sovereign,—

" He was invited to enrol himself among the Candidates for the Crown of Poland, vacant, in 1585, by the death of Stephen Bathori,—and this historical fact affords a stronger general proof of the fame of his transcendent character, than all the united testimonies even of his Contemporaries. That a young man, sprung from a Family not yet ennobled,—unemployed, save in a solitary Embassy, by his own Sovereign,—passing, perhaps, the most part of his time in literary seclusion,—should have been solicited even to be certainly unsuccessful in so glorious a race, would be utterly incredible, were it not absolutely proved. Here Elizabeth's prohibition again interfered.—She refused, says Naunton, to further his advancement, not only out of emulation, but out of fear to lose the Jewel of her times"—Lodge's *Portraits*, vol. 2.

THE ORDER OF THE WHITE EAGLE.

The early history of THIS ORDER is so little known, that it is either passed over in silence, or so obscured in fables, as to render the endeavour to throw light upon it altogether vain.

One of these fables states, that LECHUS the First, who, in 550, subjected Poland to his authority, being afterwards expelled from Sclavonia by The Wallachians and Bulgarians, was driven into a vast Plain, where he discovered a *nest* of *White Eagles*, which appeared to him to be a good omen, and induced him to found a City upon the spot,—which he named *Gnesen*, and in the Sclavonian language, *Nest*.

In commemoration of this event King LADISLAUS the fifth, in the year 1325, on the marriage of his son CASIMIR with ANNE, Princess of *Lithuania*, founded THE ORDER, and gave it a *White Eagle*, as its emblem.

That there never was a King LECHUS, SCHLÖZER in a successful Prize essay has proved satisfactorily. That LADISLAUS founded an ORDER is possible, and indeed probable, for at that period several similar Fraternities of Knights were in existence,—but the historical proofs of his having done so, are altogether wanting,—and it almost seems, as if this institution was in so little estimation, that the old Chroniclers of the Country thought nothing of it, and were altogether silent upon the subject. One fact, however, is certain, that THE ORDER of THE WHITE EAGLE of POLAND is of very remote origin.

The Kings of Poland were crowned in the City of *Gnesen*, and the *Regalia* were kept here until the year 1320, when they were removed to *Cracow*. During the independent existence of Poland, *Gnesen* was famous as The See of an Archbishop, who was Primate of Poland, and acted as Regent on the King's demise,—and who also announced the event, convened the Members of Convocation, and performed the functions of Royalty.

When AUGUSTUS the Second, Elector of Saxony, acquired the Crown of Poland, in 1697, and had to contend for it's security during many years, both with foreign and domestic foes, he renewed, in 1705, THE ORDER of THE WHITE EAGLE, and endeavoured by this Distinction to procure adherents and friends among the restless Poles, as well as to gratify his own natural inclination for dress, show, and splendour.

At it's revival THE ORDER was highly esteemed, and was conferred on The Czar PETER the Great, and Grand Duke of Russia. It also flourished under his Successor of The House of Saxony, AUGUSTUS the third.—It subsequently, however, lost it's repute, and seemed to be extinguished with the Partition of Poland in 1795,—for, it was no longer conferred, as none of the Sovereigns who then possessed parts of this Kingdom, assumed it himself, or declared himself it's Master.

In this situation it remained, until the creation of The Duchy of *Warsaw* by NAPOLEON BUONAPARTE. In the Constitutional act of The Duchy, of the 21st of July, 1807, all ORDERS of KNIGHTHOOD previously existing in Poland, were declared to exist for the future, —were at the same time renewed,—and King AUGUSTUS, of Saxony, who was at the time, Duke of *Warsaw*, as Grand Master *of all* THE POLISH ORDERS, bestowed also that of THE WHITE EAGLE. And, as POLAND is now under the Sceptre of *Russia*, it still continues like all the other Polish Orders, and is conferred by The Emperor of RUSSIA, as KING of POLAND.

It is conferred not only upon *Natives*, but also upon *Foreigners* of High Rank, as well as on Sovereign Princes. The second Polish Order, that of STANISLAUS, is inseparable from it, and every Knight of THE WHITE EAGLE receives with it that of STANISLAUS.

THE BADGE is a gold Cross, of eight points, bordered with white, enamelled transparent dark red, surmounted by a Royal crown,—between the four chief angles are golden flames, set with diamonds. In the middle of the shield of the Obverse is *The White Eagle* crowned, with outspread wings,—and, on the Reverse, the letters, A. R. *(Augustus Rex)*,—and on the four wings of the Cross, is the MOTTO, "*Pro Fide, Rege, et Lege.*" The Cross of The Grand Master bears the legend, "*Pro Rege, Grege.*"

The Badge is worn by The Knights, who consist of one Class only, on a sky blue, watered riband, from the left shoulder to the right side. On the left breast they wear also a golden STAR, of four long and four short rays, which bears a Silver Cross, with blunted points, and a red border, whereon is inscribed the MOTTO already mentioned, between the four angles of which proceed flames of fire, in silver, bordered with red.

KNIGHTS.

GEORGE, Earl MACARTNEY,—The only Official record of His Lordship having received THE ORDER of THE WHITE EAGLE of POLAND, is a Pedigree of His Family, which is recorded in THE COLLEGE of ARMS,—Died, in 1806.

Sir BENJAMIN THOMPSON, Knt., Count of THE HOLY ROMAN EMPIRE, by the Stile and Title of Count RUMFORD,—in the month of July, 1785, he was at the particular request of His Most Serene Highness The Elector Palatine, Duke of BAVARIA, decorated by His Majesty The King of POLAND, with the Royal Order of *Saint Stanislaus*,—and, upon the 6th of August, 1789, His Majesty The King of POLAND, at His Electoral Highness's request, honoured him with the Illustrious Order of *The White Eagle*."—Royal Licence, 4th of January, 1794.—Died, in 1814.

Sir WILLIAM NEVILLE HART,—*See*, The Order of St. STANISLAUS.

The Order of Saint Stanislaus.

After the decease of Augustus the third, King of Poland and Elector of Saxony, in 1763, Count Stanislaus Augustus Poñiatowsky was chosen King of Poland. Scarcely, however, had he ascended the Throne of this unquiet land, when, on the 7th of May, 1765, he founded, probably in order to establish an additional bond of Union between himself and the Nobles of the Kingdom, an Order, which, in honour to his Patron Saint whose name he bore, he entitled "*The Order of St. Stanislaus.*"

According to the Statutes ordained by him, the number of Knights was fixed at One hundred,—that the Knights of The Order of the White Eagle should receive this Order also,—but that neither they, nor the Foreigners on whom it was conferred, were included in the number of One hundred,—and that every Knight was compelled to pay yearly four ducats to *The Warsaw Hospital,* dedicated to the Infant Jesus.

This Order, however, preserved it's Dignity but a short time, as it was conferred in the most profuse manner,—and, on the division of Poland, it became entirely dormant.

By the Constitution acquired by The Duchy of Warsaw, which was founded after the peace of *Tilsit,* it received with the other Polish Orders fresh vigour, and King Augustus, of Saxony, continued to bestow it, as long as he remained Duke of *Warsaw.*

When Poland came under the Dominion of *Russia,* it was, on the 1st of December, 1815, by The Emperor Alexander, as King of

Poland, formally renewed, modified, and divided into four Classes. And, on the 16th of December, 1816, the former obligation imposed upon the Knights of paying a yearly Stipend to *The Foundling Hospital* was revived, and was fixed at the sum of four ducats annually.

THE BADGE is a Cross, of eight points, enamelled red, with small round beads on all the points,—the space between the four chief points is occupied by White, crowned Eagles, with extended wings, and the small spaces between the two points of every chief part, are filled with golden roses. The Obverse presents in the middle, the picture of *St. Stanislaus*, encircled with a wreath of laurel,—and the Reverse exhibits, on a white ground, the initials in red, S. S. (*Sanctus Stanislaus*),—with the MOTTO, " *Premiando incitat.*"

The first Class wear The Badge on a red riband, bordered with white, from the left shoulder to the right hip,—and, on the left breast, a Silver STAR.

The second Class wear it round the neck, with the same Star.

The third and fourth Classes wear it in the button-hole.

The Knights who possess THE ORDER of THE WHITE EAGLE, wear it round the neck, and without the Star on the breast.

KNIGHTS.

Sir WILLIAM NEVILLE HART, Chamberlain to His Majesty STANISLAUS AUGUSTUS PONIATOWSKY, King of Poland,—being invested with this dignity, by a special Diploma of His Majesty, dated at *Warsaw*, the 27th of December, 1794. In the following year Sir WILLIAM returned to England, and in the month of October of the same year, he received. it is said, *a Letter* from His Grace The Duke of PORTLAND, then Secretary of State for The Home Department, notifying to him, his Sovereign's most gracious permission to assume and bear the title of Knight of " THE ORDER of SAINT STANISLAUS," and to wear the *Insignia* appertaining to the same. Sir WILLIAM, soon afterwards was presented to their Majesties, and had the honour of kissing hands. He died at *Inverary Castle*, the seat of The Duke of ARGYLE, on the 23d of October, 1804.

Sir Benjamin Thompson, Knt., *Comte de* Rumford.—in 1794.

Robert Peat, D.D., Rector of Ashley *cum* Silverley, and Vicar of Kirtling, in the County of Cambridge,—One of the Chaplains in Ordinary to His Royal Highness The Prince of Wales,—to accept and wear, *in his own Country*, the Ensigns of this Illustrious Order, which was conferred upon him, on the 21st of November, 1790, (by his then description of Robert Peat, *Esquire*), by Stanislaus Augustus, late King of Poland.—Royal Licence, 2d of October, 1804.

Peter Francis Bourgeois, Member of The Royal Academy of London, —" the honour of Knighthood conferred upon him by The King of Poland, by Diploma, dated at *Warsaw*, the 16th of February, 1791, and that he may wear the *Medal* of The Order, called *Merentibus*."—Royal Licence, 16 April, 1791.

THE KINGDOM OF PORTUGAL.

See, also,

THE EMPIRE OF THE BRAZILS.

The Order of Christ.

The History of this, originally Ecclesiastical Order, is only a continuation of the History of The Templars. When the latter were extinguished, The Knights of Christ succeeded, their Institution being founded upon the ruins of The Order of The Templars, and the suppression of their Name.

This change of Name occurred in the year 1317,—and, 1319, Pope John confirmed the new Military Order of Knights of Christ, by a special Bull,—by which he retained to The Holy See the right of also appointing Knights of The Order of Christ,—whence a branch of this Order arose, as is described, among The Papal Orders.

By virtue of this Bull, The Order received again the estates of The Knights Templars for their own free disposal, were subjected to the Rule of *St. Benedict* and the Ordinances of *Citeaux,* and acquired, as the Dress of the Order, *a White Mantle,* with a *red* and *a white Cross.* These Regulations were almost exactly those of THE TEMPLARS,—the Name alone was changed,—and a Red Cross was added to the previous White one.

DIONYSIUS, the reigning King of Portugal, however, in the announcement which he promulgated, on receiving the Papal Bull, named this change or this creation of the NEW ORDER of CHRIST, *only a Reformation of* THE ORDER of KNIGHTS TEMPLARS. He restored to it, all the Estates which he had previously sequestered, annulled all the Ordinances which he had been compelled to issue against it,—and thus *The Order of Knights Templars* in his Kingdom came fully into possession of their ancient right, under the Name of THE ORDER of CHRIST."

Castro Marino was in the first instance the Seat of THE ORDER,—subsequently, in 1366, it was transferred to the Town of *Thomar,* on *The Tagus,* where a very celebrated Convent is now situate. From the three most essential of their Vows, *Poverty, Chastity,* and *Obedience,* they were by degrees in a great measure relieved. Pope ALEXANDER the Sixth absolved them from the first, on condition of devoting a third of their Revenues to the erection of the Convent of *Thomar,*—and he rendered the second, less burdensome, by permitting them to marry.

How essential a service King DIONYSIUS had rendered his Kingdom by the preservation of THE ORDER, was shown a Century later, when, under the reign of JOHN the First, PORTUGAL by means of her Discoveries established Colonies in *India,* and possessions in *Africa.* These

Enterprises were achieved solely at the expense of THE ORDER of CHRIST, to whom the Sovereigns from gratitude and with a view to excite THE ORDER to new exertions, secured to them the Proprietary of all the Discoveries which they should make, reserving only the mere Sovereignty to the Crown.

The progress made by THE ORDER, was so important, and their possessions soon became so considerable, that prudence prescribed fresh limits to their power. The Sovereigns, therefore, in the sequel declared *the Conquests* of THE ORDER to be *the property* of THE CROWN,—but granted them, the Civil government, a part of the Military authority, and the Ecclesiastical jurisdiction,—to which arrangements the Popes gave their consent. The power and dignity of THE ORDER rose even under these restrictions to a height, which the Sovereigns of Portugal conceived to be dangerous to their own security. The result, therefore, was, that in the year 1550 the GRAND MASTERSHIP of THE ORDER was for ever united with THE CROWN of PORTUGAL, and it still remains a Royal appendage at the present day.

THE ORDER continued to be, and now is, one of the richest in the world. It possesses 454 Commanderies,—which, according to the original Statutes of The Order, shall only be conferred upon those, who have fought for three years against the *Infidels*,—a people, who are now no longer to be met with. The proof, however, of a descent from an ancient *Noble Family* and of establishing the purity of his Ancestry, are indispensable in a Candidate for admission into THE ORDER.

The whole of the Knights were divided, in 1789, into three Classes, —*viz.*,—GRAND CROSSES, *six*,—COMMANDERS, 454,—and KNIGHTS, *unlimited*.

THE ORDER also includes *Priests*, who must solemnly swear to observe the three vows already mentioned.

The Admission takes place only in The Convent of *Thomar*.

For *Foreigners* on whom it is conferred, it is a mere *Honorary* Distinction. They are not comprised in the prescribed number of Members, neither are they subject to the Rules of THE ORDER, nor do they partake of it's Revenues.

The Grand Crosses wear on Ceremonial occasions and in the full

DRESS of THE Order, a Cross, *pattée, red,* charged with a Cross, *white,* pendent to a triple gold chain, round the neck. On common occasions they wear it, on a red riband, from right to left,—and, on the left breast, the STAR of THE ORDER.

The MOTTO of The Order is, "*Christiana Militia.*"

The two other Classes wear the smaller Cross, on a red riband,—THE COMMANDERS wearing it round the neck, together with a STAR on the breast,—but THE KNIGHTS wear it in the button-hole, without the Star.

KNIGHTS.

MIGUEL DE LIMAE SOUZA, of *Bombay,* in The East Indies, Esquire.—Royal Licence, 7th January, 1805.—Services rendered to The Portuguese Government, at *Goa.*

WILLIAM WYON, Esq., R. A., CHIEF ENGRAVER of The ROYAL MINT.

THE ORDER OF CIVIL MERIT OF SAINT JAMES.

THE ORDER of ST. JAMES, of *Compostella, or,* of THE SWORD, flourishes both in *Portugal* and in *Spain*,—in the latter Kingdom it is an *Ecclesiastical,* in the former a *Lay* Institution. The origin and early history of both of them, however, are the same.

The miraculous revelation of the body of THE APOSTLE, after the lapse of eight centuries from the date of his interment, and his frequent apparition in the ranks of the Christian Armies in their desperate struggles with *The Infidel,* had given so wide a celebrity to the obscure Town of *Compostella,* in Galicia, which contained the Sainted Relics, that it became the resort of Pilgrims from every part of Christendom during the Middle Ages,—and the *Escalop Shell,* the device of *St. James,* was adopted as the universal Badge of *The Palmer.* Inns for the refreshment and security of the pious Itinerants were scattered along the whole line of the route from France,—but, as they were exposed to perpetual annoyance from the predatory incursions of *The Arabs,* a number of Knights and Gentlemen associated themselves, for the purpose of their protection, with the Monks of *St. Lojo,* or *Eloy,* adopting the Rule of *St. Augustine,* and thus laid the foundation of THE CHIVALRIC ORDER OF ST. JAMES, about the middle of the Twelfth Century.

THE CAVALIERS of THE FRATERNITY, which received it's Papal Bull of approbation five years later, in 1175, were distinguished by a *White Mantle* embroidered with a *red cross,* in fashion of a *sword,* with the *Escalop shell* below the guard, in imitation of the device which glittered on the Banner of their Tutelar Saint, when he condescended to take part in their engagements with *The Moors.* The *red* colour denoted, according to an ancient Commentator, "*that it was stained with the blood of the Infidel.*"

The Rules of THE NEW ORDER imposed on it's Members the usual obligations of obedience, community of property, and of conjugal chastity, instead of celibacy. They were, moreover, required to relieve the poor, defend the traveller, and maintain perpetual war upon *The Mussulman.*—PRESCOTT.

At first THE ORDER possessed a GRAND MASTER who was chosen by themselves, and to whom was added a COLLEGE under the name of "*The Thirteen.*" Subsequently, however, THE KNIGHTS of PORTUGAL separated themselves from THE GRAND MASTER in *Castile,* and chose a GRAND MASTER of their own. Hence arose great misunderstandings between the two GRAND MASTERS, especially on the subject of Rank, which occasionally led to acts of violence. The result was, that the Kings both of Portugal and Spain gradually took upon themselves The Grand Mastership, and thus divided THE ORDER into *two Lines,* which are still in a flourishing condition. The Popes confirmed this separation, which occurred in the early part of the Sixteenth Century,—but decreed, that the Sovereigns should take no part in the *Ecclesiastical* affairs of THE ORDER, but should leave their management to the Members of THE ORDER.

This induced The Emperor CHARLES the Fifth, as King of *Spain,* to appoint a peculiar Council of THE ORDER, and to confide to it, the affairs of his other two Ecclesiastical Orders, those of *Alcantara* and of *Calatrava,*—which arrangement also received the sanction of the Popes.

Although The Knights, besides their other vows, take that of *Poverty,*—yet their ORDER, both in *Portugal* and in *Spain,* is nevertheless one of the *richest.*

In SPAIN, THE ORDER has five Posts of Honour,—viz., Two Primates, and Three Grand Commanderies,—the latter are divided into three Provinces, *Castile, Leon,* and *Arragon,*—to the first belong 43., to the second 35., and to the third 7., exclusive of Convents and Hospitals.

In PORTUGAL, THE ORDER possesses 47 places and villages, and 150 Commanderies. It also possesses near *Lisbon,* the celebrated Convent, "*Santos o Novo.*"

Both in PORTUGAL and in SPAIN the proof of *Nobility* of Sixteen Ancestors by the Father and Mother's side is required.

In SPAIN, even among the Choristers or Chaplains of THE ORDER, full investigation is made, whether their forefathers were *Jews, Saracens,* or *Heretics.* The *Novices* are required to have served six months *on board the Gallies,* and to have remained a month in one of the Convents of THE ORDER, to become acquainted with it's Rules,—these obligations may, however, be dispensed with, *on paying a sum of money.*

In SPAIN, THE ORDER of ST. JAMES still continues an *Ecclesiastical* Order, of which no further particulars need here be recorded.

In PORTUGAL, it was so only 'till the year 1789, when it was transformed by Queen MARIA into an Order of *Civil Merit,* and divided into three Classes,—*viz.,* GRAND CROSSES, 6,—COMMANDERS, 150,—and KNIGHTS, *unlimited.*

After this change THE ORDER still retained the possessions, which it had previously enjoyed.

THE BADGE is the ancient one,—a red enamelled Cross of Christ, the upper and side points of which terminate like flowers, while the lower run out straight, so as to resemble a Sword holden downwards.

THE BADGE of the two first classes is distinguished by *a red enamelled Heart,* which was added by Queen MARIA, as *The Holy Heart of Christ,* under whose protection she placed all THE PORTUGUESE ORDERS.

The Grand Crosses wear the Badge, pendent to a violet coloured riband, from right to left,—the Commanders wear it round the neck, —and the Knights, at the button-hole.

The two first Classes wear a SILVER STAR on the left breast, with the Cross of THE ORDER in the middle, on which is introduced the red enamelled Heart, worn by THE ORDER of CHRIST.

The Order of Military Merit of Saint Bento d'Avis.

The persecutions and wars of *The Moors* gave rise to many Chivalrous Fraternities,—from which, at a later period, *Ecclesiastical,* and, at length, *Lay* Orders of Chivalry were instituted, and occasioned also the Origin of "The Order of Avis."

There was formed for this purpose in Portugal, about the middle of the Twelfth Century, a union of several Noble Knights of the Kingdom, who, without assuming any definite Title, named themselves, "*The New Order of Knights.*" Alphonso Henry, first King of Portugal, very much approved this Institution. And, in the year 1162, in his presence and that of his Court, this Fraternity was, by the Papal Legate John, Abbot of *Tarnea,* formally raised to the Rank of an *Ecclesiastical* Order of Chivalry, and it's Statutes regulated in the following manner,—

That it's Members should defend the Catholick Religion, when required, with force of arms,—should be benevolent, and chaste,—in other respects should follow the Rules of *St. Benedict,* and be subject to *The Order of Cistertians.* The Grand Master was obliged to take the Oath to a Cistertian Abbot, and to promise obedience to the Pope, the King, and the Abbot General of *Citeaux.*

The first Grand Master was Don Pedro, a natural son of King Alphonso. From this period the Members of this Order named themselves "*Knights of Evora,*" which Town had been presented to them by Alphonso. When, however, in 1187, they fixed their Chief Seat at the Frontier Fortress of *Avis,* which they had defended against *The Moors,* and which Alphonso had also presented to them, they styled themselves, "*Knights of Avis.*"

In the beginning of the Thirteenth Century, THE ORDER was presented with the extensive Estates, which THE SPANISH ORDER of CALATRAVA possessed in Portugal, on condition of being subjected to the jurisdiction and visitation of THE ORDER. This arrangement lasted until the year 1385, when JOHN the first, King of Portugal, gained a victory over JOHN the first, King of Castile, assumed to himself The Grand Mastership of THE ORDER of AVIS, and forbad the Knights thenceforward to recognise the Spanish Knights of *Calatrava*, as their Superiors. The Spanish Knights, however, being dissatisfied with this injunction, brought the matter before *The Council of Basle*, which decided in their favour. But the King of Castile being too weak to enforce the decision of The Council, THE ORDER of AVIS continued to be independent, and the Grand Mastership has remained with the Crown of Portugal to the present day.

In 1789, it was changed from an *Ecclesiastical* to a *Civil* Institution. This alteration was effected by Queen MARIA, by transforming it into an Order of *Military* Merit, and assigning it's Revenues, which are very considerable, to deserving Officers of the Army.

The Members consist of,—*viz.*, GRAND CROSSES, *six*.—COMMANDERS, 49.—and KNIGHTS, *unlimited*.

THE BADGE is the ancient one, — a gold cross, *flory*, enamelled green, with Lily shaped ornaments at the four ends,—to which Queen MARIA, as in the case of *The Order* of *St. James*, added a *Heart*, enamelled red.

By the first Class it is worn, pendent to a rich green, watered riband, from right to left,—By the second Class it is worn round the neck,—and a SILVER STAR is worn by both of them, on the left breast, in the middle of which is THE BADGE, and upon it is placed a red *heart*, precisely as in THE ORDER of CHRIST.

The third Class wear the Cross, without the Heart and without the Star, in the button-hole.

GRAND CROSS.

GEORGE ROSE SARTORIUS,—*See*, THE ORDER of THE TOWER and SWORD.

KNIGHTS COMMANDERS.

JAMES LUCAS YEO, Esq., a Post Captain in the Royal Navy, K.C.B.,—His Royal Highness The Prince Regent of Portugal, dispensing with his taking the customary Oaths of the said Order, in consideration of his being of a different Communion,—" for the distinguished skill, valour, and intrepidity displayed by him, with His Majesty's Ship, *The Confiance*, under his command, at the reduction of the Island of *Cayenne*, in the months of December 1808 and January 1809."—Royal Licence, 17th of March, 1810. —Knighted, 20th of June, 1810. Died, on the Coast of *Africa*, in 1818.

Sir FREDERICK WATSON, Knt., Lieut. Colonel in The Army, " in consideration of his distinguished Services, during *The Peninsular War*."—Royal Licence, 10th of November, 1837.

Lieut. General Sir JOHN WILSON.

Sir DUDLEY ST. LEGER HILL, Knt., Colonel in The Army, Governor of Her Majesty's Colony of *Saint Lucia* and it's Dependencies, C.B., and Knight of The Royal Portuguese Military Order of The Tower and Sword, to accept and wear the Insignia of a Knight Commander of The Royal Portuguese Military Order of *St. Bento d'Avis*, " in consideration of his distinguished services, during *The Peninsular War*."—Royal Licence, 14th of June, 1838.

Sir ROBERT JOHN HARVEY, Knt., a Colonel in The Army, C.B., and Knight of The Royal Portuguese Military Order of The Tower and Sword, to accept and wear The Insignia of a Knight Commander of The Military Order of *St. Bento d'Avis*, " in testimony of Her Most Faithful Majesty's approbation of his services, during *The Peninsular War*."—Royal Licence, 16th of October, 1838.

Lieut. Colonel Sir THOMAS STEPHEN SORELL, Knt., Her Majesty's Consul General for the Austrian States in Italy, and K.H., which The Queen of Portugal hath been pleased to confer upon him, " in consideration of Services rendered to the cause of Her Most Faithful Majesty, during the Siege of *Oporto*."—Royal Licence, 28th of November, 1838.

Sir EDWARD BRACKENBURY, Knt., a Lieutenant Colonel in The Army, Knight of The Royal Portuguese Military Order of The Tower and Sword, and Knight of The Order of St. Ferdinand of Spain, of the Second class, to accept and wear the Insignia of a Knight Commander of The Royal and Military Order of *St. Bento d'Avis*, " in testimony of Her Majesty's approbation of his distinguished and important services, during *The Peninsular War*, and more particularly for his gallant conduct at the battle of *Salamanca*."—Royal Licence, 27th of December, 1838.

The Antient and Most Noble Order of The Tower and Sword.

On the 3d of May, 1808, John, Prince Regent of Portugal, afterwards King John the Sixth, in order to celebrate his Arrival in his Dominions of *Brazil,* and to reward the distinguished Services of several of the Subjects of Great Britain, who accompanied him, *revived The Antient* Order of The Sword, which had been founded by Alphonso the Fifth, surnamed "*Africano,*" in the year 1459.

On the 29th of November, 1808, he granted a Charter, and Ordinances respecting it.

This Order was, however, subsequently re-formed by Don Pedro, Duke of Braganza,—who, on the 28th of July, 1832, published the following Charter, in the name of his Daughter, Maria the Second, Queen of Portugal, in whose favour he had abdicated the Throne.

Organization of The Order.

I. The Order shall henceforward be called, " Antient and Most Noble Order of The Tower and Sword, of Valour, Loyalty, and Merit."

II. It's Degrees shall be, that of Grand Master, Principal Officers, Grand Crosses, Commanders, Officers, and Knights.

III. The Sovereign, or Regent, shall always be Grand Master. In the case of a collective Regency, the Principal Commander, and immediate Principal Officers successively shall perform the functions of The Grand Master under the Supreme direction of The Regency.

IV. The Principal Officers shall be, the Principal Commander, the Claveiro, the Principal Ensign, and the Principal Chancellor.

V. The number of Grand Crosses, Commanders, Officers, and Knights, are undetermined.

VI. There shall be moreover, Seven Minor Officers, *viz.*, The King of Arms, who shall be called, "*King of Arms, Tower and Sword*," and who shall always be a Knight of The Order,—two Heralds, and four Pursuivants, who shall have the Rank and Pay of Esquires of The Order.

Qualifications.

VII. Personal Merit, distinguished Feats of Arms, or acts of Courage, or Civic Devotion, signal Service in any Publick Career, but principally Military, shall be the only title to this Distinction.

VIII. All *Portuguese* Citizens, of whatever Class, Religious persuasion, or Profession, and all deserving *Foreigners*, shall be eligible to This Order, on proof of their Qualifications.

IX. The Qualification shall be proved before The Grand Master, in such form and manner as he shall prescribe.

X. Vague Services shall not be sufficient Qualifications.

XI. The following persons shall be exempted from proof of Services,—

1. *Military persons*, whom The Grand Master shall decorate with his own hand with the *Insignia* of The Order, for acts of Valour on the field of battle.
2. *Citizens*, on whom it shall be in a similar manner conferred, while in the act of performing some signal Act of Merit, or Civic Courage.
3. *National* or *Foreign Savans*, to whom The Grand Master shall transmit the *Insignia* of The Order, for their distinguished Merit.

Insignia of The Order.

XII. The Insignia of The Knights shall consist of a round Medal, of *Silver*, having on one side a Sword placed upon an Oaken crown, on the upper part a Tower, and around, in letters of gold, on a blue ground, the legend "*Valour, Loyalty, and Merit.*" On the reverse, the Escutcheon of The Portuguese Arms, on a Book, representing the Constitutional Charter of The Monarchy, with the inscription, "*By the King and the Law.*"

XIII. A similar MEDAL, in *Gold*, shall constitute the *Insignia* of The Officers.

XIV. Grand Crosses and Commanders may wear on the left breast, a Plate with a Sword and Tower, and the Inscription "*Valour, Loyalty, and Merit.*"

XV. On ordinary occasions THE MEDALS shall be worn, suspended from a dark blue riband. The Grand Crosses shall wear SCARFS over their dress, in the usual manner.

XVI. On Court, and Grand Gala Days, the Knights shall wear THE MEDAL suspended from a *silver* Chain, in form of a COLLAR,—Officers, from one of *gold*,—and Commanders and Grand Crosses, from a COLLAR formed of Swords and Towers.

XVII. THE INSIGNIA of The Grand Master and Principal Officers shall be the same, as the Grand Crosses.

XVIII. The Knights shall likewise be allowed to surround the Escutcheon of their ARMS with a band, with the legend of THE ORDER, "*Valour, Loyalty, and Merit.*"

XIX. THE MEDALS, *&c.*, shall invariably be conformable with the patterns, published with the Charter.

IN WHAT MANNER THE ORDER SHALL BE CONFERRED.

XX. THE ORDER shall always be conferred by The Grand Master in person, who shall dub the Recipient, *Knight*, with his own hands.

XXI. The Grand Master may not delegate his authority for this purpose, excepting in the following cases,—

1. When he shall empower a General, commanding in a Battle, to decorate with the *Insignia* of THE ORDER those, who shall have distinguished themselves in the Conflict.
2. When he shall send the *Insignia* of THE ORDER in his Name to any Native or Foreign Individual, under the circumstances stated in section 3. article XI. of this Charter.

MILITARY and CIVIC HONOURS of THE ORDER.

XXII. The Knights, Officers, Commanders, Grand Crosses, and Principal Officers of THIS ORDER, shall precede those in equal degree, of all the other MILITARY ORDERS of the Kingdom.

XXIII. A Knight of THIS ORDER shall enjoy the Rank and Honours of an Ensign in the Army, whether a simple Soldier, or exercising any Mechanical trade. The Officers shall have the Rank of Lieutenant Colonels,—Commanders, that of Colonels,—Grand Crosses, that of Brigadier,—and Principal Officers, that of Field Marshal.

ENDOWMENT of THE ORDER.

XXIV. In the Estimate of the Expenses of The State, there shall annually be inserted an *Item* to meet the charges of THE ORDER, which shall be,—

1. An *Asylum* for poor Invalids belonging to it.
2. A *College* for the Education of the Children of both Sexes, Orphans, or in extreme want, of The Knights of THE ORDER.
3. *Pensions*, which shall be granted to the Members of THE ORDER.
4. The maintenance and furnishing of *The Building*, which shall be given to THE ORDER, for the preservation of it's Archives, holding of it's Chapters, and Festivals, as shall be determined by The Statutes.

GENERAL DISPOSITIONS.

THE FESTIVAL of THE ORDER shall be celebrated annually on the 29th of April, in the manner to be determined by the Statutes.

The Antient Members of THE ORDER of THE TOWER and SWORD, who shall not have been involved in Rebellion against The King and Constitution, shall be considered "*Honorary Members*," and shall be at liberty to use the *antient Insignia*,—And those who can prove the necessary Qualifications, according to this Charter, may become effective Members, without a new grant.

KNIGHTS GRAND CROSSES.

The Right Honble. ARTHUR, Lord Viscount WELLINGTON, *Conde de Vimiera*, "for his distinguished and glorious Services, on various important occasions."—Royal Licence, 19th of October, 1811.

Sir WILLIAM CARR BERESFORD, K. C. B., *(Conde di Trancoso, in Portugal)*, "in testimony of the high estimation in which His Royal Highness THE PRINCE REGENT of PORTUGAL holds his distinguished Services."—Royal Licence, 19th of October, 1811.

Lieut. General Viscount BERESFORD, for his distinguished conduct at *Corunna, Busaco, Albuera, Badajos, Salamanca, Vittoria, Pyrenees, Neville, Nive, Orthes*, and *Toulouse*,—received, as a mark of His Majesty's gracious favour,—a Cross, and *seven Clasps*, and appointed G. C. B., K. G., Cross of The Royal Hanoverian Guelphic Order,—Grand Cross, of The Tower and Sword,—St. Ferdinand and of Merit,—and of St. Hermenégilde, and San Fernando.

PERCY-CLINTON SYDNEY, Viscount STRANGFORD, G.C.B., and G.C.H.

The Honble. EDWARD PAGET, Lieut. General of His Majesty's forces,—

Sir BRENT SPENCER, K. B., Lieut. General of His Majesty's forces,—

Sir ROWLAND HILL, K. B., Lieut. General of His Majesty's Forces,— "for their highly distinguished conduct and meritorious services, on various important occasions."—Royal Licence, 4th of May, 1812.

Sir STAPLETON COTTON, Bart., afterwards Viscount COMBERMERE, G. C. B., Lieut. General in The Army, Colonel of the 20th Regt. of Light Dragoons, "in testimony of the high sense which The Prince Regent of Portugal entertains of the great courage and intrepidity displayed by him in several Actions with the Enemy in *The Peninsula*."—Royal Licence, 11th of March, 1813.

Sir JOHN HAMILTON, Bart., Lieut. General of His Majesty's Forces, Colonel of the 2d Ceylon Regiment, and Governor of Duncannon Fort, "for the great ability and signal intrepidity displayed by him in several Actions with the Enemy, during the recent arduous Campaigns in *The Peninsula.*"—Royal Licence, 4th of April, 1816.

Admiral Sir WILLIAM SIDNEY SMITH.

Rear-Admiral Sir JOHN POO BERESFORD, Bart., K. C. B., conferred upon him by His Majesty The King of Portugal, "in testimony of his distinguished services before the Enemy."—Royal Licence, 25th of September, 1821.

CHARLES DASHWOOD, Esq., a Post Captain in The Royal Navy, late Captain of His Majesty's Ship *The Windsor Castle,* "on occasion of the visit of His Most Faithful Majesty JOHN the Sixth, King of Portugal, on board that Ship in *The Tagus,* in the month of May, 1824."—Royal Licence, 28th of March, 1825.—Knighted, 20th of April, 1825.

ADDERLEY W. SLEIGH, Esq., Acting Master in The Royal Navy, and late Captain in The Spanish and Portuguese Services, in testimony of the high sense which His Imperial Majesty Don PEDRO, Duke of BRAGANZA, entertains of the signal intrepidity and zealous conduct displayed by him, when commanding Her Most Faithful Majesty's Ship of War, "*Terceira,*" in the capture of the Island of *Porto Santo,* and destroying a Vessel under the Enemy's forts of *Cumana de Lohos* and *Porta Cruz* at the Island of Madeira,—while conducting the Blockade before *Funchal,* during the War of Liberation in Portugal, in the years 1831, 32, 33, and 34.

HONORARY KNIGHTS GRAND CROSSES.

The Right Honble. JOHN, Earl of ST. VINCENT, K. C. B., Admiral of The Red Squadron of His Majesty's Fleet, and General of His Majesty's Royal Marine Forces, to accept and wear The Insignia of an Honorary Knight Grand Cross of The Royal Portuguese Military Order of THE TOWER and SWORD, which His Majesty JOHN, King of Portugal, was pleased to confer upon him in the year 1811, "in testimony of the high sense which His Most Faithful Majesty entertained of the distinguished merits of His Lordship, and of the eminent services rendered by him to The Kingdom of Portugal."—Royal Licence, 15th of September, 1817.

Sir CHARLES STUART, K. B., and His Majesty's Envoy Extraordinary and Plenipotentiary in Portugal, "for his distinguished Services, in the performance of the important functions of one of The Governors of the Kingdom of Portugal, in the name and by the appointment of His Royal Highness THE PRINCE REGENT of PORTUGAL."—Royal Licence, 11th of March, 1813.

KNIGHTS COMMANDERS.

FRANCIS BRIAN HILL, Esq., His Majesty's Secretary of Legation at The Court of *Rio Janeiro*.—Royal Licence, 5th of June, 1810.

Sir ROBERT THOMAS WILSON, Knight of The Imperial Military Order of MARIA THERESA, and Aide-de-Camp to The King, "for his distinguished Conduct and Services."—Royal Licence, 19th of October, 1811.—*See*, AUSTRIA, Commander of MARIA THERESA.

NICHOLAS TRANT, Esq., Captain in The Royal Staff Corps, Permanent Assistant to The Quarter-Master-General of the Forces, and late Lieut. Colonel of the Queen's German Regiment, also a Brigadier-General in the Portuguese Service, and Military Governor of *Oporto*, "for his distinguished Conduct and Services."—Royal Licence, 19th of October, 1811.

THOMAS WESTERN, Esq., Rear Admiral of The Blue Squadron of His Majesty's Fleet, "in testimony of the high sense which The Prince Regent of Portugal entertains of his great Merit and Services." — Royal Licence, 26th of August, 1814. — Died, 26th of December, 1814.

Lieut. Colonel JOHN AUSTIN, Brigadier-General in the Portuguese Service, and late Military Governor and Commander of the Forces in The Kingdom of *Algarve*, the licence and authority of "*Their Excellencies the Lords Justices*," to accept and wear The Insignia of a Knight Commander of The Royal Portuguese Military Order of The TOWER and SWORD," in testimony of His Majesty The King of Portugal's approbation of the distinguished services rendered by him, while serving in the Portuguese army."—18th of October, 1821.

William Robinson, Esq., Major of The Royal Marine Artillery, C.B., and Colonel in the Service of His Sicilian Majesty, "as a signal testimony of the approbation of His Majesty The King of The Two Sicilies, of the distinguished services rendered by him in the defence of *Sicily*, while in command of *The Combined Flotilla*,"—Royal Licence, 23d of September, 1823.

Sir George Nayler, Garter Principal King of Arms, and K. H., in compliance with the earnest desire expressed by His Most Faithful Majesty John the Sixth, King of Portugal, " as an especial testimony of His Royal approbation of the manner in which he discharged the duty of investing His Most Faithful Majesty with the Ensigns of The Most Noble Order of The Garter."—Royal Licence, 5th of June, 1824.

Doctor Augustus West, "for his meritorious conduct, while in the actual service of His Most Faithful Majesty John the Sixth, King of Portugal."—Royal Licence, 28th of October, 1824.—Knighted, 24th Nov., 1824.

The Right Honourable Sir George Murray,—*See*, Austria, Order of Leopold.

Sir Galbraith Lowry Cole, G.C.B.

Sir John Campbell, Knt., K.C.B., and Knight of The Royal Portuguese Military Order of the Tower and Sword, late Lieut. Colonel in The Army, and Brigadier General in The Portuguese Service, " in testimony of the high sense which his Most Faithful Majesty John the Sixth, King of Portugal, entertains of the great courage and intrepidity displayed by him, in several Actions with the Enemy, during *The Peninsular war*."—Royal Licence, 10th of October, 1825.

William Howe Mulcaster, Esq., Post Captain in The Royal Navy, and C.B., "in testimony of the approbation of His Most Faithful Majesty John the Sixth, King of Portugal, of the signal intrepidity displayed by him at the Reduction of *The Colony of Cayenne*, in the months of December 1808 and January 1809."—Royal Licence, 10th of October, 1825.

Colonel Sir Henry Pynn, Knt., late a Brigadier-General in The Portuguese Army, " in consideration of his distinguished conduct and eminent services during *The Peninsular war.*"—Royal Licence, 16th of July, 1826.

ARTHUR-MARCUS-CECIL HILL, Esq., (commonly called Lord ARTHUR MARCUS CECIL HILL), conferred upon him in 1825, by His Most Faithful Majesty JOHN the Sixth, late King of Portugal, Brazil, and The Algarves, "in testimony of His Royal approbation of His Lordship's conduct, while actually employed in His said Majesty's service."—Royal Licence, 1st of September, 1828.

Sir CHARLES SUTTON, K.C.B.—Died, 26th of March, 1828.

Sir ROBERT HUGH KENNEDY, Knt., Commissary-General of the Forces, "for his distinguished conduct during *The Peninsular war.*"—Loyal Licence, 31st of January, 1831.

Lieutenant CHARLES SHAW,—*See*, SPAIN, ORDER of SAN FERNANDO, Third Class.

Commander JAMES WILKINSON, of The Royal Navy, "in testimony of Her Majesty The Queen of Portugal's Royal approbation of his conduct, while in Her Most Faithful Majesty's Service."—Royal Licence, 30th of October, 1837.

Lieutenant Colonel HENRY WEBSTER, "in consequence of his Services."—Royal Licence, 4th of June, 1838.

GEORGE ROSE SARTORIUS, Esq., a Captain in The Royal Navy, and Vice-Admiral in the Service of The Queen of Portugal, to accept and wear the Insignia of a Knight Commander of The Royal Portuguese Military Order of THE TOWER and SWORD,—and the Insignia of a Grand Cross of the Military Order of SAINT BENTO D'AVIS, "conferred upon him, in testimony of Her Most Faithful Majesty's approbation of his conduct in Battles on the High Seas, and Services rendered in defence of the Nation in the years 1832 and 1833."—Royal Licence, 11th of June, 1838.

Sir ROBERT JOHN HARVEY, Knt., Colonel in The Army, C. B., and Knight of The Royal Portuguese Military Order of The Tower and Sword, "in testimony of Her Most Catholick Majesty's approbation of his services, during *The Peninsular War.*"—Royal Licence, 16th October, 1838.

HONORARY KNIGHTS COMMANDERS.

DENIS PACK, Esq., Colonel in The Army, one of His Majesty's Aides-de-Camp, Lieut. Colonel in the 71st (Highland) Regt. of Foot, and a Brigadier General in The Portuguese Service,—

JOHN WILSON, Esq., Lieut. Colonel in The Royal York Rangers, and a Brigadier General in The Portuguese Service,—

ROBERT EDWARD HENRY SOMERSET, Esq., (commonly called Lord ROBERT EDWARD HENRY SOMERSET), Colonel in The Army, one of His Majesty's Aides-de-Camp, Lieut. Colonel of the 4th, (*or*, Queen's Own) Regt. of Dragoons,—

Sir RICHARD FLETCHER, Knt. and Bart., Lieut. Colonel in The Corps of Royal Engineers, and Chief Engineer with the Army in Spain and Portugal.—(*Slain at St. Sebastian's,* in 1813),—

Sir THOMAS SIDNEY BECKWITH, Knt.,Colonel in The Army, Lieut. Colonel in the 95th Regt. of Foot, and Quartermaster General of His Majesty's Forces serving in Canada,—

ROBERT ARBUTHNOT, Esq., Lieut. Colonel in The Army, and Colonel in The Portuguese Service,—

THOMAS NOEL HILL, Esq., Lieut. Colonel in The Army, and then serving with The Portuguese Army,—

WILLIAM WILLIAMS, Esq., Lieut. Colonel in the 13th, (*or*, The 1st Somersetshire) Regt. of Foot,—

GREGORY HOLMAN BROMLEY WAY, Esq., Lieut. Colonel in the 29th (*or*, The Worcestershire) Regt. of Foot,—

"for the distinguished courage and intrepidity displayed by those Officers in several Actions with the Enemy, in *The Peninsula*,"—Royal Licence, 11th of March, 1813.

JOHN MILLEY DOYLE, Esq., Lieut. Colonel in The Army, and Colonel in the Portuguese Service, "for the distinguished courage and intrepidity displayed by him in several Actions with the Enemy in *The Peninsula*."—Royal Licence, 20th of March, 1813.

WILLIAM MAC BEAN, Esq., Lieut. Colonel in The Army, and Colonel commanding the 24th Regt. of Infantry in the Portuguese Service," for the distinguished courage and intrepidity displayed by him in several Actions with the Enemy in *The Peninsula*."—Royal Licence, 24th of March, 1813.

The Honourable Sir CHARLES WILLIAM STEWART, K.B., Major-General of His Majesty's Forces, Lieutenant-Colonel of the 18th Regt. of Light Dragoons, (Hussars), and Adjutant-General to the Forces serving in Spain and Portugal, "for the great courage and intrepidity displayed by him in several Actions with the Enemy in *The Peninsula.*"—Royal Licence, 27th March, 1813.—Died, 7th of January, 1827.

GEORGE ELDER, Esq., Lieutenant Colonel in The Army, and Colonel in The Portuguese Service, "for the distinguished courage and intrepidity displayed by him in several Actions with the Enemy in *The Peninsula.*"—Royal Licence, 27th of March, 1813,—Knighted, in 1813,—and C. B.

GEORGE RIDOUT BINGHAM, Esq., Lieut. Colonel in the 53rd, (*or*, The Shropshire) Regt. of Foot, "for the distinguished courage and intrepidity displayed by him in several Actions with the Enemy in *The Peninsula.*"—Royal Licence, 30th of March, 1813.

GEORGE MURRAY, Esq., Major-General in The Army, Captain in the 3rd Regt. of Foot Guards, and Quarter-Master-General to the Forces serving in Spain and Portugal, "for the distinguished courage and intrepidity displayed by him in several Actions with the Enemy in *The Peninsula.*"—Royal Licence, 6th of May, 1813.

GEORGE ANSON, Esq., a Major-General of His Majesty's Forces, and K. C. B., "for the distinguished courage and intrepidity displayed by him in several Actions with the Enemy in *The Peninsula.*"—Royal Licence, 17th of May, 1813.

The Honourable THOMAS WILLIAM FERMOR, Colonel in the Army, and Captain in the 3d Regiment of Foot Guards, "for the distinguished courage and intrepidity displayed by him in several Actions with the enemy in *The Peninsula.*"—Royal Licence, 17th of May, 1813.

WILLIAM MAUNDY HARVEY, Esq., a Colonel in the Army, Lieut. Colonel of the 79th Regiment of Foot, and Brigadier-General in The Portuguese Service, "for the great courage and intrepidity displayed by him in several Actions with the Enemy in *The Peninsula.*"—Royal Licence, 18th of May, 1813.

JOHN HAMILTON, Esq., Lieut. General of His Majesty's Forces, and Colonel of the 2d Ceylon Regiment, "for the distinguished courage and intrepidity displayed by him in several Actions with the Enemy in *The Peninsula*."—Royal Licence, 22d of May, 1813.

ARCHIBALD CAMPBELL, Esq., Lieut. Colonel in the Army, and a Brigadier General in The Portuguese Service, "in testimony of the high sense which His Royal Highness the Prince Regent of Portugal entertains of the great courage and intrepidity displayed by him, in several Actions with the Enemy in *The Peninsula*."—Royal Licence, 2d of October, 1813.

Lieut. General Sir THOMAS PICTON, K. B., Colonel of the 77th Regiment of Foot, "in testimony of the high sense which His Royal Highness The Prince Regent of Portugal entertains of his distinguished conduct on divers important occasions."—Royal Licence, 2d November, 1813. Slain at *Waterloo*.

The Honourable Sir EDWARD MICHAEL PAKENHAM, Extra Knight of The Most Honourable Military Order of The Bath, and Major General of His Majesty's Forces, "in testimony of the high sense which The Prince Regent of Portugal entertains of the distinguished courage and intrepidity displayed by him, in several Actions with the Enemy in *The Peninsula*."—Royal Licence, 5th of January, 1814.—Slain at *New Orleans*, 8th of January, 1815.

Sir JAMES LEITH, K. B., and Lieut. General of His Majesty's Forces, "in testimony of the high sense which His Royal Highness The Prince Regent of Portugal entertains of the distinguished courage and intrepidity displayed by him, in several Actions with the Enemy in *The Peninsula*."—Royal Licence, 2d of February, 1814. *See*, FRANCE, ORDER of MILITARY MERIT.

The Honourable EDWARD STOPFORD, a Major General in the Army, "in testimony of the high sense which The Prince Regent of Portugal entertains of the great courage and intrepidity displayed by him, in several Actions with the Enemy."—Royal Licence, 10th of February, 1814.

The Honourable FREDERICK CAVENDISH PONSONBY, Lieut. Colonel of the

12th Regt. of Light Dragoons, "in testimony of the high sense which The Prince Regent of Portugal entertains of the great courage and intrepidity displayed by him, in several Actions with the Enemy."—Royal Licence, 28th of February, 1814.

Benjamin D'Urban, Esq., Colonel in the Army, Lieut. Colonel in the 2d West India Regt., and a Brigadier General in the Portuguese Service, "in testimony of the high sense which The Prince Regent of Portugal entertains of the distinguished courage and intrepidity displayed by him, in several Actions with the Enemy in *The Peninsula.*"—Royal Licence, 17th of June, 1814.

Thomas Bradford, Esq., Major-General in the Army, "in testimony of the high sense which The Prince Regent of Portugal entertains of the distinguished courage and intrepidity displayed by him, in several Actions with the Enemy in *The Peninsula.*"—Royal Licence, 24th November, 1814.

Charles Ashworth, Esq., Lieut. Colonel in the Army, and Brigadier General in The Portuguese Service, "in testimony of the high sense which The Prince Regent of Portugal entertains of the distinguished courage and intrepidity displayed by him, in several Actions with the Enemy in *The Peninsula.*—Royal Licence, 24th of November, 1814.

Sir William Robe, K. C. B., Colonel in The Army, and Lieut. Colonel in The Royal Regiment of Artillery, "in testimony of the high sense which The Prince Regent of Portugal entertains of the distinguished courage and intrepidity displayed by him, in several Actions with the Enemy in *The Peninsula.*"—Royal Licence, 14th of February, 1815.—Died, 1st of November, 1820.

Richard Blunt, Esq., Major General of His Majesty's Forces, "in testimony of the high sense which The Prince Regent of Portugal entertains of the distinguished courage and intrepidity displayed by him, in several Actions with the Enemy in *The Peninsula.*"—Royal Licence, 24th of February, 1815.

Thomas William Stubbs, Esq., Brigadier General in The Portuguese Service (late of The British Service), "in testimony of the high sense which The Prince Regent of Portugal entertains of the distinguished courage and intrepidity displayed by him, in several

Actions with the Enemy in *The Peninsula.*"—Royal Licence, 24th of February, 1815.

Major General Sir George Townshend Walker, K. C. B., "in testimony of the high sense which The Prince Regent of Portugal entertains of the great courage and intrepidity displayed by him, in several Actions with the Enemy in *The Peninsula.*"—Royal Licence, 18th of May, 1815.

The Honourable Sir Charles Colville, G. C. B., Major General in The Army, and Colonel of the 94th Regt. of Foot, "in testimony of the high sense which The Prince Regent of Portugal entertains of the great ability and signal intrepidity displayed by him, in several Actions with the Enemy during the recent arduous Campaigns in *The Peninsula.*"—Royal Licence, 23rd of May, 1815.

Sir Kenneth Alexander Howard, K. C. B., Major General of His Majesty's Forces, commanding The South West Division, "in testimony of the high sense which The Prince Regent of Portugal entertains of the great ability and signal intrepidity displayed by him, in several actions with the Enemy during the recent arduous Campaigns in *The Peninsula.*"—Royal Licence, 23rd of May, 1815.

Warren Marmaduke Peacocke, Esq., Major General in The Army, "in testimony of the high sense which The Prince Regent of Portugal entertains of his distinguished Services, whilst employed with the British Army in the defence of The Kingdom of Portugal."—Royal Licence, 29th of May, 1815.

Sir Richard Lee, K. C. B., and Rear-Admiral of The White Squadron of His Majesty's Fleet, "in testimony of the high sense which The Prince Regent of Portugal entertains of his great merit, and of the services rendered by him to the Kingdom of Portugal."— Royal Licence, 31st of May, 1815.

Major General Sir Manley Power, K. C. B., "in testimony of the high sense which The Prince Regent of Portugal entertains of the distinguished courage and signal intrepidity displayed by him, in several Actions with the Enemy in *The Peninsula.*"—Royal Licence, 20th of June, 1815.

Charles Marsh Schomberg, Esq., a Post Captain in The Royal Navy, and C. B.,—conferred upon him in 1808, "for his great merit, and the services rendered by him to *The Kingdom of Portugal.*"—Royal Licence, 23d of September, 1815.

Charles Count Alten,—*See*, Russia, Order of St. Anne.

Sir Ulysses Burgh, (now, Lord Downes), K.C.B., Lieut. Colonel in The Army, Captain in the 1st or Grenadier Regt. of Foot Guards, and Aide-de-Camp to Field Marshal His Grace The Duke of Wellington, "for the great courage and intrepidity displayed by him in several Actions with the Enemy in *The Peninsula.*"—Royal Licence, 4th of April, 1816.

Sir James Mc Grigor, Knt., M.D., and Director General of the Medical Department of His Majesty's Land Forces, "for his great merit and his services in the Field, during the recent Campaigns in *The Peninsula.*"—Royal Licence, 4th of April, 1816.

George Allan Madden, Esq., Colonel in The Army, C.B., and a Major-General in The Portuguese Service,—

John Buchan, Esq., Lieut. Colonel in The Army, C.B., and a Major General in The Portuguese Service,—

"for the great ability and signal intrepidity displayed by those Officers in several Actions with the Enemy, during the recent arduous Campaigns in *The Peninsula.*"—Royal Licence, 4th of April, 1816.

Sir Colin Campbell,—*See*, Austria, Order of Maria Theresa.

Lord Fitzroy Somerset,—*See*, Austria, Order of Maria Theresa.

James Walker, Esq., Post Captain in The Royal Navy, and C.B., "with which His Royal Highness The Prince Regent of Portugal was pleased to honour him in the year 1808, in testimony of the high sense that Prince entertains of his great merit, and of the services rendered by him to *The Kingdom of Portugal.*" —Royal Licence, 30th of April, 1816.

Major General Sir Henry Torrens, K.C.B., and Adjutant-General of His Majesty's Forces, "in testimony of His Most Faithful Majesty's approbation of the distinguished intrepidity displayed by him in Action with the Enemy in *Portugal.*"—Royal Licence, 2d of April, 1821.

Sir JOHN CROFT, of *Cowling-Hall*, in the County of York, and of *Dodington*, in the County of Kent, Bart., "in testimony of HIS MOST FAITHFUL MAJESTY's approbation of his meritorious conduct, while in the actual service of that Sovereign."—Royal Licence, 18th of December, 1821.

And His Majesty, WILLIAM the Fourth, taking into His Royal consideration the merits and services of Sir JOHN CROFT, "when employed as a Commissioner for the distribution of the Grant voted by Parliament for the relief of the distressed Inhabitants of *the ravaged Districts of Portugal*," was graciously pleased, by Warrant under His Royal Signet and Sign Manual, dated the 29th of April, 1834, to grant unto him, His Royal Licence and Authority that he, and the heirs male of his body to whom the Dignity of a Baronet shall descend, may bear and use certain *Supporters* to his and their ARMS respectively, as depicted in the Painting annexed to the said Royal Warrant.

Lieut. Colonel MICHAEL MC CREAGH, C. B., Knight of The Royal Portuguese Order of The Tower and Sword, and late a Brigadier General in The Portuguese Service, "in testimony of His Most Faithful Majesty's approbation of the distinguished services rendered by him, while serving in the Portuguese Army." —Royal Licence, 28th of December, 1821.

WILLIAM ELLIOTT, Esq., Post Captain in The Royal Navy, Captain of His Majesty's Ship *Lively*, and C. B., "on occasion of the visit of His Most Faithful Majesty JOHN the Sixth, King of Portugal, on board that Ship, in *The Tagus*, in the Month of May, 1824." —Royal Licence, 29th of March, 1825.

NEVINSON DE COURCY, Esq., Post Captain in The Royal Navy, "on occasion of the visit of His Most Faithful Majesty JOHN the Sixth, King of Portugal, on board The British Ships in *The Tagus*, in the month of May, 1824."—Royal Licence, 2d of April, 1825.

KNIGHTS.

THOMAS SEVESTRE, Esq., late Surgeon of His Majesty's Ship *The Confiance*, "for the unremitted attention paid by him to the sick and wounded of the Portuguese Forces employed in the reduction of *The Colony of Cayenne,* in the months of December, 1808 and January, 1809."—Royal Licence, 6th of October, 1810.

The Honble. Robert Lawrence Dundas, Lieut. Colonel in The Army, and Major in The Royal Staff Corps, "in testimony of the high sense which His Royal Highness The Prince Regent of Portugal entertains of the distinguished courage and intrepidity displayed by him, in several Actions with the Enemy in *The Peninsula.*" —Royal Licence, 3d of February, 1814.

John Browne, Esq., Lieut. Colonel in The Army, and Colonel in The Portuguese Service, "in testimony of the high sense which The Prince Regent of Portugal entertains of the distinguished courage and intrepidity displayed by him, in several Actions with the Enemy in *The Peninsula.*"—Royal Licence, 3d of February, 1814.

Henry Sturgeon, Esq., Lieut. Colonel in The Army, and Major in The Royal Staff Corps, "in testimony of the high sense which The Prince Regent of Portugal entertains of the distinguished courage and intrepidity displayed by him, in several Actions with the Enemy in *The Peninsula.*"—Royal Licence, 3d of February, 1814.—Killed at *Bigorre* in France, in May, 1814, three days before the Battle of *Toulouse.*

George Henry Frederick Berkeley, Esq., Lieut. Colonel in The 35th *or* The Sussex Regiment of Foot, and Assistant Adjutant General to The Forces under Field Marshal The Marquess of Wellington, "in testimony of the high sense which His Royal Highness The Prince Regent of Portugal entertains of the distinguished courage and intrepidity displayed by him, in several Actions with the Enemy in *The Peninsula.*"— Royal Licence, 26th of February, 1814.

Granby Thomas Calcraft, Esq., Knight of The Imperial Military Order of Maria Theresa, a Major General in the Army, and Lieut. Colonel of the Third *or* Prince of Wales's Regiment of Dragoon Guards, "in testimony of the high sense which His Royal Highness John Prince of Brazil, Prince Regent of Portugal, entertains of the great courage and intrepidity displayed by him, in several Actions with the Enemy in *The Peninsula.*" —Royal Licence, 9th of April, 1814.—Died, on the 20th of August, 1820.

The Honble. ALEXANDER ABERCROMBY, Colonel in The Army, and Lieut. Colonel in The Coldstream Regiment of Foot Guards, " in testimony of the high sense which The Prince Regent of Portugal entertains of the distinguished courage and intrepidity displayed by him, in several Actions with the Enemy in *The Peninsula.*"—Royal Licence, 7th of October, 1814.

HENRY PYNN, Esq., Lieut. Colonel in The Army, and Lieut. Colonel commanding the 18th Portuguese Regiment, " in testimony of the high sense which The Prince Regent of Portugal entertains of the great courage and intrepidity displayed by him, in several Actions with the Enemy in *The Peninsula.*"—Royal Licence, 17th of January, 1815.—Knighted, 23rd of February, 1815.

VICTOR VON ARENTSSCHILD, Brevet Major in the Army, and Lieutenant Colonel in The Portuguese 1st Regt. of Artillery, " as a testimony of the high sense which The Prince Regent of Portugal entertains of the great courage and intrepidity displayed by him, in several Actions with the Enemy in *The Peninsula.*"—Royal Licence, 7th of April, 1815.

SIR JOHN MACLEAN, K. C. B., Colonel in The Army, and Lieut. Colonel of the 27th (*or,* Enniskillen) Regt. of Foot,—

SIR HENRY WILLIAM CARR, K. C. B., and Lieut. Colonel of the 83d Regt. of Foot,—

Sir JOHN MAY, K. C. B., Lieut. Colonel in The Army, and Captain in The Royal Horse Artillery,—

Sir HEW DALRYMPLE ROSS, K. C. B., Lieut. Colonel in The Army, and Captain in The Royal Horse Artillery,—

Sir CHARLES BROKE, K. C. B., Lieut. Colonel in The Army, and Assistant Quarter-Master-General of His Majesty's Forces,— " in testimony of the high sense which The Prince Regent of Portugal entertains of the signal intrepidity displayed by those Officers, in several Actions with the Enemy during the recent arduous Campaigns in *The Peninsula.*"—Royal Licence, 23d of May, 1815.

JOHN FOX BURGOYNE, Esq., Lieut. Colonel in The Army, and Captain in The Royal Engineers, "in testimony of the high sense which The Prince Regent of Portugal entertains of the great courage and intrepidity displayed by him, in several Actions with the Enemy during the recent arduous Campaigns in *The Peninsula.*"—Royal Licence, 29th of May, 1815.

JOHN CAMERON, Esq., Colonel in the Army, and Lieut. Colonel of the 92d *(Highland)* Regt. of Foot, "in testimony of the high sense which The Prince Regent of Portugal entertains of the great courage and intrepidity displayed by him, in several Actions with the Enemy in *The Peninsula.*"—Royal Licence, 29th of May, 1815,—Killed at *Waterloo.*

JAMES STEVENSON BARNS, Esq., Colonel in The Army, and Lieut. Colonel in the First (*or*, The Royal Scots) Regt. of Foot, "in testimony of the high sense which The Prince Regent of Portugal entertains of the distinguished courage and signal intrepidity displayed by him, in several Actions with the Enemy during the recent arduous Campaigns in *The Peninsula.*"—Royal Licence, 28th of August, 1815.

WILLIAM COX, Esq., Lieut. Colonel in The Army, and a Brigadier General in The Portuguese Service, "in testimony of the high sense which The Prince Regent of Portugal entertains of the great courage and signal intrepidity displayed by him in several Actions with the Enemy, during the recent arduous Campaigns in *The Peninsula.*"—Royal Licence, 28th of August, 1815.—Knighted, in 1816.

Sir EDWARD BLAKENEY, K. C. B., Colonel in The Army, and Lieut. Colonel of the 7th Regt. of Foot (*or*, Royal Fusileers), "in testimony of the high sense which The Prince Regent of Portugal entertains of the distinguished courage and intrepidity displayed by him, in several Actions with the Enemy in *The Peninsula.*"—Royal Licence, 13th of November, 1815.

The Honble. Sir ROBERT LE POER TRENCH, K. C. B., Colonel in The Army, and Lieut. Colonel in the 74th Regt. of Foot, "in testimony of the high sense which The Prince Regent of Portugal entertains of the distinguished courage and intrepidity displayed by him, in several Actions with the Enemy in *The Peninsula.*"—Royal Licence, 7th of December, 1815.—Died, in March, 1824.

JAMES CAMPBELL, Esq., Colonel in the Army, Lieut. Colonel of the 94th Regt. of Foot, and C.B., "for the distinguished courage and intrepidity displayed by him, in several Actions with the Enemy in *The Peninsula.*"—Royal Licence, 11th of March, 1816.

Lieutenant-Colonel HENRY WATSON, of the 1st Portuguese Cavalry, "for the distinguished courage and intrepidity displayed by him, in several Actions with the Enemy in *The Peninsula.*"—Royal Licence, 11th of March, 1816.

WILLIAM FENWICK, Esq., Lieut. Colonel in the Army, Lieut. Governor of Pendennis Castle, and C. B., "for the distinguished courage and intrepidity displayed by him, in several Actions with the Enemy in *The Peninsula.*"—Royal Licence, 16th of March, 1816.

Sir EDMUND KEYNTON WILLIAMS, K.C.B., Lieut. Colonel in the Army, and Colonel of the 14th Regt. of Portuguese Infantry,—

Sir MAXWELL GRANT, K. C. B., Lieut. Colonel in The Army, and Colonel of the 6th Regt. of Portuguese Infantry, "for the signal intrepidity displayed by them in several Actions with the Enemy, during the recent arduous Campaigns in *The Peninsula.*"—Royal Licence, 4th of April, 1816.

ALEXANDER TULLOH, Esq., Lieut. Colonel in The Royal Artillery, C. B., and a Lieut. Colonel in the Portuguese Armies, commanding the Field Brigades of Artillery in that Service, "for the signal intrepidity displayed by him in several Actions with the Enemy, during the recent arduous Campaigns in *The Peninsula.*"—Royal Licence, 4th of April, 1816.

ARCHIBALD ROSS, Esq., Major in The Army, Captain in the 91st Regt. of Foot, and Lieut. Colonel in the Portuguese Service, "in testimony of the high sense which His Royal Highness THE PRINCE REGENT of PORTUGAL entertains of the signal intrepidity displayed by him in several Actions with the Enemy, during the recent arduous Campaigns in *The Peninsula.*"—Royal Licence, 3d of May, 1816.

Robert John Harvey, Esq., Lieut. Colonel in The Army, and late of The Portuguese Staff, " for the signal intrepidity displayed by him in several Actions with the Enemy, during the recent arduous Campaigns in *The Peninsula*."—Royal Licence, 6th of May, 1816.—Knighted, 6th of February, 1817.

Michael M'Creagh, Esq., C. B., Lieut. Colonel in The Army, and a Brigadier General in The Portuguese Service,—

William Warre, Esq., a Lieut. Colonel in The Army, Captain in the 23d Regt. of Light Dragoons, and Deputy-Quarter-Master-General of His Majesty's Forces at The Cape of Good Hope,— " for the distinguished courage and intrepidity displayed by those Officers, in several Actions with the Enemy in *The Peninsula*."—Royal Licence, 20th of May, 1816.

Dudley St. Leger Hill, Esq., Lieut. Colonel in The Army, Colonel in The Portuguese Service, and C.B., " as a testimony of the high sense which His Majesty The King of Portugal entertains of the great courage and intrepidity displayed by him, in several Actions with the Enemy in *The Peninsula*."—Royal Licence, 20th of November, 1816.—Knighted, 25th of November, 1816.

Alexander Anderson, Esq., Major in The Army, Colonel of the 11th Regt. of Portuguese Infantry, and C. B., " as a testimony of the high sense which His Majesty The King of Portugal entertains of the great courage and intrepidity displayed by him, in several Actions with the Enemy in *The Peninsula*."—Royal Licence, 6th of December, 1816.

John Rolt, Esq., Major in The Army, Colonel of the 17th Regt. of Portuguese Infantry, and C. B., " as a testimony of the high sense which His Majesty The King of Portugal entertains of the great courage and intrepidity displayed by him, in several Actions with the Enemy in *The Peninsula*."—Royal Licence, 6th of December, 1816.

Richard Armstrong, Esq., Lieut. Colonel in The Army, C. B., and Colonel of the 4th Regt. of Portuguese Infantry, " as a testimony of the high sense which His Majesty The King of Portugal entertains of the great courage and intrepidity displayed by him, in several Actions with the Enemy in *The Peninsula.*"—Royal Licence, 31st of December, 1816.

Sir Frederick Armstrong, Knt., Captain in The Army, and late Major in the Portuguese Service, "in testimony of the high sense which His Majesty The King of Portugal entertains of the great courage and intrepidity displayed by him in several Actions with the Enemy, during *The Peninsular War.*"—Royal Licence, 20th of March, 1820.

Edward Miles, Esq., Lieut. Colonel of the 89th Regt. of Foot, late Lieut. Colonel of the 38th (*or*, 1st Staffordshire) Regt. of Foot, and C. B., "in testimony of the high sense which His Majesty The King of Portugal entertains of the great courage and intrepidity displayed by him in several Actions with the Enemy, during *The Peninsular War.*"—Royal Licence, 6th of April, 1820.—Knighted, 27th of November, 1826.

Sir John Browne, Knt.,—*See*, Spain, Order of Charles the Third.

Samuel Edward Cook, a Commander in The Royal Navy, late First Lieutenant of His Majesty's Ship *The Windsor Castle*, "on occasion of the visit of His Most Faithful Majesty John the Sixth, King of Portugal, on board that Ship, in *The Tagus*, in the month of May, 1824."—Royal Licence, 2d of April, 1825.

Lieutenant William Walker, of The Royal Navy, late Lieutenant of His Majesty's Ship *The Windsor Castle*, "on occasion of the visit of His Most Faithful Majesty John the Sixth, King of Portugal, on board that Ship, in *The Tagus*, in the month of May, 1824."—Royal Licence, 18th of April, 1825.

Lieutenant John Somerville, of The Royal Navy, late Lieutenant of His Majesty's Ship *The Windsor Castle*, "on occasion of the visit of His Most Faithful Majesty John the Sixth, King of Portugal, on board that Ship, in *The Tagus*, in the month of May, 1824."—Royal Licence, 18th of April, 1825.

Lieutenant Charles Robert Dashwood, of The Royal Navy, late Lieutenant of His Majesty's Ship *The Windsor Castle*, "on occasion of the visit of His Most Faithful Majesty John the Sixth, King of Portugal, on board that Ship, in *The Tagus*, in the month of May, 1824."—Royal Licence, 18th of April, 1825.

Lieutenant EDWARD PITTS, of The Royal Navy, late Lieutenant of His Majesty's Ship *The Windsor Castle,* "on occasion of the visit of His Most Faithful Majesty JOHN the Sixth, King of Portugal, on board that Ship, in *The Tagus,* in the month of May, 1824." —Royal Licence, 18th of April, 1825.

Lieutenant JOHN *de* COURCY DASHWOOD, of The Royal Navy, "on occasion of the visit of His Most Faithful Majesty JOHN the Sixth, King of Portugal, on board His Majesty's Ship *The Windsor Castle,* in *The Tagus,* in the month of May, 1824."—Royal Licence, 18th of April, 1825.

ROBERT PURKIS HILLYAR, M.D., late Surgeon of His Majesty's Ship *The Windsor Castle,* "on occasion of the visit of His Most Faithful Majesty JOHN the Sixth, King of Portugal, on board that Ship, in *The Tagus,* in the month of May, 1824."—Royal Licence, 18th of April, 1825.

Lieutenant HENRY KNOLLES, of The Royal Navy, late Lieutenant of His Majesty's Ship *The Windsor Castle,* "on occasion of the visit of His Most Faithful Majesty JOHN the Sixth, King of Portugal, on board that Ship, in *The Tagus,* in the month of May, 1824."—Royal Licence, 21st of April, 1825.

FREDERICK WILLIAM RICHARD SADLER, late Master of His Majesty's Ship *The Windsor Castle,* "on occasion of the visit of His Most Faithful Majesty JOHN the Sixth, King of Portugal, on board that Ship, in *The Tagus,* in the Month of May, 1824."—Royal Licence, 21st of April, 1825.

Lieutenant JOHN SEATON, of His Majesty's Ship *Lively,* "on occasion of the visit of His Most Faithful Majesty JOHN the Sixth, King of Portugal, on board that Ship, in *The Tagus,* in the month of May, 1824."—Royal Licence, 21st of April, 1825.

Lieutenant RICHARD CROZIER, of His Majesty's Ship *Lively,* "on occasion of the visit of His Most Faithful Majesty JOHN the Sixth, King of Portugal, on board that Ship, in *The Tagus,* in the month of May, 1824."—Royal Licence, 21st of April, 1825.

EBENEZER JOHNSTON, Esq., Surgeon of His Majesty's Ship *Lively,*

" on occasion of the visit of His Most Faithful Majesty John the Sixth, King of Portugal, on board The British Ships, in *The Tagus*, in the month of May, 1824."—Royal Licence, 21st of April, 1825.

George Voller Oughton, Gentleman, Purser of His Majesty's Ship *Lively*, " on occasion of the visit of His Most Faithful Majesty John the Sixth, King of Portugal, on board, that Ship, in *The Tagus*, in the month of May, 1824."—Royal Licence, 21st of April, 1825.

Edward Augustus Parker, Lieutenant of His Majesty's Marine Forces on board *The Windsor Castle*, " on occasion of the visit of His Most Faithful Majesty John the Sixth, King of Portugal, on board that Ship, in *The Tagus*, in the month of May, 1824." —Royal Licence, 20th of September, 1825.

David Barry, M.D., Surgeon to His Majesty's Forces, and late First Surgeon to The Portuguese Army, " in consideration of his distinguished conduct in the Field, while in the actual service of His Most Faithful Majesty John the Sixth, King of Portugal." —Royal Licence, 26th of September, 1825.

Stephen Giles, First Lieutenant of His Majesty's Marine Forces on board *The Windsor Castle*, " on occasion of the visit of His Most Faithful Majesty John the Sixth, King of Portugal, on board that Ship, in *The Tagus*, in the month of May, 1824."—Royal Licence, 11th of October, 1825.

Thomas Peacocke, Esq., a Major in The Army, and Lieut. Colonel in The Portuguese Service, " in consideration of his distinguished conduct, while in the actual service of His Most Faithful Majesty John the Sixth, King of Portugal."—Royal Licence, 7th of November, 1825.

Robert Sharpe, a Lieutenant in The Royal Navy, and late Lieutenant of His Majesty's Ship *The Windsor Castle*, " on occasion of the visit of His Most Faithful Majesty John the Sixth, King of Portugal, on board that Ship, in *The Tagus*, in the month of May, 1824."—Royal Licence, 11th of April, 1826.

Major Benjamin Orlando Jones, " for his Services rendered to The

Portuguese Government, during *The Peninsular War.*"—Royal Licence, 17th of December, 1827.

Lieutenant GEORGE READ, of The Royal Navy, late Lieutenant of His Majesty's Ship *Lively,* "on occasion of His Most Faithful Majesty JOHN the Sixth's, King of Portugal, visit on board that Ship, in *The Tagus,* in the month of May, 1824."—Royal Licence, 17th of April, 1828.

THOMAS STAUNTON ST. CLAIR, Esq., Lieutenant Colonel in The Army, "in testimony of approbation of his conduct, while employed in the Military Service of His Most Faithful Majesty."—Royal Licence, 23d of February, 1831.

Lieutenant THOMAS BALDOCK, of The Royal Navy, "in consideration of the services rendered by him and the great skill he displayed, whilst conveying Her Most Faithful Majesty's August Consort, His Royal Highness THE PRINCE FERDINAND of PORTUGAL, to LISBON."—Royal Licence, 1st of May, 1837.

Lieutenant LIONEL CHARLES HENRY WILLIAM FITZ-GERALD, "in approbation of his conduct, especially at the Battle of *Valongo,* on the 18th of August, 1833, while in Her Most Faithful Majesty's actual Service."—Royal Licence, 13th of May, 1837.

JAMES JOHNSTON, Esq., Lieut. Colonel of Her Majesty's 99th (*or,* Lanarkshire) Regt. of Foot, "in testimony of the approbation of Her Majesty THE QUEEN of PORTUGAL of his services, during *The Peninsular War.*"—Royal Licence, 11th of September, 1838.

ROBERT PAGE, of *Madeira.*

The Military Order of Our Lady of The Conception of Villa Viçosa.

This Order was founded by John the Sixth, on the 6th of February, 1818, being the day of his Accession to The United Kingdom of Portugal, Brazil, and Algarve.

On the 10th of September, 1819, the following Statutes were framed at *Rio de Janeiro*, for the regulation of The Order,—

Article 1. This new Order shall be called, The Military Order of Our Lady of The Conception of Villa Viçosa, and shall consist of a Grand Master, Actual and Honorary Grand Crosses, Commanders, Knights, and Servants.

2. The King or Queen of Portugal shall enjoy the Rank of Grand Master, and the inalienable right of admitting persons into The Order.

3. All Members of The Royal Family, of either Sex, shall always be Actual Grand Crosses of The Order.

4. This Order shall comprise, 12 Honorary Grand Crosses,—40 Commanders,—100 Knights,—and 60 Servants,—together with the Commanders and Knights *Natos* (or, *ex officio*), nominated for the Service of The Order, and who shall be regarded, as *Extraordinary*. The number of Members to be fixed,—unless The Founder of The Order should see reason to exceed it.

5. The Honorary Grand Crosses shall be conferred upon persons, possessing *Titles*,—the rank of Commander, on those who shall have been made *Hidalgos* in the Royal Palace,—and the rank of Knight, on persons of *Noble* birth and Functionaries, who, by their services, may have merited the Royal consideration.

6. The Insignia of this Order shall consist of a Great Star, of nine points, of white enamel,—and rayed with gold, with nine small Stars of the same enamel, placed upon rays between each of it's points, and decorated with the Royal crown upon the superior point. In the centre, on a field of dead gold, shall be represented The Divine Salutation, in a cypher of bright gold, and around it, on an enamelled face of light blue, shall be inscribed the legend, "*Padroeira do Reino,*" *(Patroness of The Kingdom).*

7. This *Insignia* shall be greater or less, according to circumstances. Grand Crosses and Commanders shall wear the larger size,—and they may likewise wear the smaller, on days which are not Festival days. The Knight shall wear the smaller size,—and the Servants shall wear *Insignia* with the same characters and of the same size as that of the Knights, but made in *Silver*, without either gold or jewels.

8. The *Insignia* shall be worn, attached to a light blue riband of *Chamalote (silk and camel's hair)*, bordered with white,—The Grand Crosses shall wear a broad riband, passed over the right shoulder,—The Commanders shall wear a similar riband, round the neck,—The Knights and Servants shall wear the *Insignia*, suspended from the left button-hole. The Grand Crosses and Commanders shall likewise wear the *Insignia* in a plate of gold, or embroidered on the left side of the coat or outer garment, as is customary in other Orders.

9. In order to preserve uniformity in the *Insignia* of The Order, the use of *Insignia*, which shall not be conformable in size, ornament, and character to the plate above-mentioned, are absolutely forbidden.

10. The Grand Crosses, Commanders, and Knights of this Order shall enjoy all the honours, rights and privileges which are enjoyed by The Grand Crosses, Commanders and Knights of the other Military Orders of the Kingdom,—And the maintenance of these privileges, and the cognizance and decision of all matters relating to The Order, to be entrusted to the two Courts of *Conciencia y Ordens* of The United Kingdom.

11. The Titles of The Grand Crosses shall be issued by The Secretary of State for The Interior, according to custom,—Those of The Commanders and Knights shall be comprised in Patents,—Those of the Servants, in Warrants issued by The Courts of *Conciencia y Ordens,*

and signed by The Grand Master. The Courts of *Conciencia y Ordens* of *Lisbon* and *Rio de Janeiro* shall cause these Titles to be issued to those who shall produce Grants of the favour, signed by the Secretary of State for the Affairs of The United Kingdon, without any process of qualification.

12. The Chapel of Our Lady of The Conception of Villa Viçosa, in the Province of *Alemtejo*, and the Royal Chapel of The Court, to be the Heads of the NEW ORDER,—THE FESTIVAL of THE PATRONESS of THE ORDER to be annually celebrated in both of these Chapels,—that of *Villa Viçosa*, on the customary day,—and that of *The Court*, on the 8th day of The Conception.

13. All the Honorary Grand Crosses, Commanders, Knights and Servants, who may be within one league from the head of The Order, shall be obliged to be present at the celebration of The Festival of The Patroness.'

14. The two Royal Institutions of THE OFFICERS and SLAVES (*of The Virgin*) erected in The Church of Our Lady of The Conception of *Villa Viçosa*, shall be incorporated with THIS ORDER, continuing the obligations of their own Statutes, and at the same time complying with the requisitions of THIS ORDER.

15. The Dean of the Royal Chapel of *Villa Viçosa* shall be Commander *Nato* of THIS ORDER,—The Canons of the same Royal Chapel, or the Prior and Beneficiaries of The Church of Our Lady of The Conception of *Villa Viçosa*, and those of The Court of The Royal Corporation of The Slaves, erected in the same Church, shall be likewise perpetual Knights *Natos*.

16. THIS ORDER shall be independent of *Religious* Profession.

17. Persons rewarded with THIS ORDER, shall present themselves personally or by proxy before The Dean of The Royal Chapel of *Villa Viçosa*, and swear, in his hands, to defend The Mystery of The Immaculate Conception of The Virgin Mary.

18. A book shall also be kept in The Royal Chapel of THE COURT, in which shall be matriculated all the Honorary Grand Crosses, Commanders, Knights, and Servants, possessing titles passed by The Court of *Conciencia y Ordens* of *Rio de Janeiro*.

19. All persons matriculated in The Royal Chapel of The Court

shall afterwards take the Oath before The Dean of The Royal Chapel of *Villa Viçosa,* and matriculate themselves in the Book of The Royal Corporation of The Slaves, and shall then present to The Court of *Conciencia y Ordens* the Certificates of the Oath and Matriculation within two years from the day, on which their titles shall have been passed by The Chancery.

20. Relates to the emoluments of the Functionaries,—and

21. Charges certain persons to watch over the observance of these Statutes.

The Royal Order of Saint Isabella.

This Order was founded on the 2d of May, 1804, by the reigning Queen, Charlotte, and confirmed by The King.

The Queen is Grand Mistress, and confers The Order, on *ten* Members, who must be of the highest Rank.

The Badge of The Order consists of a gold Medal, with the effigy of *St. Isabella,* Queen of Portugal, which is worn, pendent to a rose coloured riband, with white borders, passed across the shoulder like a Scarf.

The Motto is, " *Pauperum Solatio.*"

Crosses of Honour.

Cross for The Campaign of The Peninsula,—Instituted by King John the Sixth, on the 28th of July, 1816. The Peninsular War being divided into six Campaigns, those of the Officers who have taken part in them, possess the right of wearing on the left breast a Roman Cross, with six laurel branches, in *gold*,—and, in the centre of the Cross, which is attached to a blue and red riband, the No. 6., indicative of the Six Campaigns. Those Officers who have only been present at one, two, or three Campaigns, wear the same Cross, but in *silver*, with the legend on the reverse, "*War of The Peninsula.*"

Cross of Command,—Instituted, on the 28th of July, 1816, by King John the Sixth,—This is a Roman Cross, with Stars equal in number to the Battles at which each individual commanding a Regiment, or Battalion has been present, and having the name of the possessor in the Exergue. It is worn at the button-hole, attached to a blue and red riband.

Cross of Fidelity of Transmontana,—Instituted, on the 24th of July, 1823, by King John the Sixth, for all those Officers who composed the Corps of the Army of Lieut. General The Marquess de Chaves at the period of the Abolition of The Constitutional Cortes at Lisbon. The Cross bears the effigy of the King,—and, on the reverse, "*Heroic Fidelity,—Transmontana*, 1823. It is attached to a green and white riband.

Cross of Fidelity to King and Country,—Instituted, on the 24th of July, 1823, by King John the Sixth, and conferred on all those persons who followed the King to *Villa*, from the 30th of May to the 5th of June,—and, on the Military men who followed The Infant D. Miguel to *Santarem*. The Cross bears the effigy of the King,—and, on the reverse, "*Fidelity to King and Country*, 1823," It is worn attached to a riband of green and white, and red in the centre.

Cross of Volunteers of Monte Video,—instituted, on the 25th of July, 1824, by King John the Sixth. On one side is the effigy of the King,—and, on the reverse, "*Monte Video*, 1822 *and* 1823." It is worn, attached to a sky blue riband with a red and blue border.

Cross of Emigration from 1826 to 1828,—instituted, on the 21st of September, 1828, by Don Miguel. On one side is the effigy of the Founder,—and, on the other, the date of Emigration, and the number of battles in which the Possessor has taken part, from the publication of the Charter of *Brazil* to the 7th of March, 1827. It is worn, attached to a white and red riband.

THE KINGDOM OF PRUSSIA.

The Order of Generosity.

The Order of Generosity was instituted in the month of May, 1667, by Frederick III., Elector of Brandenburg, and afterwards King of Prussia, whilst he was only Electoral Prince.

The Knights wore a gold cross, of eight points, enamelled, *azure*, having on the upper part the letter F., in gold,—and, in the centre, this Motto, " *La Générosité*," pendent to a rich black, watered riband.

Voltaire was a Knight of this Order.—*See*, The Orders of The Black Eagle, and of Military Merit.

The Order of The Black Eagle, *or*, of The Prussian Eagle.

Frederick the First, King of Prussia, having founded, when quite a youth and whilst he was only Electoral Prince, The Order of Generosity, intended thereby to create a chivalrous Distinction, both for *Civil* and *Military* persons. The Badge was worn, suspended from the neck, by a rich black, watered riband, two fingers wide.

Having attained The Royal Dignity, His Majesty instituted this celebrated and famous Order at *Königsberg*, on the 18th of January, 1701, the day previous to his Coronation,—with the intent, that the Knights, by appearing in their Installation Robes, might render that Ceremony more brilliant on the following day.

The Order of The Black Eagle, which consists but of one Class, is The Chief Order in the State.

By the Statutes of The Order—which, with a few alterations, were confirmed on the 18th of Jan., 1810,—the number of Knights, exclusive of The Members of The Royal Family, was limited to *Thirty*, who were all to be admitted into The Order of Generosity previous to their reception into this, unless they were Sovereign Princes,—but the number is now unlimited.

The reigning King of Prussia is always Grand Master,—and the Princes of The House of Brandenburg are born Knights of this Order, but they cannot be formally admitted, until after they have received confirmation.

All the Knights, without distinction, enjoy the Rank of Lieutenant-General of the King's forces,—and have a right to surround the Escutcheon of their FAMILY ARMS with THE COLLAR of THE ORDER.

It is only conferred on Sovereigns, Princes, and Persons of high Rank, or on meritorious Statesmen. On Sovereigns it is bestowed without regard to age,—but, on other persons, not before the age of *Thirty*. No one receives it, who does not already possess THE PRUSSIAN *Red Eagle*, of the first Class. If such is not the case, he receives the latter at the same time, but only wears it's Cross on a smaller riband, round the neck.

THE ENSIGN of THE ORDER is a gold Cross, of eight points, enamelled blue, having at each angle a Spread Eagle enamelled black, and charged in the centre with a Cypher of the letters F. R., the initials of the Founder's name. The Knights commonly wear this Cross suspended from a rich broad, watered, Orange coloured riband, worn sashwise over the left shoulder,—and a SILVER STAR, of eight points, embroidered on the left side of their outer garment, whereon is an Escutcheon, containing an Eagle displayed, *black*, on a red ground, holding in his dexter claw a chaplet of Laurel, and, in the other, a Thunderbolt, with this MOTTO, round it, "*Suum cuique.*"

It is stated, that The King chose *The Black Eagle*, being the ARMS of PRUSSIA,—and the *Colour* of the riband, as a particular compliment to his Mother, who was a PRINCESS of *Orange*.

On days of Ceremony the Knights wear THE BADGE, pendent to a rich COLLAR of gold, composed of round pieces of gold, each enamelled with four cyphers of the letters F. R.,—in the centre of the piece is set a large diamond, and over each cypher a regal crown, all richly chased, intermixed with Eagles displayed, enamelled black, alternately, and holding in their claws Thunder-bolts of gold.

GRAND CROSS.

ARTHUR, Duke of WELLINGTON, K. G.

The Honble. Sir CHARLES WILLIAM STEWART,—*See*, RUSSIA, ORDER OF ST. GEORGE, of the Fourth Class.

The Order of Military Merit.

This Order was instituted by Frederick *the Great*, King of Prussia, on his Accession to the Crown in 1740,—who then changed The Order of Generosity into that of Merit.

Persons who were decorated with the *antient* Cross, returned it, when they received the *new* one,—and those who had not received the latter, were allowed to wear The Order of Generosity. until their death. After this alteration, the extinction of The Order of Generosity took place by the demise of the last possessor, who had not in lieu of it received The Order of Merit.

There are no Statutes in existence.

Frederick, during the first years of his reign, conferred this Order indiscriminately on such as distinguished themselves either in Arms or in Arts, without distinction of Birth, Religion, or Country,—and, among others, he honoured his friend Voltaire, by presenting him with it.

No pensions are associated with it,—and there were some persons, who thought their Merit not sufficiently recompensed by this new Title,—but His Majesty was not very ready to grant pecuniary Rewards. To those who were most in his favour; he sometimes presented *snuff-boxes*, on which was inscribed, "*Amitié augmente le prix.*"

This Order which consists but of one Class, and which at first was termed "*Ordre pour le mérite,*" received from Frederick William the Third, the name of "*Ordre pour le mérite Militaire.*"—and, since that time, it has been exclusively conferred upon Military men.

Three golden *leaves of Oak*, on the ring of this Cross, are a peculiar distinction, by which the King honours eminent degrees of Merit.

The Badge of The Order is a gold Cross, of eight points, enamelled blue, without a medallion,—on the upper point is the letter F., surmounted by a royal crown,—and, on the three other points, the Motto of the Order, "*Pour le Mérite*,"—between each of the angles is a golden eagle, with expanded wings.

It is worn round the neck, pendent to a rich black, watered riband, edged with silver.

Knights.

Charles Wood, Esq., Captain in the 10th (*or*, The Prince of Wales's Own Royal) Regt. of Light Dragoons (Hussars), "in testimony of the approbation of The King of Prussia, of the distinguished Military services rendered by him to the common cause, during the last Campaign."—Royal Licence, 1st of March, 1815.

Lieut. Colonel Charles Gordon, of the 41st Regt. of Foot,—Had the Royal Licence,—but it does not appear, that it was ever recorded in the Books of The College of Arms.

Henry Frederick Cooke, Esq.,—*See*, Russia, Order of St. George.

Thomas Noel Harris, Esq.,—*See*, Russia, Order of St. Wladimir, of the Fourth class.

The Order of The Red Eagle.

The time of the Institution of this antient Order is uncertain.

In 1705, it bore the appellation of, "The Order of Sincerity, *or*, of The Red Eagle," being then under the protection of The Hereditary Prince George William, of Anspach and Bayreuth, who completely organised it in 1712, being the year of his Accession to the Crown. But various alterations have been made under successive Grand Masters.

In 1734, The Order was revised by The Margrave Frederick, who named it "*The Order of Bayreuth*,"—and, in 1759, he added Grand Crosses. In 1777, the Order was again re-organised by The Margrave Christian-Frederick-Charles-Alexander of Brandenburg, Anspach and Bayreuth.

King Frederick-William the Second, by Letters Patent, dated the 12th of June, 1792, declared himself Grand Master of The Order, and ordained it to rank next to that of The Black Eagle.

King Frederick-William the Third added to The Order, by a decree of the 18th of January, 1810, a *second* and *third* Class,—and, in the same month and year, a *fourth*, and a *Jewel* for the most antient Members of the second Class.

Those Knights who have passed through the third Class, wear *three golden leaves of Oak,* on the rays of the *Star* at the top,—And the Knights of the first Class, who have passed through the third and second, wear the ring with *three golden leaves of Oak* on the superior ray of the *Jewel.*

THE BADGE of THE ORDER is an eight pointed, white enamelled Maltese Cross, of gold,—on one side, is a shield of white enamel, on which is depicted a *Red Eagle,* bearing the *Hohenzollern Arms* upon it's breast, and resting upon a laurel crown,—on the reverse, is the Royal cypher F. W., on a white ground, surmounted by the Royal crown.

THIS BADGE is suspended from a broad, rich, watered, white riband, with an orange coloured edge, which is worn scarfwise over the left shoulder.

THE STAR is in all respects the same, as that of THE ORDER of THE BLACK EAGLE of PRUSSIA,—except that *The Red Eagle,* bearing the *Arms of Hohenzollern* upon it's breast, and holding a Laurel crown in it's claws, is embroidered in the centre of the Star. This Eagle appears in the middle of an escutcheon of silver, which is surrounded with a fillet of silver tissue, and thereon is embroidered the MOTTO of The Order, "*Sincere et constanter,*" in letters of gold. This STAR is worn upon the left breast of the upper apparel.

The Knights of THE BLACK EAGLE are Members of the First Class of THE RED EAGLE, and wear the Cross suspended from the neck by a narrower riband. The King, and the Princes of the Blood, wear the Decoration of the third Class at their Coat button-hole.

Although it is usually bestowed upon General Officers, it is, by the Statutes, not more appropriated to *Military* men, than to those of *Civil* employments.

GRAND CROSS.

The Right Honble. Sir GEORGE MURRAY,—*See,* AUSTRIA, ORDER of LEOPOLD.

Sir ROBERT THOMAS WILSON,—*See,* AUSTRIA, Commander of MARIA THERESA.

The Honble. Sir CHARLES WILLIAM STEWART.—*See,* RUSSIA, ORDER of ST. GEORGE, of the Fourth Class.

THIRD CLASS.

CHARLES BROWN, Esq., M.D., of *Margaretta Farm*, in the County of Norfolk, " in testimony of the regard and esteem and of the high sense entertained by FREDERICK WILLIAM the Third, King of PRUSSIA, of his important services, whilst actually employed in the service of His Majesty during a long series of years, as his first Physician, and as first Physician to His Majesty's Court and Army."—Royal Licence, 23d of May, 1817.—Died, on the 11th of May, 1827, *æt.* 80.

FOURTH CLASS.

BENJAMIN HEBELER, Esq., His Prussian Majesty's most excellent Consul General for GREAT BRITAIN and IRELAND.

THE ORDER OF SAINT JOHN.

THE PRUSSIAN ORDER of SAINT JOHN was instituted by FREDERICK WILLIAM the Third, on the 23d of May, 1812, in honour of the memory of THE ORDER of THE ANTIENT GREAT PRIORY of ST. JOHN at *Brandenburg*, which was suppressed on the 23d of January, 1811.

The King is Sovereign Protector of THE ORDER, and names THE GRAND MASTER who is dependent upon himself. His Majesty nominates also THE KNIGHTS, who form but one Class of an unlimited number. They receive THE ORDER, either as a token of Merit, or as a mark of Royal favour, but they are required to be of *Noble* descent.

THE BADGE is a gold Cross, of eight points, enamelled white,—in the four angles are black Prussian Eagles, crowned, and with extended wings. It is worn round the neck, pendent to a black riband. The Knights also wear a Cross of the same shape, without Eagles, embroidered with white silk, on the left breast of the coat.

The Grand Master is distinguished by wearing both Crosses of a larger size.

The Knights are authorized to wear THE CROSS of THE ORDER in their ARMORIAL ACHIEVEMENTS.

THE DECORATIONS of THE BLACK EAGLE and of THE RED EAGLE of the First Class, as well as THE PRUSSIAN ORDER of SAINT JOHN, are ornamented with *Diamonds*. They are a particular and extraordinary mark of Distinction, with which The King honours the highest degree of Merit.

KNIGHT.

JOHN CHARLES STAPLETON, Esq., late an Officer in The Regiment of Cuirassiers of The Royal Prussian Guard, " in testimony of the approbation of His Majesty The King of PRUSSIA of his distinguished conduct, while in His Prussian Majesty's actual service."—Royal Licence, 6th of July, 1822.

The Order of The Iron Cross.

Of all THE ORDERS which have ever been founded, there is none so expressive, none destined for so high and holy an object, none so simple in itself, so devoid of ostentation, and yet so important, and so well conceived, as THE ORDER of THE IRON CROSS. The idea first originated with THE FOUNDER himself,—a circumstance which none will doubt, who have already acknowledged the tender sensibility and deep feeling of this Sovereign in so many relations and important affairs.

When, at the close of the year 1812, an overruling PROVIDENCE had inflicted on the icy plains of *Russia* a fearful judgement on the daring Oppressor of all Human rights, and the hope dawned that at length the period was approaching, when the yoke of Foreign rule and oppresssion so burdensome to EUROPE might be cast off, there prevailed in the PRUSSIAN people but one mind and one power, which were as rare as they were estimable. In the true sense of the word, it might be said, there was but one voice among them, during the first months of the year 1813, "*War for Freedom, and Justice or Death.*" When their Sovereign determined upon war, animated alike with the heroick spirit of his people, and trusting in the help of THE ALMIGHTY,—and that help was given.

PRUSSIA had ORDERS and DISTINCTIONS, but these

Distinctions were gained under common place circumstances,—they had in fact been given to those, who had been compelled to fight by force for the Foreign Oppressor, and had personally distinguished themselves in this relation. How, then, were splendid actions in the great approaching Cause, and in so extraordinary a time to be recompensed by these Orders!—And, yet, it was of the greatest importance to give an outward sign of the brilliant innate enthusiasm, in order to acknowledge and reward heroick deeds, and to unite together those who distinguished themselves in these glorious and decisive days.

Under these glowing sentiments King FREDERICK WILLIAM the Third instituted THE ORDER of THE IRON CROSS, at *Breslau*, on the 10th of March, 1813,—and, on the same day, it was announced in the following manner,—

"In the present great emergency, on which the existence of our Father-land, as a Nation depends, the powerful feeling which raises the Native so high, deserves to be honoured and perpetuated by peculiar Monuments,—That the fortitude with which the people bore the intolerable evils of an iron time did not descend to pusillanimity, is proved by the exalted courage which now animates every breast, and which can only be continued by a reliance on Religion and by faithful attachment to King and Country.

"WE have, therefore, determined to distinguish Merit, which may be displayed in the War about to commence, whether in actual Contest with the enemy, or otherwise in the field, or at home,—but ever in relation to the great struggle for Freedom and Independence, and we have resolved not to bestow it after the close of this War."

THIS ORDER, the distribution of which has ceased with the War that occasioned it's origin, consists of,—*viz.*, Grand Crosses,—Knights, of the first,—and Knights, of the second Class.

The Cross is of *cast Iron*, bordered with silver,—the obverse is smooth,—but the reverse displays on the upper portion of the Cross, the initials of The Founder, F. W., surmounted by the Royal crown,—beneath, are extended into the side parts of the cross, three leaves of oak, and on it's lower portion is inscribed, the date of the year 1813.

The Grand Crosses wear a Cross of double the dimensions of that which is worn by the other two classes, pendent to a black riband, with a white border, round the neck. It was only given for obtaining a decisive battle, or for the taking of a place of importance. It was also conferred upon the Commander of a Fortress, should he have defended it with glory.

The two other classes, whose Cross is of equal size, wear it in the button-hole, and on a similar riband, if it was acquired in action against the enemy, and on a white riband with a black border, when this was not the case.

The Knights of the first Class wear also a similar Cross, of black riband with a white border, or of white riband with a black border, placed in the fashion of a Jewel on the left breast.

In the bestowal of this Order, neither rank nor condition were regarded,—but to be a Prussian was an indispensable requisition,—hence, *Foreigners* could not receive it.

By a Royal Ordinance of the 5th of May, 1813, there was suspended in the Church of every Garrison, a Tablet on which were inscribed the names of those, who had died in the field of Battle, whilst performing deeds worthy of being rewarded with The Iron Cross.

A Royal Ordinance of the 12th of March, 1814, declares, that, after the decease of a Member, the Cross should pass, as by inheritance, to some other individual who should have distinguished himself in the same Campaign.

All the old Colours of the *Infantry* and Standards of the *Cavalry*, which were borne in this War, are embroidered as a peculiar distinction, with *an Iron Cross* at the Point.

THE ORDER OF LOUISA.

THE ORDER of LOUISA was created by King FREDERICK WILLIAM the Third, in the month of August, 1814, for LADIES, who had given during the last sanguinary wars the greatest proofs of Patriotism and of Humanity.

The reason why the King named it "THE ORDER of LOUISA," is not mentioned in the announcement,—but who can doubt, that he wished by it's means to establish a lasting Memorial of a faithful, beloved, and alas! too early lost CONSORT in life,—that he instituted it in constant remembrance of one of the noblest of Women, whom PROVIDENCE in the time of the deepest affliction and most depressing sorrows, as a still severer trial took from his side. And how could he have chosen a purer name than that one, who, in every respect, was revered, beloved, and honoured as a *Woman*, and a *Queen?*—And, if it was the inscrutable decision of THE MOST HIGH, that she should not survive to see happier times, her remembrance will live for Centuries in THE INSTITUTION which arose out of this propitious Æra, and the model of all who wear THE ORDER devoted, consecrated to her, shall, ever be, Queen LOUISA!

THE DECORATION of THE ORDER is a small gold Cross, enamelled black, which is of the same size for all who possess it. On both sides, in the middle, is a round light blue shield, which bears on the obverse, an L., encircled by the significant emblem of a wreath of Stars,—and, on the reverse, the dates 1813 and 1814. It is worn on the white and black riband of THE IRON CROSS, which is so intimately connected with it, in a loop on the left breast, without any further ornament.

THIS ORDER is composed of One Hundred and Sixty-one Ladies, married or single,—and no regard is paid either to birth or rank for their admission. A Princess of The Royal Family is at the head of THE ORDER, as far as the usual details are concerned.

The King decides as to the forfeiture of THE ORDER, on receiving the opinion of The Chapter.

This Cross is most highly and deservedly esteemed by THE LADIES of PRUSSIA,—in fond recollection of the interesting, high-spirited, and beauteous Consort of the excellent and patriotick FREDERICK WILLIAM, on whose rectitude and honour, even the misconduct of his Ministers, during the disastrous War, had not brought the slightest unkind observation.

The Order of Concord.

The Order of Concord was instituted by Christian Ernest, Margrave of Brandenburgh, on his return from Spain in the year 1660, to distinguish the part which he had taken in restoring Peace and Union to many of the Princes of Europe.

The Badge of The Order is a gold Cross, of eight points, enamelled white,—in the centre, is a medal bearing two Olive branches passing saltierwise through two crowns, and circumscribed with the word, "*Concordans.*" On the Reverse, are the cypher of The Founder, and the date of the Institution of The Order.

The Cross is surmounted with an Electoral Crown, and it is worn, pendent to an orange-coloured riband.—Clark, *vol.* 2, *p.* 133.

The Order of Favour.

There subsisted for some time in Prussia an Order, called, "The Order of Favour,"—which, according to it's denomination, was conferred with very little distinction.—Johnson's Life of Frederick *the Great.*

Marks of Honour and of Distinction.

By a Royal Ordinance of the 14th of June, 1793, King Frederick William the Second, granted to the Military, who, during the Campaign of 1793, had deserved well of their Country, a Medal, in *gold*, for Subalterns,—and another, in *silver*, for the Privates,—to be suspended by a black riband from the button-hole of the coat.

King Frederick William the Third, by a Royal Ordinance of the 30th of September, 1806, founded a gold Medal, as a Mark of Military Honour of the *First* Class,—and, a Silver Medal, as a Mark of Honour of the *Second* Class, to recompense such of his Soldiers as should distinguish themselves by particular instances of valour.

The *Gold* Medal could not be acquired, without having previously obtained the *Silver* one.

The Silver Medal is worn with a black riband, bordered with white, at the button-hole of the coat. It bears on the Reverse this inscription, "*Services rendered to the State.*"

By an order of The King, of the 30th of September, 1814, the distribution of the Gold Medal has ceased,—and, in lieu of it, a Silver Cross is to be bestowed, as a mark of Military honour of the First Class. It bears on the Reverse, the initials of the King, with a Crown.

On the 18th of January, 1810, King Frederick William the Third instituted a mark of Civil Honour, suspended from the button-hole of the coat, to recompense Merit and Services rendered to the State. On the Reverse is this inscription, "*Services towards the State.*"

King Frederick William the Third also founded a Brass Medal to be distributed among all Military Men, who took part in the Campaigns of 1813 and 1814, without having given any cause of complaint. It bears on the Reverse this inscription, "*For duty and fidelity during the War,"*—Above, are the initials of the King, with the crown. Round it, are these words, " *God was with us, Glory be to him.*"

By an Ordinance of the 18th of January, 1832, Frederick William the Third, King of Prussia and Prince of *Valengin*, founded, in memory of The Principality of *Neufchâtel* being rescued from the hands of Rebels, who intended to destroy the Government, a Badge of Honour for all those, who had taken an active part in the Military operations against the Insurgents, or who had spontaneously taken up arms for the maintenance of tranquillity in their own Districts.

This Decoration consists of a Silver Medal, which is worn at the button-hole of the coat, pendent to a riband with the *Prussian* colours and those of *Neufchâtel*,—on one side, are the initials of the King's name, with the inscription, " *Fidelity to duty, and to our Country,"*—and, on the other side, the Arms of The Principality of Neufchâtel.

By an Ordinance of the 18th of June, 1825, the day of the Battle of *Belle Alliance*, King Frederick William founded marks of Services and of Distinction, in order to commemorate the glorious achievements in the years 1813, 1814, and 1815.

The Cross of Service and of Distinction is granted to Officers for *twenty-five* years service, and is suspended from the button-hole of the coat. One Campaign is reckoned two years, and the medal of the Cross bears on the Reverse, the Roman numerals XXV.

The Mark of Service and of Distinction, or *The First Class*, is granted to Military men for *twenty-one years* of service, or for three Campaigns, and is attached to the left breast of the coat.

The Mark of Service and of Distinction, or *The Second Class*, is conferred on Military men for *fifteen years* of service, or for two Campaigns or one, and is attached to the left breast of the coat.

The Mark of Service and of Distinction, or *The Third Class*, is granted to the Military for *nine years* of service, and is also attached to the left breast of the coat.

THE POPEDOM OF ROME.

The Order of The Golden Spur.

It is a disputed point among the old Writers, which of the Popes first conceived and executed the idea of founding The Order of The Golden Spur. Many wish to recognise Pope Pius the Fourth, as it's Founder, and adopt the year 1559, as the date of it's establishment. His Predecessor, Pope Paul the Third, may, nevertheless, be regarded with probability, as it's author. There certainly is no historical proof of it's having existed prior to the reign of this Pope. In the Letters of Institution, the Knights were named, "*Knights of The Golden Militia,*" and not, "*Knights of the Golden Spur,*" which title is only employed in common conversation. At an earlier period they bore the name of "*Counts Palatine of The Lateran,*" which was expressly mentioned in the original document. They enjoyed numerous rights and privileges, and held rank before *The*

Knights of The Maltese and *German Orders.* These, however, have all been lost in the course of time, as The Order has suffered many alterations from the constant changes of Government,—and the Ordinances are now, merely an empty form, which declare, that the Knights shall enjoy all the advantages and privileges, which are due to them from right and custom,—for there are now no *benefits* connected with them.

It was bestowed, as an Order of Merit, on Official persons, learned men, artists, and other persons who had rendered service to The State, or whom the Holy Father selected for such distinction, and even upon *Foreigners.* Neither rank nor station, but simply the profession of *The Roman Catholick Religion,* is required as a qualification to receive it.

The Pope usually confers it, on the Anniversary of his Election to the Papal Chair, upon those who attend him in the quality of Pages, as he goes in Procession from The Vatican to the Church of The Lateran.

Pope Benedict the Fourteenth gave The Badge that form which it has retained without alteration, and still possesses. This is a golden Maltese Cross, enamelled white, with a small gold Spur, hanging to both points of the lower wing. All the Knights, forming as they do only one Class, wear it on a narrow red riband in the left button-hole.

A peculiarity of this Order, which occurs in no other, consists in the circumstance that it is not bestowed by The Superior alone. The Papal *Nuntios,* the *Prelates* who are Members of the highest Papal Court of Justice, and some other *Romish Prelates,* possess the right of *each* naming *two* Knights of The Golden Spur. The same right is possessed by The Ducal House of Sforza Besarini, on whom as early as 1539, almost from the commencement of The Order, Pope Paul conferred the same power by a special Privilege, which has constantly been confirmed by succeeding Popes,—and which, in it's Diploma of The Order, styles the Knights according to the form then in use, "*Counts Palatine of The Lateran.*" The Badge which *they* bestow, differs in some particulars from that which is conferred by The Popes.

The irregularities which have existed, and do stil exist, from this diversified bestowal of THE ORDER,—the circumstance of it's not having always been bestowed with due care, and the infrequency of the distribution,—together with the fact which is asserted, *that it is to be obtained for money,* have long been injurious to it's Dignity, and will still operate to it's Prejudice.

The Order of Christ.

This *Second* Papal Order of Merit, which stands in higher repute than that of "The Golden Spur," is properly a Portuguese Order,—hence it's history is described more circumstantially under the Kingdom of Portugal, and the following details may be sufficient to be mentioned here.

Pope Clement the Fifth having abolished, in 1312, *The Order of The Knights Templars,* King Dionysius of Portugal resisted this decision, and suffered The Order throughout his Dominions to remain in almost undisturbed possession of their property. The misunderstandings which arose in consequence between the two Courts, led, under Pope John the Twenty-second, the Successor of Clement, to this conclusion,—that the Order of The Templars, under the altered title of "*Knights of Christ,*" continued in Portugal,—the Pope also, in the year 1319, confirmed this, in some measure, New Order, but reserved to himself and to his Successors the right of likewise appointing Knights of "*The Order of Christ.*" From that time to the present, the Popes have exercised this right, and have conferred The Order of Christ, as an Order of Merit, both upon *Natives* and *Foreigners* of The Catholick Religion, of every rank and condition,—nor is it required, that the person who receives it, should be of *noble* descent,—though such obligation is imposed upon the Knights of *The Portuguese Order of Christ.*

In the Seventeenth Century a gross abuse appears to have taken place in the distribution of The Order,—

which SALVATOR ROSA ridicules in some very severe expressions in his Satires on Painting.

THE BADGE of THE ORDER is a gold Cross, enamelled red, and edged with gold. Formerly it was worn, pendent to a gold chain,—but it is now worn, attached to a scarlet riband round the neck. It is occasionally, adorned with diamonds and pearls, bestowed as a peculiar mark of distinction.

Medals, and Marks of Honour.

The Medal of Military Merit was founded by Pope Gregory the Sixteenth, in the summer of the year 1832, by a Decree which has not been made publick.

The first Class is of *gold,*—the second, of *silver.* Both of them bear the effigy of the Sovereign Pontiff, and are ornamented with the Papal crown, and the Keys,—always of the same *metal,* as the *Medal,*—or else are surrounded with Laurel branches, in green enamel.

The Medal surrounded by Laurel branches is conferred on Subaltern Officers in *gold,* and, on Under-Officers, in *silver.*

A similar Medal, which dates from Pope Pius the Seventh, and bears his effigy and those of his respective Successors, is still distributed in great quantities, at present with the effigy of Gregory the Sixteenth. It is generally of *bronze* for Private Soldiers, and of *bronze* or *silver* for Under-Officers.

In the year 1816, Pope, Pius the Seventh, instituted a Cross of Honour, as a reward for those Military men who had distinguished themselves by contributing to free the Ecclesiastical States from *The Banditti!* who infested them. The decoration is of silver, gilt, and enamelled white, and bears the inscription, "*Latronibus fugatis, Securitas restituta.*" In what a condition must that State be, where *distinctions* are conferred *for such services as these!*

Pope Pius the Seventh also instituted a Decoration, to be conferred on all Presidents, actual and future, of The Academy of Painting of *St. Lucas,* with the authority to wear it even after their Presidency shall have expired. It is named, "*Ordine del Moretto.*" Artists, such as Thorwaldsen, Camuccini, *&c.,* who have been Presidents of this

Academy have, therefore, the right to wear this Mark of Honour,—but which ought no more to be regarded as *an Order of Knighthood,* than that which is borne by The Presidents of the fourteen Divisions of the City, on the centre of the Arms of these different Quarters.

Neither can the Decoration which is worn by *The Chapters of Loreto* by styled *an Order of Knighthood,* although it is the remains of an ancient Order of Knighthood, called "*Ordine Lauretano,*" or "*Order of the Defenders of the Holy Church of Loreto.*"

THE EMPIRE OF RUSSIA.

General Remarks.

The Emperor of Russia is Grand Master of All the Russian Orders. His Majesty appoints all the Members, either from his own inclination or at a recommendation of a Chapter, which, for all Orders, consists of a Chancellor, a Chief Master of the Ceremonies, and a Treasurer.

No determinate number of Members is fixed for any of the Orders.

Neither proof of *Nobility*, nor a certain age, nor any particular Profession of Religion are required, for any of The Orders. They all confer upon a Citizen, the Russian degree of *Nobility*,—hence, in the list of The Orders in The Court Calendar of *St. Petersburg*, a "*de*" is prefixed to all the Names. Even if a Member makes no use of it, his children and grand-children may avail themselves of it, for The Empress Catharine ordered, that every one should obtain the Rank of *Nobility*, who could prove that his father or grand-father had been *a Knight of The Russian Order*.

All the Orders are borne together. They are all worn over the

Coat. He who has several, commonly wears only THE ORDER of *St. Andrew*, and under this, THE ORDER of *St. George*, or of *Wladimir*.

No one is allowed to wear any one of THE ORDERS set with *Brilliants*. With this embellishment they are bestowed, as a peculiar mark of honour or favour from THE EMPEROR, and only on such persons as previously possess THE ORDER, *without* brilliants. THE ORDERS of *St. George* and of *Wladimir* are, however, never adorned with *Diamonds*.

Revenues are associated with some of THE ORDERS.—but *Foreigners* do not participate in them.

Of THE FESTIVALS of THE ORDERS, those of *St. Andrew*, *St. Anne*, *St George*, and *St. Wladimir*, are celebrated at COURT.

Of the interest of the Capital of 200,000 roubles, which is preserved by THE CHAPTER of THE ORDER,—10,000 roubles a year are intended for the support and instruction of the daughters of Knights, who are destitute of property,—the daughters of Knights of *St. George* and of *St. Wladimir* of the 3d and 4th class, and of *St. Anne* of the 2d and 3d class, but they must not possess more than 100 *peasants*. Those who are entirely *Orphans*, are specially provided for,—their Education is conducted in the Publick Female Institutions, of which the Empress mother has the direction.

To the Capital from which invalid Generals, Staff, and Superior Officers receive provision, contributions have been made since 1816, among other means, on bestowals of THE ORDER. He who receives THE BADGE, in *brilliants*, pays for the ORDER of *St Andrew* 2000, for that of *St. Alexander Newski* 1500, for that of *St. Anne* of the first class 500, and of the second 100 roubles.

Besides ORDERS of Knighthood, Crosses of Merit, and MEDALS, there are in RUSSIA, MARKS of HONOUR in various shapes. During the last forty years, great exploits of Generals and Officers have been rewarded with *Swords* or *Sabres*, enriched with Gold or Diamonds. These ARMS of HONOUR usually bear the inscription, "*For Courage*." Sometimes the acts for which these honourable distinctions are conferred, are specified upon them.

THE MAIDS of HONOUR of the Empress wear the PORTRAIT of HER MAJESTY, ornamented with *Diamonds*,—And THE LADIES of THE

Palace, bear Her initials on a Medal, set also with *Diamonds*, and suspended by a watered, silk riband from the left breast.

The Five *Original* Orders of The Russian Empire are,—

Saint Andrew,
Saint Catharine,
Saint Alexander Newski,
Saint George,—and
Saint Wolodimir,

and they were instituted by the Two most remarkable Princes, who ever filled The Imperial Throne.

Peter *the Great* was the Founder of those of *Saint Andrew, Saint Catharine*, and *Saint Alexander Newski*,—and those of *Saint George*, and *Saint Wolodimir*, were instituted by The Empress Catharine the Second.

The Imperial Order of Saint Andrew.

The Imperial Russian Order of Saint Andrew, was instituted on the 30th of November, O. S., 1698, by The Czar, Peter *the Great*, who wished to follow the example of other Sovereigns, by founding an Order of Knighthood, for recompensing persons who should signalize themselves by services to the State.

He is said to have chosen St. Andrew for it's Patron, as this Apostle, according to tradition, was the Founder of Christianity among the *Muscovites*.

The first opportunity that offered, was the Revolt of *The* Strelitz,—as soon as it was quelled, all those who had particularly distinguished themselves in suppressing the commotion, were decorated with the New Order.

It is the highest Order of Russia, comprising but one Class,—and is conferred only on Sovereign Princes, Russian Nobility of the first Rank, or *Foreigners* of the highest Distinction. Those on whom it is conferred, wear also the Decorations of The Orders of Saint Alexander Newski and Saint Anne.

The Knights of The Order of Saint Andrew have the Rank of Lieutenant-Generals.

The Badge of The Order is the image of *St. Andrew* on the Cross, richly chased and enamelled, upon an Imperial Eagle of gold, the heads of the Eagle ducally crowned, and over both the heads an Imperial Crown. The letters S. A. P. R. on the cross, are the initials of the words, " *Sanctus Andreas Patronus Russiæ.*" This, on ordinary occasions, is worn pendent to a broad light blue riband, sashwise over the right shoulder.

The Jewel is a Star, of eight points, embroidered in silver, on the left breast of the outer garment, with a circle of a light blue colour, on which is the Motto " *Pour la Foi et la Fidélité,*" in the Russian language,—in the centre, the Imperial eagle, upon a golden ground, surmounted by a crown.

On days of Ceremony The Badge is worn, pendent to a Collar of gold, composed of square chains, and Imperial Eagles, ducally crowned, and over both heads an Imperial crown, in the right claw a Sceptre, and in the left a Mound, upon the breast a red shield charged with *St. George* slaying a Dragon, — ovals charged with a cross *saltier*, blue,—and shields with the letters G. G. in a cypher, on each side of every one of them two banners, and on the top an Imperial crown.

The Knights are dressed in a splendid costume, prescribed by The Empress Anne,—and the Members who reside in *St. Petersburg*, are bound to be present at the annual Festival of The Order, which is celebrated on the 30th of November, O. S., under the penalty of 50 Roubles.

The Decoration of The Order of St. Andrew, enriched with *Diamonds*, is granted as an extraordinary mark of Honour, conferred by The Emperor on the highest degree of Merit.

Knights.

The Right Honble. William Shaw Cathcart, Viscount Cathcart, K.T., Ambassador Extraordinary and Plenipotentiary to His Majesty The Emperor of All the Russias, a General in the Army, *&c.*, to accept and wear the Insignia of The Imperial Russian Order of *St. Andrew*, and The Cross of The Imperial Military Order of *St. George*, of the Fourth class, " which His Imperial Majesty has been pleased to confer upon His Lordship, as a testimony of the high sense which that Sovereign entertains of his services, and of the devotion and zeal displayed by His Lordship in the Field on several occasions, during the late Campaign on *The Continent*."—Royal Licence, 6th of May, 1814.

The Most Noble WILLIAM SPENCER, Duke of DEVONSHIRE, K.G.—" Service, as Ambassador Extraordinary and Plenipotentiary to *St. Petersburgh*, on occasion of the Coronation of The Emperor NICHOLAS."—Royal Licence, 19th of March, 1828.

THE ORDER OF SAINT CATHARINE.

THE ORDER of SAINT CATHARINE was instituted by The Czar PETER *the Great*, on the 25th of November, O. S., 1714, in honour of the Empress CATHARINE the First,—to whose ability and good conduct His Majesty in a great measure owed his delivery from the perilous situation in which he was involved on the banks of the river *Pruth*, in 1711.

THE CZAR declared THE EMPRESS, Sovereign of the Order,—and, although at it's first institution *both Sexes* were admitted of it, it is now exclusively confined to *Ladies*, and those on whom it is conferred, are persons of the highest Distinction in Europe.

It is divided into two Classes,—*viz.*, GRAND CROSSES,—and CROSSES, of the second Class.

THE BADGE is a Medallion, of gold, enriched with Diamonds, and charged with the image of *St. Catharine*, pendent to a broad, watered riband of a deep red colour, bordered with silver, worn sashwise over the right shoulder. On this riband are embroidered, in the Russian language, the words, "*For Love and our Native Land.*" On the Reverse, is represented a nest of Eaglets upon the top of an old Tower, at the foot of which two old Eagles are returning to their young, holding serpents in their beaks,—above the figures, are the Latin words, "*Aequat munia comparis.*"

On the left side of the Stomacher, is embroidered a SILVER STAR, of eight points, on the centre of which is a Cross, with this MOTTO, "*Pour l'Amour et la Patrie.*"

The Cross of the second Class is attached to a Rosette, of the same coloured riband, placed on the left breast of their upper vestment. It was founded by the Emperor PAUL, in the year 1797.

The Empress is The Grand Mistress of The Order,—the Festival of which is solemnized on the 25th of November, O. S., on the 7th of December, of the present style.

Since the year 1796, there has been a SEMINARY of The ORDER of ST. CATHARINE, so named, because it's Revenues are partly derived from The Commanderies of this Order, and because, on the first admission of Pupils, the Ladies of The Order received the privilege of filling up the nominations in The Seminary with poor children of their own Selection.

Previous to her Marriage with His Serene Highness the reigning Duke of WURTEMBERG, Her Royal Highness CHARLOTTE AUGUSTA, Princess Royal of GREAT BRITAIN, was, by the then reigning Empress of Russia, created a Lady of THIS ORDER, and, at the Ceremony of Her Nuptials, Her Royal Highness wore the *Insignia.*

The Order of Saint Alexander Newski.

The Imperial Order of Saint Alexander Newski was instituted by The Czar Peter *the Great*, in 1722, in honour of Alexander Jaraslowitsch, Prince of *Novogorod*, one of the Heroes and Saints of the Russian Empire,—who, in 1240, gained considerable advantages in battles against the Swedes, and on the borders of the *Neva*, from whence the Surname of *Newski* is taken. He died a Monk, in 1263, at *Wolodimir*, under the name of Alexis.

This Order consists but of one Class,—and is intended to recompense, both *Civil* and *Military* Services. It was conferred for the first time after the death of The Czar, by the Empress Catharine the First, on Prince Menzikof, on the 8th of April, 1725. Officers who are decorated with this Order, must at least have the Rank of Major General. The Order of Saint Andrew is only conferred upon those, who have received that of Saint Alexander Newski.

The Badge of The Order is a large gold Cross, of eight points, enamelled dark red,—between the angles, are Imperial eagles, of gold, and Imperial crowns,—and, in the centre, enamelled white, is *St. Alexander* on horseback, to whom a hand from the clouds extends a wreath of laurel,—the reverse displays only in the centre, on a white ground, the letter A. (*Alexander*), several times intertwined.

It is worn, pendent to a broad, watered riband of a deep red colour, sashwise over the left shoulder,—with a silver Star, of eight points, on the left breast. In the centre of this Star, on a silver surface, are the two intertwined letters S. A. (*Sanctus Alexander*), covered by a red Ducal Hat, round which, on a poppy coloured circle, in the Russian language, is the MOTTO, "*For Labour and our Country.*"

THE DECORATION of THE ORDER of ST. ALEXANDER NEWSKI, enriched with *Diamonds*, is granted as a very high mark of Honour, bestowed by the Emperor on Merit of an eminent degree.

KNIGHTS of THE ORDER of SAINT ANDREW wear the Badge of THIS ORDER suspended from the neck.

The Festival is celebrated on the 30th of August, O. S.

This Order is distinguished by the appellation of "*The Red Riband.*"

KNIGHTS.

Sir SAMUEL GREIG, Vice-Admiral in The Imperial Service of Russia, conferred upon him, the 7th of July, 1776.—Registered in The College of Arms, in 1778. — *See*, THE ORDER of SAINT GEORGE.

The Right Honourable Sir GEORGE MURRAY,—*See*, AUSTRIA, THE ORDER of LEOPOLD.

Sir GORE OUSELEY, Bart.

In 1797, the Right Honourable ADAM Lord DUNCAN, Viscount DUNCAN of *Camperdown*, gained a most signal Victory over the Dutch Fleet, in which two Admirals were made Prisoners,—and some Russian ships of war having been present in that Engagement, His late Imperial Majesty, PAUL the FIRST, was pleased to confer the Ensigns of this Order upon His Lordship,—and to accompany the same with a Letter, highly expressive of the esteem and friendship which he entertained for that distinguished Naval Commander.

On the 12th of September, 1797, Lord DUNCAN received the Royal Licence to accept and wear the *Insignia*.

The Imperial Military Order of Saint George.

The Imperial Military Order of Saint George was instituted by The Empress Catharine the Second, on the 26th of November, O.S., 1769, to reward the Officers in her Service, both *Naval* and *Military*.

It is divided into Five Classes.—Knights of the first Class receive 700 roubles *per annum*,—those of the second Class, 400,—and those of the third, 200. A hundred of the oldest Members of the fourth Class receive 100 roubles. The Widow of a Knight receives for one year, the pension of her husband.

Knights of the first and second Classes enjoy the Rank of Major Generals.

In order to be admitted into the first Class, it is necessary to have gained, as General in Chief, a great battle, and count 25 years of effective service, or 18 campaigns at sea.

To be received into The Order, it is necessary to have taken a ship from the enemy,—or a battery, or some strong position,—to have displayed extraordinary courage in defence, or by skilful tactics to have obtained or contributed to a Victory—to have offered as a Volunteer on a perilous enterprise, and executed it,—to have mounted the first in the breach in the storming of a fortress,—or, finally, when landing on the Enemy's shore, to have alighted the first upon the beach.

The Badge of The Order is a gold Cross, enamelled white,—in the centre of the obverse, is the Armorial bearing of The Grand Principality of Moscow, *St. George* in full armour, on horseback, slaying a Dragon,—the cypher on the reverse, is that of *St. George*.

This beautifully formed Cross is worn by the first two Classes of equal size.

The first Class wear it on a riband consisting of three black and two alternate yellow stripes, from right to left,—and on the left breast, the STAR embroidered in gold,—in the centre, upon a red ground, are the initials of *St. George*, surrounded by a blue circle, on which is inscribed the MOTTO, "*For Service and Bravery*," in the Russian language.

The second Class wear it, on a narrow similar riband, round the neck, and with the same Star on the left side.

The third Class wear a smaller Cross, suspended in the same manner as the second, but without a Star.

The fourth Class wear the same Cross, at the button-hole on the left side.

And the fifth Class, which is granted to Subalterns and Soldiers, is a Silver Cross, and is also worn at the button-hole on the left side.

THE FESTIVAL of the Order is celebrated on the 26th of November, O. S., in every year. No particular Costume is prescribed. Generals are obliged to attend in their Uniform without embroidery.

THE EMPEROR is Sovereign GRAND MASTER, and this Dignity is for ever annexed to The Imperial Crown.

Forty Thousand roubles are the fixed annual Revenue.

The *Insignia*, when a Knight dies, must always be restored to THE ORDER.

KNIGHTS OF THE FIRST CLASS.

ARTHUR, Duke of WELLINGTON, K. G.

Sir ROBERT THOMAS WILSON, a Major General of His Majesty's Forces, Knight Commander of The Imperial Military Order of Maria Theresa, and of the Royal Portuguese Military Order of The Tower and Sword, also Knight of The Imperial Russian Military Order of St. George, of the Third Class, and of The Red Eagle of Prussia, *&c.*, to accept and wear the Insignia of The Imperial Russian Order of St. Anne, of the First Class, "as a signal testimony of His Imperial Majesty's high sense of the great zeal, distinguished intrepidity, and unremitting exertions displayed by him, in the cause of Europe, during the late arduous campaigns."—Royal Licence, 21st of July, 1814.

KNIGHTS OF THE SECOND CLASS.

Sir SAMUEL GREIG, Vice-Admiral in The Imperial Service of Russia, the three following Orders of Knighthood conferred upon him by Her Imperial Majesty, CATHARINE the Second,—*viz., the first,* THE MILITARY ORDER of ST. GEORGE, of the *Second Class,* conferred upon him, on the 22d of September, 1770, "for his particular courage and brave behaviour shown at the glorious Victory obtained over the Enemy's Fleet on the Coasts of *Asia,* as likewise for the means given by him towards the destruction of the same."—*Secondly,* THE SLESWIG-HOLSTEIN ORDER of ST. ANNE, conferred upon him, on the 25th of March, 1774, "for the distinguished zeal with which he manifested his service, and thereby made himself worthy of Her Imperial Majesty's particular favour."—*Thirdly,* THE RUSSIAN ORDER of ST. ALEXANDER NEWSKI, conferred upon him, on the 7th of July, 1776, by Her Imperial Majesty on board of The Fleet at anchor in *The Road of Cronstadt,* at the time when she most graciously honoured the same with Her presence, "for his pains and zealous services."—Given by The Imperial College of Admiralty in *St. Petersburg,* the 15th of January, 1778,—And registered in The College of Arms, the same year.

HENRY, Marquess of ANGLESEY, K. G.

General Lord HILL, G. C. B.

GEORGE ACKLAM, Esq., a Post Captain in The Royal Navy, and Commander of His Majesty's Ship *Ranger,* "in testimony of the high sense which The Emperor of Russia entertains of his long and distinguished services before *Dantzic.*"—Royal Licence, 3d March, 1814.

NEIL CAMPBELL, Esq., Colonel in the Army, and Major of the 54th (*or,* West Norfolk) Regt. of Foot, C.B., to accept and wear the Insignia of The Imperial Russian Order of St. Anne, of the Second class,—and the Cross of The Imperial Military Order of St. George of Russia, of the Fourth class, "in testimony of the high sense which The Emperor of All the Russias entertains of his military services, in the several Actions with the Enemy during the recent campaign."—Royal Licence, 2d of June, 1814.

—Knighted, 15th of October, 1814. Died, Governor of Sierra Leone, in 1827.

The Honble. FREDERICK CATHCART, a Captain in the 2d (*or*, Royal North British) Regt. of Dragoons, " in testimony of the high sense which The Emperor of All the Russias entertains of his military services in the several Actions with the enemy, during the recent Campaign."—Royal Licence, 2d of June, 1814.

ALEXANDER MACDONALD, Esq., Lieut. Colonel in The Army, and Captain in The Royal Horse Artillery, " in testimony of His Imperial Majesty's approbation of the distinguished services rendered by him, during the siege of *Dantzic*."—Royal Licence, 25th of February, 1815.

Sir GORE OUSELEY, Bart.

Sir HUDSON LOWE, Major General, K.C.B.

Sir EDWARD CODRINGTON,—*See*, FRANCE, ORDER of ST. LOUIS.

CHARLES BECKWITH, Esq., Lieut. Colonel in The Army, and on the Half-pay of The Rifle Brigade, C.B., " in testimony of the high sense which The Emperor ALEXANDER entertains of the highly distinguished conduct and intrepidity displayed by him, in the memorable Battles in *The Netherlands*, in 1815." — Royal Licence, 3d of March, 1822.

KNIGHT OF THE THIRD CLASS.

Sir ROBERT THOMAS WILSON, Knight of The Imperial Military Order of Maria Theresa, Knight Commander of The Royal Portuguese Military Order of The Tower and Sword, and a Major General in The Army, " which His Imperial Majesty The Emperor of Russia was pleased to present to him in his Camp in front of *Jauer*, on the Morning of the 27th of May, 1813, in the presence of his Army, as a memorial of His Imperial Majesty's high esteem and regard, and as a distinguished testimony of that Sovereign's approbation of the great gallantry and zeal displayed by him, in a series of eminent services throughout the whole of the war between Russia and France."—Royal Licence, 31st of December, 1813.

KNIGHTS OF THE FOURTH CLASS.

Sir COLIN CAMPBELL,—*See*, AUSTRIA, ORDER of MARIA THERESA.

Lord FITZROY SOMERSET,—*See*, AUSTRIA, ORDER of MARIA THERESA.

FREDERICK THESIGER, Esq., Captain in The Royal Navy, and late a Captain of the Second Rank in the Fleet of Russia, having, during his services in The Navy of Her late Imperial Majesty CATHARINE Empress of All the Russias, had the good fortune to receive Her said Majesty's gracious approbation of his conduct and to be honoured, in consequence thereof, with the Cross of The Military Order of *St. George*, of the Fourth Class,— And, that he had since in Her Majesty's service had the honour to be actively employed, during the late War, and particularly as Aid-de-Camp to Vice-Admiral Lord Viscount NELSON at the attack of the line of defence before *Copenhagen*."—Royal Licence, 27th of October, 1803.—Died, 2d of September, 1805.

The Right Honble. WILLIAM SHAW CATHCART, Viscount CATHCART, —*See*, RUSSIA, ORDER of ST. ANDREW.

The Honble. Sir CHARLES WILLIAM STEWART, K. B., (now, Marquess of LONDONDERRY, G. C. B., and G. C. H.,) His Majesty's Envoy Extraordinary and Minister Plenipotentiary to The Court of Prussia, Lieut. General in The Army, and Knight Commander of the Royal Portuguese Military Order of The Tower and Sword, to accept and wear the Insignia of The Imperial Russian Military Order of St. George, of the Fourth Class, and also the Insignia of a Knight Commander and Grand Cross of The Royal Swedish Military Order of The Sword, and the Decorations of The Grand Prussian Orders of The Black and Red Eagle, which their Majesties The Emperor of Russia, The King of Sweden, and The King of Prussia, have been respectively pleased to confer upon him, " in testimony of the high sense which those Sovereigns entertain of his signal services in the cause of Europe, and of his distinguished gallantry and conduct in the Battle of *Culm*, on the 30th day of August, 1813, (wherein he was wounded), and in the Battles fought before Leipsig on the 18th and 19th days of October, 1813."—Royal Licence, 21st of March, 1814.

On the 8th of May, 1837, The King was pleased to give and grant to General The Marquess of LONDONDERRY, G. C.B., His Royal Licence and permission, to accept and wear THE MEDAL "instituted by THE EMPEROR of RUSSIA, in commemoration of *The Capture of Paris* in the year 1814, which His Imperial Majesty had been pleased to confer upon him.

JOHN MARSHALL, Esq., Post Captain in The Royal Navy, and commanding His Majesty's Ship *Shamrock*, to accept and wear the Insignia of The Imperial Russian Order of St. George, of the Fourth Class,—and of a Knight of The Royal Swedish Military Order of the Sword, "in testimony of the high sense which The Emperor of Russia and The King of Sweden entertain of his bravery and services during the Siege of *Gluckstadt*, and at the Blockade of *Hamburgh* and *Haarburgh*."—Royal Licence, 10th of January, 1815.

THOMAS SYDENHAM, Esq., to accept and wear the Insignia of The Imperial Russian Military Order of St. George, of the Fourth Class,—and of a Knight of The Royal Swedish Military Order of the Sword, which their Imperial and Royal Majesties The Emperor of Russia and The King of Sweden have been respectively pleased to confer upon him, "in testimony of the high sense entertained by those Sovereigns of the distinguished services rendered by him to the common cause, in the presence of the Enemy, while serving in the Field in the Combined Army under the command of His Royal Highness The Prince Royal of Sweden, and the bravery displayed by him on the field of Battle, during the last Campaign."—Royal Licence, 21st of September, 1814.—Died at Lisbon, 28th of August, 1816.

HENRY FREDERICK COOKE, Esq., Lieut. Colonel in The Army (Colonel on The Continent of Europe), K.C.G., and C. B., to accept and wear The Insignia of The Imperial Russian Order of St. GEORGE, of the fourth Class,—Of The Royal Prussian Order of MILITARY MERIT,—and of a Knight of The Royal Swedish Military Order of THE SWORD,—with which their Majesties The Emperor of RUSSIA, The King of PRUSSIA, and The King of SWEDEN had respectively been pleased to honour him, "in testimony of the high sense which those Sovereigns entertain of his distinguished services before the Enemy, while serving upon *The Continent*, during the late Campaigns."—Royal Licence, 15th of March, 1822.

CROSS.

SAMUEL BENTHAM, Esq., Colonel of the Regiment of *Rajsk* in the Service of Her Imperial Majesty of All the Russias, having been entrusted with the command of the first Division of the Flotilla employed on the *Liman* under His Serene Highness NASSAU SIEGEN, had the good fortune to meet with Her Imperial Majesty's most gracious approbation of his conduct, in consequence of which he was honoured with the Cross of THE MILITARY ORDER of ST. GEORGE.—Royal Licence, 25th of July, 1789.

HENRY FANSHAWE, Esq., late Lieut. Colonel of His Majesty's 83d Regt. of Foot, now Colonel of The Regiment of *Elotzkoy* in the Service of Her Imperial Majesty of All the Russias, having been entrusted with the command of a Squadron of Gallies and Gunboats making part of the *Liman* Flotilla under His Serene Highness Prince NASSAU SIEGEN, had the good fortune to meet with Her Majesty's most gracious approbation of his conduct in the several Actions of the 7th, 17th, and 18th of June, and 1st of July, in consequence of which he was honoured with the Cross of THE MILITARY ORDER of ST. GEORGE.—Royal Licence, 28th of July, 1789.

Major General WILHELM *de* DORNBERG, Colonel-Commandant of The Duke of BRUNSWICK OELS' Corps of Cavalry, "for his great merit and services."—Royal Licence, 20th of May, 1813.

The Imperial Order of Saint Wolodimir, *or*, Wladimir.

The Imperial Order of Saint Wolodimir was founded by The Empress Catharine the Second, on the 4th of October, 1782, the Anniversary of her Coronation, in memory of Wolodimir *the Great*, the First Sovereign of All the Russias in 976, who introduced the Christian Religion into Russia, and acquired the appellation of "*Like unto the Apostles.*"

This Order is intended, as a recompense for peaceful civil virtue, and to assure merit of it's just reward, in any Class of the Community,—the military, citizens, and artists may equally obtain it.

It consists of four Classes,—and it is not necessary to pass through the lowest to arrive at the first. Civil Servants of Government have a right to the Decoration, after an uninterrupted and faithful service of thirty-five years.

Knights of the *first* class receive a Pension of 600 roubles,—those of the *second* class, 300,—those of the *third*, 200,—and those of the *fourth*, 100.

The Knights of the first Class wear the Cross, pendent to a wide riband of a ponceau colour, with a black border, across the right shoulder,—on the reverse of the cross, in Russian characters, is the date of the Foundation, *viz.*, 22d September (4. October), 1782. On the left breast of the coat they wear the Jewel, on which are four Russian letters, implying, "*The Holy Prince Wolodimir, like unto the Apostles,*"—and around the centre is, in Russian, "*Utility, Honour, and Glory.*"

The Knights of the second class wear the same Cross, suspended from the neck, and the same Jewel, as those of the first.

Knights of the third class have a smaller cross, suspended in the same manner, but without a Jewel.

Members of the fourth class wear the Cross at the button-hole of the coat. The persons who obtain it for Military services, put a Rosette on the riband.

The Anniversary of The Order is celebrated on the 22d of September, O. S.

This Decoration is never to be set with *Diamonds*.

Second Class.

Henry Fanshawe, Esq., Knight of The Russian Military Order of St. George, 18th of July, 1789,—and who, by His Britannick Majesty's gracious permission, holds a distinguished Command in the Russian Army.—Royal Licence, 20th of March, 1807.—*See*, Order of St. George.

John, Lord Strafford.

James Wylie, Esq., Counsellor of State, and First Surgeon to His Imperial Majesty The Emperor of Russia, and Inspector General of His Armies and Military Hospitals for the Medical Department. — Royal Licence, 20th of March, 1807. — Created a Baronet, 2d of July, 1814.

Dr. Alexander Crichton, First Physician in Ordinary to The Emperor of All the Russias, and to The Imperial Family, and also one of His Imperial Majesty's Actual Counsellors of State, "in testimony of the high sense which His Imperial Majesty entertains of his merit and conduct, while in his actual service."—Royal Licence, 4th of July, 1820.

Third Class.

Lieutenant General Sir John Lambert, K. C. B., Colonel of the 10th (*or*, North Lincolnshire) Regt. of Foot,—Royal Licence, dated the 15th of September, 1828, to accept and wear the Insignia of The Imperial Russian Order of St. Wladimir, of the Third Class,—and of a Commander of The Royal Bavarian Order of

MAXIMILIAN JOSEPH, which their late Majesties The Emperor of Russia and The King of Bavaria were pleased to confer upon him, "in testimony of the high sense which those Sovereigns entertained of his Services before the Enemy in *The Netherlands*, in the year 1815."

Captain EDWARD CURZON.—Battle of *Navarino*.—*See*, FRANCE, ORDER of ST. LOUIS.

JOHN A. OMMANEY, Captain in The Royal Navy, C. B.—Battle of *Navarino*.

Sir ARCHIBALD WILLIAM CRICHTON,—*See*, RUSSIA, ORDER of ST. ANNE, of the Second Class.

FOURTH CLASS.

The Honble. GEORGE CATHCART, Lieutenant in the 6th Regt. of Dragoon Guards, "in testimony of the high sense which The Emperor of All the Russias entertains of his military services in the several Actions with the enemy, during the recent Campaign."—Royal Licence, 3d of June, 1814.

The Honble. HERBERT GARDNER, Captain in the Royal Regiment of Artillery, to accept and wear the Insignia of The Imperial Russian Order of St. Wolodimir, of the Fourth Class, which His Majesty The Emperor of All the Russias hath been pleased to confer upon him, at the recommendation of His Most Serene Highness CHARLES AUGUSTUS Duke of SAXE WEIMAR, "in consideration of the distinguished intrepidity displayed by him before *Maubeuge*."—Royal Licence, 22d of July, 1814.

THOMAS NOEL HARRIS, Esq., Captain in His Majesty's 36th (*or*, Herefordshire) Regt. of Foot, to accept and wear the Insignia of The Imperial Russian Order of St. Wolodimir, of the Fourth Class,—and of the Royal Prussian Order of Merit, "in testimony of the sense entertained by The Emperor of Russia and The King of Prussia of the distinguished conduct displayed by him, during the memorable Campaigns of 1813 and 1814."—Royal Licence, 2d of January, 1815.

GEORGE MACLEOD, Captain of the 35th Regt. of Foot (late a Lieutenant in The Royal Regiment of Artillery), " in testimony of His Imperial Majesty's approbation of the distinguished services rendered by him, during the siege of *Dantzic.*"—Royal Licence, 23d of February, 1815.

Lieutenant WILLOUGHBY MONTAGU, of The Royal Regiment of Artillery, " in testimony of His Imperial Majesty's approbation of the distinguished services rendered by him, during the Siege of *Dantzic.*"—Royal Licence, 28th of April, 1815.

Captain GEORGE STORY, late of the Staff of His Excellency General Count *de* LANGERON, " in testimony of His Imperial Majesty's approbation of the distinguished services rendered by him in the Field."—Royal Licence, 20th of May, 1815.

Lieutenant JAMES EDGECOMBE, of The Royal Navy, " in testimony of the high sense which The Emperor of Russia entertains of the signal intrepidity displayed by him, during the Blockade of *Hamburgh* and *Haarburgh.*"—Royal Licence, 29th of September, 1815.

Lieutenant and Adjutant ROBERT GILBERT, of The Royal Marine Artillery, " in testimony of the approbation of His Imperial Majesty The Emperor of All the Russias, of the signal intrepidity displayed by him in command of a Corps of The Royal Marine Artillery (employed as a Rocket Corps), at the Siege of *Dantzic,* in the year 1813."—Royal Licence, 29th of September, 1817.

CHRISTOPHER STRACHEY, Esq., Post Captain in The Royal Navy, " in testimony of the approbation of His Imperial Majesty The Emperor of All the Russias, of the distinguished intrepidity displayed by him in Action with the Enemy before *Dantzic,* in the year 1807."—Royal Licence, 20th of March, 1820.

JOHN HARE, Esq., Lieut. Colonel in the 27th Regt. of Foot, and C.B., " in testimony of the high sense which His Imperial Majesty The Emperor of Russia entertains of the highly distinguished conduct and intrepidity displayed by him in the memorable Battles in *The Netherlands,* in the year 1815."—Royal Licence, 16th of March, 1822.

Sir RICHARD DOBSON, Knt., conferred upon him by The Emperor

ALEXANDER of Russia, and graciously permitted by His Majesty King WILLIAM the FOURTH to wear the Decoration. *See*, DENMARK, ORDER of THE DANNEBROG, of the fourth Class.

Sir GEORGE ADAM WOOD,—*See*, AUSTRIA, ORDER of MARIA THERESA.

Sir NEIL DOUGLAS,—*See*, AUSTRIA, ORDER of MARIA THERESA.

JOSEPH CHAPPELL WOOLLNOUGH, Esq., K. H., Commander in The Royal Navy, " for the bravery and high military talents displayed by him, in the blockade of *Hamburg* and *Haarburg*, in 1814."

Sir GORE OUSELEY, Bart.

HENRY JOHN CODRINGTON, Midshipman in The Royal Navy.

The Imperial Order of Saint Anne.

The Imperial Order of Saint Anne was instituted, at *Kiel*, on the 3d of February, O. S., 1735, by Charles Frederick, Duke of *Schleswick-Holstein-Gottorp*, in honour of the Empress Anne of Russia, and of his Consort Anne-Petrowna, the eldest daughter of Peter *the Great*.

In 1762 His Imperial Highness Peter, Duke of *Schleswick-Holstein-Gottorp*, and Grand Duke of Russia, succeeding to The Imperial Dignity, upon the death of The Empress Elizabeth, this Order then became the *Fourth* of that Empire, and so it remains at present.

At the time of it's Institution, this Order consisted but of one Class, of *Six Knights*. But, when Paul the First ascended the Throne of Russia, in 1796, he declared it to be a *Russian* Order, divided it into three classes, and constituted it to be a recompense for Merit, both in his own Dominions and abroad,—and, he finally ordained, that whoever in future should receive The Order of Saint Andrew, should, at the same time, be decorated with that of Saint Anne.

In this state it remained until the year 1815, when The Emperor Alexander added a Second Class to it, to which Military men only could be admitted, and those of that Class have the Cross enamelled on the hilt of their Sword.

Persons admitted into the First Class, must have at least the Rank of Major General. The oldest Members of this Class receive an annual Pension.

THE BADGE of THE ORDER is a Cross, composed of four large Rubies set in gold, the angles between the Cross set with Diamonds, and on the centre is a Medallion, enamelled with the figure of *St. Anne*. It is worn by the Knights of the first class, pendent to a broad, watered crimson riband, edged with yellow, passed scarfwise over the left shoulder.

THE STAR of The Order is of silver, with eight points or rays,—on the centre, a red cross on a gold ground, with the following MOTTO, which is also inscribed on a red ground, "*Amantibus Justitiam, Pietatem, Fidem.*"

The Knights of the second class wear the Cross suspended from the neck.

The Decoration of The Order of the first and second class, enriched with *Diamonds*, or with an Imperial Crown of gold, enamelled, attached to the ring of the Cross, and on the ray at the top of the Star, is a particular distinction, with which The Emperor, NICHOLAS the first, honours eminent Merit.

The Knights of the third class wear the Cross at the button-hole of their coat.

Those of the fourth class have the Cross enamelled on the hilt of their sword.

The fifth class, which is granted to Subalterns and Soldiers, is a gold MEDAL, on which is enamelled a red cross,—the whole surmounted with a crown,—is worn at the button-hole of the coat.

On solemn FESTIVALS, the Knights of the first class appear in the full HABIT of The Order.

THE ANNIVERSARY of The Order is celebrated on the 3d of February, O. S.

FIRST CLASS.

General Sir TOMKYNS HILGROVE TURNER,—*See*, OTTOMAN EMPIRE, Superior Class.

CHARLES Count ALTEN, K. C. B., Lieutenant-General in The Army, also Lieutenant-General in the Hanoverian Service, Commander of the Hanoverian Troops serving in The Netherlands and in France, Colonel of the First Light Batallion of The King's

German Legion, and Knight Grand Cross of The Royal Hanoverian Guelphic Order,—His Majesty's Royal Licence, dated 23rd of March, 1816, that he might accept and wear the Insignia of The Imperial Russian Order of St. Anne, of The First Class,—of The Royal Order of Wilhelm of The Netherlands, of The Third Class,—and of an Honorary Knight Commander of The Royal Portuguese Military Order of The Tower and Sword, which their Imperial and Royal Majesties The Emperor of Russia and The King of The Netherlands, and His Royal Highness The Prince Regent of Portugal, had been respectively pleased to confer upon him, "in testimony of the high sense which they entertain of the highly distinguishĕd intrepidity and conduct displayed by him throughout *The Peninsular War*, and at the memorable Battles fought at *Quatre Bras* and *Waterloo*, on the 16th and 18th of June, 1815."

Major General Sir Robert Thomas Wilson.

Francis Charles Seymour Conway, Esq., (commonly called, Earl of Yarmouth, afterwards Marquess of Hertford), "in testimony of His Imperial Majesty's approbation of His Lordship's zeal, while in The Emperor's actual service, during his residence in this Country."—Royal Licence, 5th of July, 1821.

Sir Thomas Tyrwhitt, Knt., Gentleman Usher of The Black Rod of The Most Noble Order of The Garter, "as an especial mark of the favour of The Emperor Alexander, and in testimony of his approbation of the manner in which he discharged the duty of investing His Imperial Majesty with the Ensigns of The Most Noble Order of the Garter, in the year 1813."—Royal Licence, 11th of September, 1824.

Lieutenant-General Sir Frederick Adam, K. C. B., and Lord High Commissioner in and to The United States of The Ionian Islands,—Royal Licence, dated the 15th of September, 1828, to accept and wear the Insignia of The Imperial Russian Order of Saint Anne, of The First Class,—and of a Knight of The Imperial Austrian Military Order of Maria Theresa, which their Imperial Majesties the late Emperor of Russia and The Emperor of Austria were pleased to confer upon him, "in testimony of the high sense which those Sovereigns entertained of his Services before the Enemy in *The Netherlands*, in the year 1815."

Second Class.

Sir Samuel Greig, Vice-Admiral in The Imperial Service of Russia, conferred upon him, on the 25th of March, 1774.—Registered in The College of Arms, in 1778.—*See*, Order of St. George.

Sir John Ross, Knt., Captain in The Royal Navy, and C. B., to accept and wear The Insignia of The Second Class of The Imperial Russian Order of Saint Anne, *in Diamonds*,—and those of a Knight Commander of The Royal Swedish Military Order of The Sword, "for his Services in *The Baltic*, in the year 1812." —Royal Licence, 24th of December, 1834.

Sir William Congreve, Bart., conferred by The Emperor of Russia, on account of the effect of *The Congreve Rockets*, at the battle of *Leipsic*, in 1813.—Died, in 1828.

Francis Banks, Esq., Commander in The Royal Navy, and commanding His Majesty's Ship *Blazer*, to accept and wear the Insignia of The Imperial Russian Order of St. Anne, of the Second class,—and of a Knight of The Royal Swedish Military Order of The Sword, "in testimony of the high sense which The Emperor of Russia and The King of Sweden entertain of his bravery, and of the services rendered by him in the blockade of the river *Elbe*, during the operations before *Hamburgh*, and at the siege of the Fortress of *Gluckstadt*."—Royal Licence, 24th of November, 1814.

Sir Archibald William Crichton, Knt.,—*St. Anne*, of the Second Class, *St. Wladimir*, and *The Legion of Honour*.

Captain Richard Dickinson.—*See*, France, Order of St. Louis.

Captain The Hon. Frederick Spencer, of the Royal Navy.—Battle of *Navarino*.

Thomas Fellowes, Captain in The Royal Navy, C. B.—Battle of *Navarino*.

Gawen W. Hamilton, Captain in The Royal Navy, C. B.—Battle of *Navarino*.

Hon. I. A. MAUDE, Captain in The Royal Navy, C. B.—Battle of *Navarino.*

LEWIS DAVIES, Captain in The Royal Navy, C. B.—Battle of *Navarino.*

Hon. WILLIAM ANSON, Captain in The Royal Navy, C. B.—Battle of *Navarino.*

Rt. Hon. HENRY Viscount INGESTRIE, Captain in The Royal Navy, C. B.—Battle of *Navarino.*

GEORGE B. MARTIN, Commander in The Royal Navy, C. B.—Battle of *Navarino.*

ALEXANDER MACDONALD, Esq., Lieut. Colonel in the Army, and Captain in The Royal Horse Artillery, "in testimony of the high sense which His Imperial Majesty entertains of the distinguished services rendered by him, during the siege of *Dantzic.*"—Royal Licence, 17th of February, 1815.

Sir JOHN DICK, of *Braid,* a *Nova Scotia* Baronet, was created a Knight of ST. ANNE, by The Empress CATHARINE the Second. The essential Services which this Gentleman rendered to the Russian Fleet, when it anchored at *Leghorn,* during the first war which subsisted between that Power, and the Turks, entitled Sir JOHN to this honourable Distinction. He was then the British Consul-General in Tuscany, and His Britannick Majesty was graciously pleased to permit him to accept and wear the Decorations of this Illustrious Order. GEORGE the Third conferred upon him the honour of K. B.,—He died in 1805.

A Dispatch from Field-Marshal The Duke of WELLINGTON, dated *Head-Quarters, Paris!*—the 21st of August, 1815, contains the following List of Officers, upon whom His Imperial Majesty THE EMPEROR of RUSSIA had conferred Decorations of different Classes of *The Orders of St. George, Anne,* and *Wladimir* respectively, "in testimony of His Imperial Majesty's approbation of their services and conduct, particularly in the late Battles fought in *The Netherlands,*"—

Second Class, St. George.

Lieut. General The Marquess of Anglesey.
Lieut. General Lord Hill.

Third Class, St. George.

Lieut. General Sir H. Clinton.
Major General Cooke.
Major General Kempt.

Second Class, St. Wladimir.

Major General Sir O. Vandeleur.
Major General Sir J. Byng.
Major General Sir D. Pack.

Third Class, St. Wladimir.

Major General Lord E. Somerset.
Major General Sir J. Lambert.
Major General Sir C. Grant.
Major General Maitland.
Major General Sir H. Vivian.
Colonel Mitchell.

Fourth Class, St. George.

Colonel Sir J. Elley, Assistant-Adjutant General.
Colonel Reynell, 71st Regt.
Colonel Sir A. Barnard, 95th Regt.
Colonel Honble. A. Abercromby, Assistant-Quarter-Master General.
Colonel Sir J. Colborne, 52d Regt.
Colonel Woodford, Coldstream Guards.
Colonel Honble. Sir F. Ponsonby, 12th Light Dragoons.
Colonel Hervey, Acting Military Secretary.
Lieut. Colonel Sir R. Hill, Royal Horse Guards.
Lieut. Colonel Lord F. Somerset, Military Secretary.
Lieut. Colonel Lord Saltoun, 1st Guards.

Fourth Class, St. Wladimir.

Colonel Hepburn, 3d Guards.
Colonel Sir G. Wood, Royal Artillery.
Colonel Muter (now Straton) 6th Dragoons.
Colonel Carmichael Smyth, Royal Engineers.
Lieut. Colonel Macdonnell, Guards.
Lieut. Colonel Sir H. Bradford, 1st Guards.
Lieut. Colonel Lord Greenock, Assistant-Quarter-Master General.
Lieut. Colonel Cooke, 1st Guards.
Lieut. Colonel Sir C. Broke Vere, Assistant-Quarter-Master General.
Lieut. Colonel Sir H. G. Berkeley, Assistant-Quarter-Master General.
Lieut. Colonel Ross, 95th Regt.
Lieut. Colonel Sir G. Scovell, Assistant-Quarter-Master General.
Lieut. Colonel Dick, 42d Regt.
Lieut. Colonel Douglas, 79th Regt.
Lieut. Colonel Nixon, 28th Regt.
Lieut. Colonel Lygon, 2d Life Guards.
Lieut. Colonel Hare, 27th Regt.

First Class, St. Anne.

Lieut. General Charles Count Alten.
Major General Sir E. Barnes, Adjutant-General.
Major General Adam.

Another Dispatch from Field-Marshal The Duke of Wellington, dated *Head-Quarters, Paris,* the 8th of October, 1815, contains the following List of Officers, upon whom His Majesty The Emperor of Russia had conferred Decorations of *The Order of St. Anne,* " in testimony of His Majesty's approbation of their services and conduct, particularly in the late Battles fought in *The Netherlands,*"—

Second Class of St. Anne.

Colonel Frederick von Arentschildt, 3d Hussars, K.G.L.

Lieut. Colonel Robert Torrens, West India Regt.
Lieut. Colonel John Waters, A.A.G.
Lieut. Colonel Charles Beckwith, 95th Foot.
Lieut. Colonel William Campbell, A.Q. M.G.
Lieut. Colonel Colin Campbell, Royal Scots.
Lieut. Colonel Arthur Clinton, 1st Dragoons.
Lieut. Colonel John Hicks, 32d Foot.
Lieut. Colonel William Elphinstone, 33d Foot.
Lieut. Colonel Henry Mitchell, 51st Foot.
Lieut. Colonel A. G. Norcott, 95th Foot,
Lieut. Colonel A. Cameron, 95th Foot.
Lieut. Colonel J. B. Clarke, 2d Dragoons.
Lieut. Colonel Sir John May, K.C.B., Royal Artillery.
Lieut. Colonel Sir Hew Ross, K.C.B., Royal Artillery.
Lieut. Colonel Sir Robert Gardiner, K.C.B., Royal Artillery.
Lieut. Colonel Sir William Gomm, K.C.B., Royal Artillery.
Lieut. Colonel John Bull, Royal Artillery.
Major Edward Kelly, 2d Life Guards.
Major A. M'Donald, Royal Artillery.

Medals, and Marks of Honour.

The Cross of Ismaël.—This Cross was instituted, on the 2d of December, 1790, by The Empress Catharine the Great, in memory of the capture of *Ismaël.* The Russian inscription upon it signifies, "*Ismaël was taken,* 2d *December,* 1790."

The Cross of Saint George.—The Emperor Alexander, in 1807, founded the Cross of St. George, which he declared, could only be acquired in time of war, by land or sea,—and which conferred on it's possessor a pecuniary right, equal to one-third of his pay. A Decree, dated the 22d of October, 1814, declares, that the Crosses conferred on the Under-officers and Privates in the Campaigns of 1813 and 1814, shall pass, after their deaths, to individuals in the same Corps, who at that time were proposed for the honour of this Distinction without obtaining it.

The Cross of Saint George may be considered, as *The fifth Class* of The Order of Saint George. It is worn at the button-hole.

Medal, in gold, for the Riband of The Order of Saint George.—This Medal, founded by The Emperor Alexander, is intended for Officers, and is worn at the button-hole. It reckons to it's possessors for three years service, as well for obtaining pensions as for acquiring The Order of St. George.

Medals of 1807.—The Officers and Privates of the *Landwehr* (*Militia*) disbanded in the month of September of this year, who had been present at a Battle, wear,—the former, a Medal, of gold,—the latter, one, of silver,—attached to the riband of The Order of St. George. Those Officers, who have not taken part in an Engagement, wear the MEDAL, attached to the riband of THE ORDER of ST. WLADIMIR.

THE CROSS of BAZARDGICK.—The Emperor ALEXANDER the First granted to the Russian Soldiers this CROSS, as a reward for the valour which they displayed in the capture of *Bazardgick.* The Russian inscription upon it signifies, "*For the Assault and Conquest of Bazardgick,* 22d *May,* 1810." On the other side of the Cross is inscribed, "*To distinguished valour.*"

MEDALS of 1812.—In memory of the year 1812, The Emperor ALEXANDER instituted a Medal of Silver, which, on the obverse, displays in an angle encircled with rays, the Eye of Providence, with the date 1812,—on the reverse is inscribed, in the Russian language, "*Not unto us, not unto us, but unto thy Name,*" be ascribed Praise, Honour, and Thanksgiving. This Medal is worn in the button-hole, on a light blue watered riband by all Military men, who took part in the Campaign of 1812,—as well as by all Ecclesiastics and Medical men, who exerted themselves in battles and under the fire of the enemy. In 1814, this Medal, in bronze, was given to the oldest of every Noble family,—and, in 1816, it was also allowed to the oldest female member of every Noble family, to wear them on the riband of THE ORDER of ST. WLADIMIR. Civil persons wear it on the riband of THE ORDER of ST. ANNE.

MEDALS of 1814.—The Emperor ALEXANDER, in memory of the Capture of Paris, in 1814, founded a MEDAL, of silver, which is worn, attached to a riband, one half of which is blue, and the other half the same as the riband of ST. GEORGE. On the Medal is inscribed, "*For the Capture of Paris,* 19th *March,* 1814."

MEDAL, FOR THE PERSIAN CAMPAIGN.—This MEDAL is of silver, and is worn, attached to a riband, one half of which is like that of THE ORDER of ST. WLADIMIR, and the other half like that of THE ORDER of ST. GEORGE. It was founded by The Emperor NICHOLAS the First, in order to recompense the courage of the Russian Soldiers, who were present at The Persian Campaign. On the Reverse, are the words, "*For the war of Persia.*"

MEDAL, FOR THE CAMPAIGN OF TURKEY.—THIS MEDAL was distributed among the Russian Soldiers, who were present at The Campaign of Turkey. It was instituted by The Emperor NICHOLAS, is of Silver, and bears, on the Reverse, the words, "*For the Campaign of Turkey.*"

THE MEDAL OF SAINT ANNE.—The Emperor NICHOLAS instituted this MEDAL, for Military men of long and faithful service. It may be considered, as *The fifth Class* of THE ORDER of ST. ANNE. On the Reverse, is the number of the present Possessor. The Emperor has granted it, surmounted by a Crown, to various Foreigners,—for example, to the entire Company of the Under-officers of the Guard at Berlin.

MEDAL, FOR THE CAPTURE OF WARSAW.—The Emperor NICHOLAS founded this SILVER MEDAL, in memory of the submission of *Warsaw*, during the memorable Campaign of the Insurrection of THE POLES. It was granted to all those, who were present in the ranks, and fought at this period. On one side are the words, "*For the Capture of Warsaw*, 26th *September*, 1831,"—and, on the other, "*Utility, Honour, and Glory.*"

MARKS OF DISTINCTION OF SERVICE.

The Emperor NICHOLAS has conferred upon his faithful Servants, MARKS of DISTINCTION,—on which, reckoning from the fifteenth year of service, the years of service, from five years to five years, are noted in Roman numerals. This Mark of Distinction is worn by *Military men*, attached to the riband of THE ORDER of ST. GEORGE,—and, by *Civilians*, to the riband of THE ORDER of ST. WLADIMIR, on the left breast of the coat.

There are also conferred upon MERCHANTS, in Russia, GOLD MEDALS of a moderate size, for different causes. They are worn, suspended from the neck,—and bear on one side, the Portrait of The Emperor,—and, on the other, the following Inscriptions,—

1. "*For Zeal and Services.*"—This is granted to those, who have been employedby The Electoral Assemblies, and who are in the service of the Crown, and have signalized themselves by their zeal.

2. "*For Utility.*"—This is granted, for success in the promotion of Manufactures, and improvements in Industry.

THESE MEDALS are worn successively attached to the ribands of THE ORDERS of ST. ANNE, ALEXANDER NEWSKI, ST. WLADIMIR, and ST. ANDREW. When an individual, who has worn them attached to all the above-named ribands, again renders himself worthy a Recompense, he obtains one of these Medals, with the same Inscription, *in Diamonds.*

ARTISANS and PEASANTS can obtain, by similar Merit, SILVER MEDALS, resembling the preceding, and bearing the same inscription. They are worn successively attached to the same ribands, but are not subsequently decorated with Diamonds.

GOLD and SILVER MEDALS of a moderate size are also conferred, as Military Distinctions, upon Functionaries and armed Persons of *The Mahometan Religion,* who have served against the Enemy. They are worn, suspended from the neck by the riband of THE ORDER of ST. GEORGE,—and bear, inscribed upon them, the words, "*For Courage.*"

THE NOBILITY, FUNCTIONARIES, and MERCHANTS, receive GOLD MEDALS, with the inscription, "*For having saved the unfortunate,*" as a recompense for their efforts in saving drowning persons,—or, for assistance in case of Fire. Under-officers, Soldiers, Artisans, and Peasants, obtain similar Medals, in *silver*, for the like actions. They are worn at the button-hole, attached to the riband of THE ORDER of ST. WLADIMIR or of ST. ANNE.

To the Fund from which Invalid Generals, Staff and Superior Officers receive a PENSION, *fees,* since the year 1816, have been paid by those also who receive MEDALS,—*viz.*, from one, set with *brilliants,* 500 roubles,—from one, in *gold,* with a blue riband, 200,—on the riband of *St. Alexander Newski* and *Wladimir,* 150,—on the riband of *St. Anne,* 100,—and from *Silver Medals,* 25 roubles.

THE KINGDOM OF SARDINIA.

The Royal Sardinian Supreme Order of The Annunciation.

This Order, which was originally styled, "*Ordre Militaire du Lac d'Amour*,"—subsequently, "*The Order of The Collar*,"—and now usually, "*The Order of The Annunciation*," was founded by Count Amadeus the Sixth, of *Savoy*, either in the year 1355 *or* 1362. It was probably instituted by motives of Piety, connected with the wish to preserve the remembrance of the valiant deeds of his Predecessor in the Government, Amadeus the Fifth, or, *The Great*, who had distinguished himself by his victory over The Turks at the Siege of *Rhodes*, in 1310.

Some persons, however, seek the reason of it's origin in the attachment of Amadeus to a Lady from whom he received a *Bracelet*, woven with the tresses of her hair in love-knots, which he wished to perpetuate in the form of a Decoration. This last opinion accords in some measure with the earlier name of THE ORDER, —but historical proofs are wanting for all these suppositions.

For the same reason the four letters upon The Badge, F. E. R. T., have been variously interpreted. ANDREW FAVIN calls it "*The Order of The Snares of Love,*" in regard to the favour of the *Bracelet,* and that these letters signify, "*Frappés, Entrés, Rompés, Tout.*" But FRANCIS CAPRE pronounces this to be an error.

Other writers explain these Gothick characters by the words, "*Fortitudo ejus Rhodum tenuit,*" in memory of the glorious Action of AMADEUS *the Great,* when he compelled The Turks to raise the Siege of *Rhodes.*

But ASHMOLE says, that *Mons.* GUICHENON, in his *History of The Royal House of Savoy,* has fortunately cleared this mistake from all obscurity. For the Device of The House of Savoy, consisting of these four Gothick letters, appears to be more ancient than the time of AMADEUS *the Great,* in regard that LEWIS *de Savoy,* Baron *de Vaud,* who died in 1301, always wore this Device, as is apparent on some of his Coins. And further, that on the Monument of THOMAS *de Savoy,* who died on the 20th of January, 1233, is a *Dog,* on the collar of which is the word "*Fert,*" without points, or separation of the letters,—being a strong proof, that this was but one word. To strengthen his assertion, GUICHENON exhibits another convincing testimony from a *brass Coin* of the same Earl THOMAS, on the Reverse of which are two Knots, of the fashion already described, and the word "*Fert,*" in the centre. He likewise gives us the figure of a *Silver Coin* of PETER *de Savoy*,—who abode for some time in England, in the reign of King HENRY the Third, and built that stately Structure in *The Strand*, called "*The Savoy*," recently pulled down, —and on which was represented the Device "*Fert*," in Gothick characters.

From all these evidences it is apparent, that this *Word*, as also the *Knot*, and *Cross* were used for the Device, Badge, and Arms of THE HOUSE of SAVOY, long before AMADEUS *the Great* so gloriously relieved The Isle of *Rhodes*, against the power of the Turks, or the time of the Institution of THE ORDER.

In any case THE ORDER is of very ancient origin,—and it is stated to have continued under the denomination of THE ORDER of THE COLLAR, until the time of CHARLES the third, surnamed *The Good*, Duke of Savoy, who changed it's title to that of, "*The Annunciation*," from the picture of the history of The Blessed Virgin Mary, by him first appended to THE COLLAR, in 1518.

When Duke VICTOR AMADEUS the second, of Savoy, in the year 1713, by a treaty with France, received the Kingdom of Sicily, he assumed the title of KING,—but five years afterwards he exchanged with The Emperor CHARLES the Sixth the possession of *Sicily* for that of *Sardinia*, to which he succeeded in 1720. By this means "THE ORDER of THE ANNUNCIATION," of The Dukedom of *Savoy*, was raised to an ORDER of The Kingdom of *Sardinia*,—THE FESTIVAL of which is celebrated on the 25th of March, being that of *The Annunciation of The Virgin*.

The most ancient STATUTES are those of AMADEUS the Eighth, dated the 30th of May, 1410, the original whereof is lodged in the Archives at *Turin*,—these he augmented in 1434,—and both of them are printed by SANSOVIN.

Duke CHARLES the Third made new Statutes, on the 11th of September, 1518,—these were enlarged by Duke EMANUEL PHILIBERT, in 1568, and were published in the year 1577.

THE KING of SARDINIA is GRAND MASTER of THE ORDER, and appoints the Knights,—whose number, with the exception of the Sovereign and the Heir Apparent, was first fixed at 15., then at 20.,—but is now undetermined, although those above 20., are named " *Supernumeraries*."

At present their number is small,—but it is, however, on the most respectable footing, and is counted among the most distinguished ORDERS in Europe.

The Knights of THE " *Supreme*" ORDER of THE ANNUNCIATION, as it is usually termed,—and who form but one Class, must at least be of the Ancient *Nobility*, and must already possess the *second* Sardinian Order, that of *St. Maurice and St. Lazarus*, whose Badge is also worn with it. He who has not yet The Order of *St. Maurice and St. Lazarus*, is, previous to his reception, created a Knight by three strokes of the Sword,—in which case, however, he does not receive THE BADGE of THE ORDER.

All the Knights acquire the title of, " *Excellency*."

THE BADGE is a round Medallion, of gold, enamelled white, on which is depicted The Annunciation of The Virgin Mary, surrounded by true-lovers' knots, of gold,—Of such knots, alternating with the initials F. E. R. T., also consists the gold COLLAR, to which is appended The Badge, round the neck, close to the Collar of their Habit, and in this manner AMADEUS the tenth, Duke of Savoy, is represented in his portrait at full length, formerly hanging in the Gallery at *Whitehall*,—to which COLLAR, within a circle formed of three love-knots, is suspended the figure of *The Annunciation*, as related by ST. LUKE.

The Knights are likewise decorated on the left breast, with a STAR, in the form of a Sun,—in the middle of which, the figure of *The Annunciation* is pourtrayed, surrounded by a single circle of mosaic, embroidered in silver, on which is the word " FERT."

The Riband is sky blue, and watered.

On the death of a Knight, the COLLAR is returned. But, from this observance, by a special understanding The Prince Abbot of *St. Gall* was always exempted. It appears, that a former Abbot having, on some occasion, rendered an important service to THE HOUSE of SAVOY, by sending Auxiliaries to it's assistance, was hence appointed a Knight of The Order of The Annunciation,—and that his Successor did not send back THE COLLAR, but begged permission to wear it likewise, which was granted, both to him and to his Successors,—so that every Prince Abbot of *St. Gall* was a Knight of this Order, although he was not reckoned among the Statutory number of Twenty.

THE ORDER has five Officers,—*viz.*, a CHANCELLOR, who is a Bishop or Archbishop,—a SECRETARY, who is the Minister of Foreign Affairs, for the time being,—an ALMONER, who is the Almoner to the King, for the time being,—a TREASURER,—and a HERALD at ARMS, who is the first Brigadier of the King's Body-guard, for the time being.

The first four Officers wear THE BADGE, not pendent to THE COLLAR, but to a sky blue riband, round the neck, and have also THE STAR.

The HERALD wears a CROSS,—in the middle of which is delineated The Annunciation,—appended by a sky blue riband in the button-hole, and has no Star.

The other Officers have peculiar Dresses of Ceremony.

GRAND CROSS.

Field Marshal The Duke of WELLINGTON, K.G.

The Right Honble. EDWARD, Viscount and Baron EXMOUTH,—*See,* SPAIN, ORDER of CHARLES the Third.

The Order of St. Maurice, of *Savoy*.

The Order of Knights Hermits of St. Maurice had their origin on the retreat of Amadeus the Eighth, Duke of *Savoy*, to his castle in the Desert of *Ripaille*, near the charming banks of the Lake of *Geneva*, and was bestowed by him, in the year 1434, upon those *Ten* Knights of his Court, who had retired into seclusion with him,—as well to preserve the memory of *St. Maurice*, the titular Patron of *Savoy*, as of his *Lance* or *Sword*, and *Ring*, the essential symbols of Knighthood, passing from hand to hand, from the Kings of *Burgundy* to the Princes of *Savoy*,—by whom they have been preserved, with great honour and veneration, and are, as a pledge of Dominion, delivered to them with great Solemnity, at their Inauguration.

Nine years after the Institution of this Order, The Founder was prevailed upon to accept the *Popedom*, on the deposition of Pope Eugenius the Fourth, by the Council of *Basil*, on the 5th of November, 1439, by the name of "Felix *the Fifth*,"—but, on the 6th of April, 1449, he freely resigned the Chair to Pope Nicholas the Fifth, and immediately retired again to his solitude in *Ripaille*.

His resignation appears to have been coupled with the conditions, that he should be the *First* Cardinal in The Conclave,—that the Pope should always receive him *standing*, and offer him his *mouth* to kiss,—that he should be perpetual Cardinal-Legate in the States of *Savoy* and *Switzerland*, and in the Archbishoprics of *Geneva*, *Sion*, *Bress*, &c.

And lastly, that all the *Cardinals* of his creation should be recognised by the Pope.

After he had made a Peace so acceptable to the Church, and so honourable to himself, he spent the remainder of his life with great devotion at *Ripaille*, where he died, on the 7th of January, 1451, with an extraordinary reputation of sanctity, and lies buried in the Cathedral Church of *Lausanne*.

This Order of St. Maurice was made subject to the Rule of *St. Augustine*,—And the Knights had assigned to them for their Habit, a grey or ash-coloured gown, and chaperon, a vest of the same, girded with a golden girdle, the chaperon and sleeves being faced with red camlet,—and, on the breast of the mantle, a cross, *pomettée*, made either of white taffeta, or white cloth,—But the Cross, which The Master or Chief of The Order wore, was made of gold, or rich embroidery.

Soon after the death of The Founder, the Order, which was never very flourishing, fell into obscurity and disuse.

The Royal Piedmontese Military Order of St. Maurice and St. Lazarus.

When the Reformation advanced with rapid progress, threatening destruction to the Roman Catholick faith, diffusing itself far and wide, and acquiring a prodigious number of adherents, many foundations and associations were established, for the preservation and defence of the threatened Communion, and for checking the increase of *The Protestant Religion.* Calvin, who, with his friends, had established themselves in *Geneva,* acquired many disciples in the neighbouring Country of *Savoy.* To oppose this establishment Emanuel Philibert, Duke of *Savoy,* in 1572, founded or revived "*The Order of St. Maurice,*" and imposed upon the Knights, the obligation of the preservation and defence of *The Catholick Church,* as their principal duty.

This establishment was very agreeable to Pope Gregory the Thirteenth, who confirmed it, on the 16th of September, 1572,—and, he shortly afterwards, by his Bull, dated the 13th of November in the same year, for their greater honour, united "*The Order of St. Maurice,*" which had existed only a few weeks to "*The Order of St. Lazarus,*" which he had recently suppressed in *Italy,*—and he ordained, that this new Institution should thenceforth be called, "The Order of ST. MAURICE and ST. LAZARUS." He also constituted the said Duke, and his Successors, Dukes of *Savoy*, GRAND MASTERS thereof,—and he obliged them to equip two Gallies, for the service of the Papal See, to be employed against *Pirates*.

The history of *The Order of St. Lazarus* has already been described under the Kingdom of FRANCE.

The altered circumstances of the times have, however, changed the original Institution of THE ORDER, have transformed it altogether, and it has become from an *Ecclesiastical* and *Military* Order, one of *Civil* Distinction, which is conferred on *Civilians* as well as on *Military* men, and is connected with essential advantages. THE ORDER has preserved the earlier possessions of it's COMMANDERIES. THE KING of SARDINIA confers them, as Grand Master General of THE ORDER, on the Members for their use, and reserves to himself a part only of the Revenues, which he bestows in *Pensions*, on those, who have not received *Commanderies*.

The attainment of THIS ORDER, therefore, is an object of importance, —but the proof of descent from eight *Noble* Ancestors is a necessary condition,—unless the party is absolved by The King. Such is the case also with the vow of Obedience to the Grand Master, of the obligation to attend on *Lepers*, and of Connubial fidelity,—so that a Knight can neither form a second Marriage, nor can he marry a Widow, without the consent of the Pope.

THE OFFICERS of The Order are,—

1. THE CHIEF CONSERVATOR, who is President of the Administration of the Estates of The Order.
2. THE GRAND MASTER of The Hospitals, who exercises the superintendence over the Hospitals.
3. THE AUDITOR, who transacts the justice and police of The Order.

These three have the Grand Cross.

4. THE SECRETARY, who also usually has the Grand Cross.

5. The Fiscal and Patrimonial Advocate has the smaller Cross.
6. The Herald at Arms has no decoration.

The other Officers have likewise Uniforms and Dresses of Ceremony.

No Member can attain the *First* Class, that of Grand Crosses, unless he has previously belonged to the *Second*, that of Knights.

The *White* Cross of The Badge, is the original one of *The Order of St. Maurice*, to which the *Green* one of *St. Lazarus* was added, when their union took place. It is worn by The Grand Crosses, pendent to a green riband, round the neck,—by the Knights it is worn, of a smaller size, in the button-hole,—where it is also worn by the Knights of *The Order of the Annunciation*, who are always Knights of this Order.

The Star which is of silver, presents the obverse of the Badge.

On occasions of Ceremony all the Members appear in a peculiar State Dress. In common, they wear a green Military Uniform.

Grand Crosses.

The Right Honble. Edward, Viscount and Baron Exmouth,—*See*, Spain, Order of Charles the Third.

The Right Honble George, Viscount Keith, G.C.B., Member of The Imperial Ottoman Order of The Crescent, Admiral of The Red Squadron of His Majesty's Fleet, *&c.*, "in testimony of the approbation of His Majesty The King of Sardinia, of the services rendered by His Lordship at the Bombardment of *Genoa*, in the year 1809."—Royal Licence, 3d of August, 1822.

Commanders.

Lieut. Colonel Gideon Gorrequer, Member of the Imperial Ottoman Order of The Crescent, Knight of The Royal Hanoverian Guelphic Order, and Knight Commander of The Royal Sicilian Military Order of Saint Ferdinand and of Merit, "for his distinguished conduct at the attack and reduction of *Genoa*, in 1814."—Royal Licence, 3d of December, 1833.

Knights.

Lieutenant Colonel Charles Ashe à Court,—*See*, The Two Sicilies, Order of St. Ferdinand and of Merit.

John Benjamin Heath, Esq., Consul General in London, for Sardinia and Genoa.

John Shearman, Esq., late a Captain in the 60th Regt. of Foot, now a Captain in the 13th Regt. of Foot, "in consideration of his Services in the Field, while Brigade Major of *The Italian Levy*."—Royal Licence, 4th of April, 1816.

Alexander Maclachlan, Esq., Captain in the Royal Regiment of Artillery, "in testimony of the high sense which His Sardinian Majesty entertains of his services in the Field, while commanding a Corps of Artillery *of The Italian Levy*, on the Eastern Coast of Spain, in 1813 and 1814."—Royal Licence, 6th of September, 1816.

Richard Roberts, Esq., Lieut. Colonel in the Army, and Captain in the 62d, (*or*, The Wiltshire) Regt. of Foot, "as a testimony of the high sense which His Majesty Victor Emanuel, King of Sardinia, entertains of the distinguished services rendered by him at the siege of *Genoa*."—Royal Licence, 17th of June, 1817.

George Bentham, Esq., Post Captain in The Royal Navy, "in testimony of the high sense which His Majesty The King of Sardinia entertains of the distinguished intrepidity displayed by him in command of His Majesty's Sloop *Heron*, in action with the Enemy at the memorable attack upon the City of *Algiers*, on the 27th of August 1816."—Royal Licence, 25th of June, 1817.

Richard H. Fleming, Esq., Lieutenant (*now*, Commander) in The Royal Navy, as a signal testimony of his distinguished merit and intrepidity, when commanding The *Invincible* Battering Vessel, and, subsequently, The *Explosion* Vessel, in the Battle of *Algiers*.—*See*, The Two Sicilies, Order of St. Ferdinand and of Merit.

Edward Brace, Esq., Post Captain in The Royal Navy, C.B., Knight of The Royal and Distinguished Order of Charles the Third of Spain, and of The Royal Order of Wilhelm of the Netherlands, "in testimony of His Sardinian Majesty's approbation of his distinguished Services before the Enemy at the Capture of *Genoa*, and the Bombardment of *Algiers*."—Royal Licence, 12th of April, 1825.

THE MILITARY ORDER OF SAVOY.

Under this denomination VICTOR EMANUEL, King of SARDINIA, founded, on the 14th of August, 1815, a NEW ORDER, which he destined exclusively for THE MILITARY.

It consists of four Classes.

THE BADGE of THE ORDER is a gold Cross, enamelled red, upon which is another Cross enamelled white, surrounded with a wreath of laurel, and surmounted by a Crown, of gold. It is worn by the Grand Crosses, pendent to a broad light blue, watered riband, from the right shoulder to the left side.

THE STAR, which is of Silver, contains in the centre, upon a blue ground, the cypher of the Founder, V. E., crowned, and surrounded with the MOTTO, "*Al merito ed al valore.*" If a Grand Cross be a Knight of the highest ORDER of THE ANNUNCIATION, the Star is worn on the coat *below* that of the higher Order.

KNIGHTS, if they are not Officers, enjoy a yearly *Pension* of 120 *lire.*

The Civil Order of Savoy.

This Order was founded by Charles Albert, King of Sardinia, by an Ordinance, dated at *Turin*, the 29th of October, 1831, to recompense those, who had shown Fidelity and Attachment to the Sardinian Throne,—and especially those, who, by arduous study and great exertions have become the Ornament of The State, and have rendered important services,—more particularly those. who have devoted themselves to Publick Education and have produced results to which may be attributed the greater value, since, on the good Education of Youth, depends the welfare of Individuals, of Families, and of the Kingdom.

It consists of one Class only.

Pensions are attached to The Order to the amount of 30,000 *lire*.

It is conferred, free of all expense.

The Badge is a Cross, of gold, enamelled blue, and in the centre, on a white ground, is the cypher of the Founder, C. A., in letters of gold, —and, on the reverse, within a circle, is the Motto, "*Al merito civile*, 1831." It is worn, pendent to a blue riband, with two white stripes, on the left side of the coat.

Military Medal.

Charles Albert, King of Sardinia, in order to recompense feats of Military Valour, instituted, on the 26th of March, 1833, for the Land and Sea Forces, a Mark of Honour, consisting in a Medal, of gold, and silver, displaying on one side a Cross, surmounted by a Crown, with the Motto "*For Military Valour,*" — and, on the reverse, two branches of Laurel, between which is inscribed the name of the person, who is honoured with the Decoration,—on the Exergue, are stated the Action rewarded, and the date.

This Medal is conferred on all Military persons, whether of the Land or Sea service, from the rank of General, and the possessors enjoy the same honours and advantages, as The Knights of The Military Order of Savoy. The possessors of the *gold* Medal enjoy a Pension of £ 100.,—those who have the *silver* Medal, receive annually £ 50. The Medal and Pension descend to the Family of a Possessor. The Widow receives the Pension so long as she continues single. If there be only Children, these latter enjoy the income until the youngest attains the age of 15 years.

THE
DUCHY OF SAXE-COBURG-GOTHA.

The Saxe-Ducal Order of Ernest.

This Order was instituted in the year 1690, by Frederick the First, Duke of Saxe-Gotha and Altenburg, eldest son of Duke Ernest *the Pious*. It was styled "*The Order of German Integrity*," the Motto being "*Fideliter et Constanter*."

It was restored in the year 1833, in honour of the memory of the said Duke Ernest *the Pious*, Progenitor of the Joint Saxe-Ducal House, and in remembrance of that Branch of it which became *extinct* in 1825.

The renewal of it was attended with certain modifications, which existing circumstances rendered desirable. The original Motto of "*Fideliter et Constanter*" was retained,—and it was determined, that The Order, independently of being a Badge of Honour for the several Members of The Saxe-Ducal Houses, should moreover be bestowed, as a Reward and a mark of

Distinction, upon such State Functionaries and Subjects whose achievements, integrity, talents, fidelity, services, attachment and devotion to their Prince and their Native land, were of a nature to entitle them to the gratitude of The State.

THE ORDER is common to the Three Branches of The Ernestine Saxe-Ducal House of the Line of *Gotha*.

It's GRAND MASTERS are,—

1. The reigning Duke of SAXE-ALTENBURG.
2. The reigning Duke of SAXE-COBURG-GOTHA.
3. The reigning Duke of SAXE-MEININGEN-HILDBURGHAUSEN.

It is conferred by mutual consent, by these THREE BRANCHES jointly, at the recommendation of any one of them. To each of the Three separate Courts, a limited number of Decorations are allotted, subject to The Statutes, and that number is not allowed to be exceeded.

THE ORDER consists of Four Classes,—*viz.*, GRAND CROSSES,—COMMANDERS of the *first* Class,—COMMANDERS of the *second* Class,—and KNIGHTS.

Besides these Classes, a SILVER CROSS, and a SILVER MEDAL of MERIT, are attached to THE ORDER.

THE INSIGNIA consist of a Cross, of eight points, enamelled white, surmounted by a crown of gold, round which is a golden border,—and, on each of the points, is a golden ball. The obverse displays a round shield of gold, on which is a bust, in gold, of Duke ERNEST, surrounded with blue enamel, bearing the MOTTO, "*Fideliter et Constanter*,"—this circle is again surrounded with a green coloured wreath of oak, ornamented with gold,—but THE ORDER when given to *Foreigners*, does not exhibit this latter embellishment. The reverse of The Cross shows the Saxon Arms, round which is a circle of blue enamel, with an inscription, in letters of gold, denoting the date when THE ORDER was founded,—this circle is likewise surrounded by an oak-wreath, as above described.

THE CROSS of THE ORDER is worn, pendent from a rich deep red, watered riband, edged with green, over the left shoulder of The Grand

Crosses — but the Commanders wear it round the neck,—and the Knights suspend it from the button-hole.

THE STAR consists of eight rays, of gold and silver,—and, in the centre, the green wreath of oak leaves, round the blue circle, containing the MOTTO, and within, upon a shield of gold, a Crown, in green enamel.

On the 12th of November, 1826, ERNEST, then Duke of *Saxe-Coburg-Saalfeld,* took the Title of SAXE-COBURG-GOTHA.

MEDALS, founded by Duke ERNEST, of *Saxe Coburg-Gotha,* as Duke of *Saxe-Coburg-Saalfeld.*

THE DUCAL CAMPAIGN MEDAL of SAXE-COBURG was instituted, in 1816, by Duke ERNEST, and distributed among all the Military, who had taken part in the Campaign of 1814. The letters, "*Ernest* H. Z. S. C. S.," signify, "ERNEST, *Duke of Saxe-Coburg-Saalfeld.*" This Medal is worn attached to the button-hole.

In 1814, MEDALS, of *gold* and *silver*, were instituted by Duke ERNEST, for distribution among the Saxon Military men under his command. These Medals were only granted for brilliant Actions, and on very particular recommendation, in compliance with the decision of a Commission appointed for this purpose. They are given of *gold* or *silver*, according to the degree of merit, and are worn at the button-hole.

In 1814, *Iron* MEDALS were instituted by Duke ERNEST, as General-in-Chief of the 5th Corps of The German Army, for Volunteers. This Medal has been distributed among 1617 individuals,—*viz.*, A Volunteer Battalion of Chasseurs of *Berg*, a Volunteer Squadron of Chasseurs of *Berg*, a Volunteer Corps of Chasseurs of *Nassau*, a Volunteer Company of Chasseurs of The Duchy of *Saxony*, a Volunteer Company of Chasseurs of *Lippe-Detmold, Lippe-Schaumbourg*, and *Waldeck*.

The Corps of Saxon Volunteers consisted of 6 Companies of Infantry,—2 Squadrons of Hussars,—2 Squadrons of Cavalry Chasseurs, —one Company of Sappers,—2 Squadrons of Dragoons,—and one Battery of six.

A Decree of THE KING of SAXONY authorizes the wearing of this mark of Honour, under the sole condition, that it's possessor shall not wear the riband *without* the Medal. The Saxons wear it attached to a riband of two white and three green stripes.

This Medal is worn at the button-hole,—on one side is the inscription, "*Union renders powerful,—Love of Country, invincible,*"—and, on the other, "*To the Volunteer Defenders of their Country of the 5th Corps of The German Army, their General-in-Chief,* ERNEST, *Duke of Saxony.*"

THE
DUCHY OF SAXE-GOTHA-ALTENBURG.

The Saxon Ernestine Order.

See, Saxe-Coburg-Gotha.

Medal.

The Ducal Medal of Saxe-Gotha-Altenburg was instituted in the year 1816, by the late Duke Emilius Leopold Augustus, of *Saxe-Gotha-Altenburg*, for his Federal troops, who had made the Campaigns of 1814 and 1815.

This Medal, which is worn at the button-hole, was distributed, in *bronze*, to the Soldiers,—and, in *bronze, gilt*, to the Officers and Persons of the same Rank. On one side, in the centre, is a Crown, of gold, with the inscription, "*For Bravery in the Field*,"—and, on the other, a rose, of gold, surrounded with an ornamental border, — in the Exergue, are the words, "Duchy of Gotha-Altenburg, MDCCCXIV, MDCCCXV." It is worn, pendent to a dark green riband, edged with a border of black and yellow.

THE DUCHY OF SAXE-MEININGEN-HILDBURGHAUSEN.

THE SAXON ERNESTINE ORDER.

See, SAXE-COBURG-GOTHA.

MEDAL.

THE MEDAL of SAXE-MEININGEN was instituted in 1816, for the troops of *Meiningen*, by The Duchess LOUISA ELEANORA, Mother of the reigning Duke, and destined for all those who had taken part in the Campaign of 1814 and 1815. One side of this Medal is exactly similar to THE CAMPAIGN MEDAL of *Saxe-Coburg*, representing a Maltese Cross, surrounded by an oaken crown,—the other side bears the inscription, "*To the Defender of his Country*," and, above, in a circle, "LOUISA ELEANORA, *Duchess Dowager of Saxony, first Governess and Regent of the Country*." The Medal is worn at the button-hole.

THE GRAND DUCHY OF SAXE-WEIMAR-EISENACH.

The Order of Vigilance, *or*, The White Falcon.

Charles Augustus, Grand Duke of Saxe-Weimar-Eisenach, *&c.*, keeping in remembrance the recovery of Independence, recently effected and secured to the whole German Empire by native energy and valour,—and being desirous on his part to bestow upon those, who either by word or by deed had signalized themselves in contributing toward it's achievement, some mark of Distinction worthy of their merits, the results of which had proved specially beneficial to His Grand Ducal House and to His Territories, had determined to restore, in it's most material features, but with such modifications as the altered circumstances of the times rendered desirable, The Order of Vigilance, *or*, The White Falcon, instituted on the 2d of August, 1732, with the approval of His then Imperial Majesty, and since worn by several Princes and other high State Personages, for the encouragement of Fidelity and for the reward of Patriotick feeling on the part of the Nation and her Chief,—Did accordingly renew the said ORDER, adding to it's original intent and meaning His resolve, that it should thenceforth be conferred upon such of his State Functionaries and Subjects, in recompense for faithful Services rendered to Him, as from their qualifications and merits he then considered, or might thereafter deem worthy of his especial favour,—

He thereupon ordained, at *Weimar*, on the 18th of October, 1815, that THE DUCAL ORDER of VIGILANCE, *or*, THE WHITE FALCON of SAXE-WEIMAR, instituted on the 2d of August, 1732, was from this day restored,—and should continue to be, the only GRAND DUCAL ORDER of SAXE-WEIMAR.

It consists of three Classes,—

The first Class comprehends The Grand Master, who is always to be the reigning Grand Duke of Saxe-Weimar, for the time being,—the Princes of The Grand Ducal House, and *Twelve* Knights Grand Crosses. No one in his Territories shall be eligible for The Grand Cross, until he shall have attained to the rank of a Supreme Privy Counsellor of State, or to that of Major General in the Army.

The second Class consists of *Twenty-five* Commanders,—for which no one shall be eligible, who has not attained to the rank of a Privy Counsellor of Government, to that of a Counsellor of State, of a Counsellor of Justice, of a Colonial Counsellor, or to the rank of a Major in the Military Service.

The third Class consists of *Fifty* Knights.

The principal duties of every Member of THE ORDER are,—

1. Fidelity and Devotion to his own Native Fatherland, and to the duly constituted National Authorities.
2. The constant cultivation and developement of National feel-

ing, National Art, and National Manners,—improvement in the general system of Legislation, of the Government, of the Constitution and of the Administration of Justice, as well as the diffusion of Truth and Knowledge upon sound principles, corresponding with the firmness of the German national character,—all this is to be the aim of every Member of The Order.

3. Every Member shall be bound to render practical Assistance to such of his oppressed Fellow citizens, as are suffering from the effects of War,—but more especially to those, who have been wounded in their Country's cause, and to the Survivors of those Warriors who have fallen in the struggle for their Native land.

As a Symbol suitable to the times, admonitory of the discharge of these obligations and actively urging on to *Vigilance* for the German weal, The Badge of The Order shall consist of a golden, white enamelled Falcon, with golden beak and talons, upon a golden octagonal, green enamelled Star,—between which is placed a red quadrangular Star, of smaller dimensions, with white enamelled points. Above the Star is a Royal Crown of gold,—and, on the reverse, is the octagonal Star in white, and the quadrangular Star in green, enamel.

In the middle is a blue enamelled shield, bearing the Motto, "*Vigilando ascendimus.*" This shield is surrounded with a golden laurel-wreath, for *Civilians*,—but, for the *Military*, with the emblems of war, surmounted by a Royal Crown, of gold.

The Silver Star belonging to The Badge, is of White Silver,—it is to be worn on the left breast. In the middle of it, is a flying Falcon, upon a golden ground, surrounded with the Motto, "*Vigilando ascendimus,*" in blue enamel. This golden circle rests upon the green enamelled Star, which latter rests upon the large silver one.

The Badge is worn by the Knights Grand Crosses, suspended over the right shoulder from a broad dark red, watered riband. The Star of The Order is attached to the left side of the breast.

The second Class wear The Badge, suspended from a narrower red riband, round the neck, whence it descends to the breast.

The Knights of the third Class wear a similar small Badge, suspended from the button-hole by a red riband.

THE RESTORATION of THE ORDER of VIGILANCE is celebrated annually, on the 18th of October, as a National Festival, commemorative of the Liberation of GERMANY from the miseries of a Foreign yoke. And, on this occasion, every Knight shall actively devote his energies, to ameliorate the sufferings of his Fellow Citizens.

THE ORDER has a Chancellor, and a Secretary. The Chancellor is always to be the Principal Grand Ducal Minister of State of Saxe-Weimar.

Immediately after the decease of a MEMBER of THE ORDER, his Successors are to send the *Insignia* of it to The Chancellor.

THE FALCON was received into ARMORIAL BEARINGS, and, in fact, into THE GREAT SEAL itself by The Emperor HENRY the Fourth. The *White* Falcons from *Iceland* and *Norway* were particularly esteemed. The attachment for this bird lasted for several ages, and even so late as the middle of the Eighteenth century,—so that it could not fail to be honoured, like it's companion the *Eagle,* and to be chosen as the Symbol of an ORDER of KNIGHTHOOD. Hence it was so greatly honoured, that Laws were enacted to secure it against snares and plots, as is now done in Germany, in the case of *Nightingales,*—for surely none, *but the vilest Epicure,* would think of *eating* these charming SONGSTERS.

MEDALS,

MEDAL FOR FAITHFUL WARRIORS.—Instituted by The Grand Duke CHARLES AUGUSTUS, of *Saxe-Weimar-Eisenach*. It is of *bronze*.

MEDALS FOR CIVIL MERIT.—The Grand Duke CHARLES AUGUSTUS, whilst in *Paris*, in 1815, caused a MEDAL to be struck, bearing on one side his effigy, and, on the other, "*Carolus Augustus Magnus Dux Saxoniæ*," or, "*Mitescunt aspera secula*." These Medals were distributed, as Medals of Civic Merit, in *gold*, *silver*, and *bronze*,—and are worn by express permission, attached to the red riband of THE ORDER of THE WHITE FALCON. This Medal becomes the property of the heirs of the possessors.

A smaller Medal, bearing on the reverse the words, "*Doctorum frontium præmia*," was also struck off at *Paris*, and distributed by THE DUKE, amongst the distinguished men of LEARNING.

The Grand Duke CHARLES FREDERICK, since the year 1829, conferred the small Medal of Civic Merit, with the effigy of the Grand Duke CHARLES AUGUSTUS on the obverse, and the words, "*Meritis nobilis*," surmounted by an oaken crown on the reverse. This Medal is distributed also in *gold* and *silver*, and is worn at the button-hole, attached to a riband of National colours.

THE KINGDOM OF SAXONY.

The Royal Saxon Military Order of Saint Henry.

The Royal Saxon Military Order of Saint Henry, which was destined to recompense the Merit of Officers of the Saxon Army, in time of war, was founded, on the 7th of October, 1736, at *Hubertsburg*, by Augustus the Third, King of Poland, in celebration of his Fortieth birth-day.

On the same day he decorated himself with The Order, and also The Prince Royal and several Generals, without any regard to the difference of their Rank. The distribution, however, rested there.

In 1763, the Royal Founder died, and his son the late Elector, having survived him only a few months, The Order became *dormant*.

On the 4th of September, 1768, His Royal Highness Prince Xavier, of Saxony, Tutor and Administrator of The Electorate, deemed it expedient to renew this

ORDER, as a reward for brave Actions, and as a proper recompense for the Officers who served in the troops of The Elector. The DECORATION was then changed, which the Prince and the Elector assumed for the first time, and distributed several Grand Crosses, as well as Crosses of Commanders, and those of an inferior Rank.

No grant of THIS ORDER, however, took place before the year 1796, at which period it may be considered to have been really instituted, by a distribution of seven Crosses for Knights,—and, ever since 1807, it has been granted in it's three Classes.

On the 23d of December, 1829, ANTHONY, King of Saxony, added to THIS ORDER, a new Class,—that of *Commanders* of the Second Class,—and gave to it STATUTES, in 15 paragraphs, to the following effect,—

THE ORDER bears the name of The Saxon Emperor, HENRY *the* SAINT, who reigned from 1002 to 1024,—and the collective Members of The Order are named " KNIGHTS of THE ROYAL SAXON MILITARY ORDER of SAINT HENRY,"

The Dignity of Grand Master belongs to, and remains inseparably attached to the person of The King of Saxony,—

The Members of THE ORDER are divided into Four Classes,—*viz.*, GRAND CROSSES,—COMMANDERS of the First Class,—COMMANDERS of the Second Class,—and KNIGHTS, whose number is indefinite.

All the Commanders nominated up to the date of the Statutes, belong to the First Class.

THE BADGE consists of a gold, eight pointed Cross, of the Maltese form, having a broad border of white enamel. In the centre is an Escutcheon of yellow enamel, and thereupon appears a Pedestrian figure of The Emperor SAINT HENRY, completely armed, and decorated with the ornaments and attributes of The Imperial Dignity, and his name, S. HENRICUS. On a blue circle, which surrounds the Escutcheon, are these words, "FRID. AUG. D. G. REX SAXONIÆ INSTAURAVIT." On the Reverse, is likewise a blue enamelled Escutcheon, and thereon is displayed on a shield, *per* fess *sable* and *argent*, two Swords placed crosswise. These Swords are surrounded with a crown of Laurel, and the whole is encircled with a fillet or border, of blue enamel, on which appears the MOTTO of The Order, "*Virtuti in Bello.*" The four principal angles of this CROSS are filled up with green branches, and garlands of *Rue*, which are the antient Devices of The Electoral House of Saxony.

There are three kinds of Decorations,—the largest, for the Grand Crosses,—a smaller, for Commanders.—and the smallest, for Knights.

The Grand Crosses wear the Decoration suspended by a sky blue silk riband, with a lemon coloured edge, four inches wide, over the right shoulder. They wear besides on the left breast, a radiated STAR, of eight points, embroidered in gold, four inches in diameter,—on the centre of which is depicted the obverse of the Badge, with the inscription "*Virtuti in Bello.*"

Commanders wear the Cross from a riband three inches wide, round the neck,—And the Commanders of the First Class bear the Jewel already mentioned, of a reduced size, with a narrower riband.

Knights wear their Cross, attached to a riband two fingers broad, at the second button-hole, with a rosette.

The Members of The Order are to wear the Decoration constantly. The King, as Perpetual Grand Master, and the Princes of the Blood, alone enjoy the right to have their Decorations enriched with precious stones, unless this favour should have been expressly granted.

The King alone nominates the Members of The Order. He takes, however, into consideration, the representations which may have been made to him by The Generalissimo of The Saxon troops, during a Campaign.

THIS ORDER, purely MILITARY, is exclusively destined to Officers serving in The Saxon Army, and is conferred without respect to the difference of Religion, noble Birth, or years of Service,—Merit alone, resting upon distinguished Actions in the field, joined to true Allegiance to King and Country, can open the way to the attainment of it.

But no Officer is allowed to ask for it,—Should officers, however, in the service of Foreign States render themselves deserving by Military Actions in the field, as respects the person of The King, the Royal Saxon Territory or Troops, in such case they are honoured with The Order.

According to the Regulations, one must be a Lieutenant General, and have commanded a Corps in time of War, to be nominated a Grand Cross. To be a Commander of the First Class, it is necessary to be either Lieutenant General or Major General, and to have commanded a Brigade against the Enemy. In fine, it is only after having attained the rank of a Superior Officer, and made, as such, several Campaigns, that the Cross of Commanders of the Second Class can be obtained. However, eminent Services rendered to the King's person, or *Fifty* years of honourable labour in the service of the State, make an exception to the usual order of Promotion.

The tie of Allegiance and of the most inviolable Fidelity to King and Country is rendered stronger to every Knight on his reception into this Order. It is, indeed, rendered imperative to him, to promote and sustain the Renown of the Army by his own example and zealous co-operation.

Should any Knight, contrary to the Royal expectation, through cowardly or dishonourable behaviour, render himself unworthy of The Order, inasmuch as the loss of The Order is not without being accompanied by a *legal Penalty*, according to circumstances,—A Commission of Knights of the various Ranks shall be constituted, upon whose Report the King himself shall decide.

The Members receive their Diploma, with the Statutes, signed by The King, and counter-signed by The Chancellor of The Order, and confirmed by The Great Seal, without any charge whatever.

The Nomination to this Order confers the privilege of annexing this Dignity to their Titles, and of joining the *Insignia* of it to their Achievement, in the following manner,—

1. For Grand Crosses, the Escutcheon is out-bordered on every side by the rays of the Jewel which lays underneath, the whole surrounded by the riband of The Order, on which is the Motto, and to which the Decoration is appended,—

2. For Commanders of the First Class, the Escutcheon is

placed on the Decoration, and surrounded by the narrower riband of The Order,—

3. For Commanders of the Second Class, the small riband of The Order, to which the Cross is attached, surrounds the Escutcheon.

4. For Knights, the Cross attached to a Rosette, is suspended from the Escutcheon.

On the death of a Member of The Order, the *Insignia*, together with the *Statutes*, are to be returned to The Chancery of The Order,—Also on Promotion to a higher Class of The Order, the Insignia previously received, are in like manner to be returned.

COMMANDERS OF THE FIRST CLASS.

The Right Honble. Sir GEORGE MURRAY.—*See*, AUSTRIA, ORDER of LEOPOLD.

Colonel E. KERSLEY.

Colonel HARVEY.

Colonel FREMANTLE.

The Royal Domestick Order of the Rue Crown of Saxony.

The Royal House of Saxony until the year 1807, possessed only "The Order of Saint Henry," which is confined to the Military. There was, therefore, no means in the hands of the reigning Prince, to confer Honorary Distinctions on deserving Statesmen of the higher Class, in the manner in which Sovereigns grant marks of approbation, by the conferring of a Domestick Order. For this reason, then, Frederick Augustus, King of Saxony, founded "The Order of The Green Crown," *or*, "Crown of Rue."

This was done on the 20th of July, 1807, just at the time when Napoleon Buonaparte, then Emperor of France, after the conclusion of the Peace of *Tilsit*, halted a few days in *Dresden*, on his return to Paris.

King Frederick Augustus was influenced also in his establishment of this Order, by gratitude for the Protection of Divine Providence experienced by Saxony, during the latter years of the War, to which the Motto, "*Providentiæ Memor*," bears allusion.

The Order consists only of one Class, and is conferred gratuitously, according to merit.

The Statutes are neither printed, nor published.

The first distribution of this Order took place on the 20th of July, 1807. The King is always The Grand Master,—and the Princes of the Blood, together with the Nephews of the reigning Monarch, are Knights of THE ORDER by birth.

The nomination to this Order, both for the Members of The Royal Family and other reigning Sovereigns, as well as for Foreign Dignitaries, is the exclusive right of The Grand Master.

THE DECORATION of THE ORDER consists of a Cross, of eight points, of a light green colour, with a white enamelled border,—the Escutcheon in the middle, on both sides, is surrounded by a wreath of *Rue*, of sixteen leaves,—on the obverse, this wreath encircles the initials, F. A., of the Founder, with a crown,—and, on the reverse, is the MOTTO, "*Providentiæ Memor.*"

This Cross is worn, pendent to a broad dark green riband, over the right shoulder. And, on the left breast, is embroidered a silver STAR, in the centre of which are also the words "*Providentiæ Memor,*" surrounded by a wreath of *Rue*.

KNIGHT.

ARTHUR, Duke of WELLINGTON, K.G.

The Royal Saxon Order of Civil Merit.

The Order of Civil Merit was instituted on the 7th of June, 1815,—after the conclusion of the Peace at *Presburg*, and the return of King Frederick Augustus to his Dominions,—in honour of the proofs of Fidelity, Attachment, and Zeal, which he had received from his Servants and Subjects, during his absence,—caused by the events that took place between the years 1813 and 1815.

In the Preamble to The Statutes, which are dated at *Dresden*, the 12th of August, 1815, His Majesty declares, that he had for a very long period entertained the intention of establishing an Order, for the reward of distinguished Civil Services,—and, during his absence from his Dominions, he experienced from his Servants and Subjects so many fresh proofs of loyal attachment, and prudence in the performance of their services, that he could no longer refrain from honourably and publickly distinguishing those, who had thus acquired peculiar claims to his consideration and gratitude,—whilst, at the same time, he encouraged, by a permanent Institution, an emulation in the service of his Country and of His Royal House. He had, therefore, decided upon establishing an Order of Civil Merit, and making the following Regulations respecting it,—

1. The right of conferring The Order, and of promoting in it, shall belong to the King exclusively.

2. Every Native, who shall have rendered beneficial services to the State, or otherwise distinguished himself by social virtues, may be admitted into The Order.

3. We reserve to ourselves the right of conferring The Order upon *Foreigners* also, if they have acquired a claim to Our gratitude, and to that of The State.

4. The Order shall be composed of Three Classes,—*viz.*, GRAND CROSSES,—COMMANDERS,—and KNIGHTS.

The *fourth* Class shall comprehend those, to whom THE CIVIL MEDAL has been granted.

5. THE ENSIGN of THE ORDER shall be an octagonal cross of gold, on white enamel, having in the middle of it, a round white Shield with a golden edge,—and, on the Obverse side, the Saxon Arms, with the inscription, " FREDERICK AUGUSTUS, *King of Saxony*, the 7th *June*, 1815,"—and, on the Reverse side, a wreath of Oak, surrounding the MOTTO, "*For Merit and Fidelity.*"

Note. His Majesty King ANTHONY has deemed it proper to direct, that peculiar Decorations bearing the MOTTO, "*For Merit*," instead of the above, should be prepared for those who may be nominated *Foreign* Members of THE ORDER.

THE CIVIL MEDAL shall contain Our Effigy on the Obverse side, with the same inscription as on the Obverse side of the shield of The Order. The Reverse side of The Medal shall resemble entirely that of the Decoration of The Order.

We have fixed the day of Our Return as that of the Establishment of The Order, with the view of founding a permanent Memorial of the cordial Love, with which We have been received by Our Subjects, and on the same day in future the conferring of The Order, and Promotion in it, shall take place as a rule.

6. THE CROSS of the FIRST CLASS shall be suspended from the right shoulder by a white, watered riband, with two grass-green stripes, four inches broad,—and there shall be worn besides on the left breast, a sexagonal radiating STAR of silver, with the Oak-wreath and Inscription, as above.

THE COMMANDERS shall wear the same Decoration round the neck, on a similar riband, three inches broad.

THE CROSS of the KNIGHTS, being of somewhat smaller dimensions, shall be fastened at the top to a hole by a similar loop, two inches broad.

THE MEDAL of MERIT shall be worn in a loop-hole, on the same riband.

7. These Decorations of The Order shall be borne by the Members continually, and with every Dress.

8. The Members of The Order shall add to their Titles the degree of Honour conferred on them by admission into the same, and shall be charged to annex the Insignia of The Order to their Arms,—that is to say, in the first Class, *the Star*,—in the second, *the Cross on the riband*,—and in the third, *the Cross in the loop*.

9. Admission into The Order shall impose upon all Members of it, an additional obligation to observe the most inviolable Fidelity to The Prince, and the Country,—and shall be an incitement to contribute, with all their power and according to the best of their judgement, to the Honour and Welfare of SAXONY. But, if the unhoped for circumstance should occur, that a Member of The Order becomes unworthy of the Distinction conferred upon him, a Council of The Order, to be summoned by Us for the purpose, shall deliberate upon his exclusion, and make a Report to Us of their opinion thereupon.

10. As We have reserved to Ourselves the nomination of a Common Chancellor for all our Orders, we will, therefore, commit to the same the direction of the Council of The Order of Civil Merit,—which Council shall be composed of two Grand Crosses, four Commanders, and a Secretary of The Order, besides The Chancellor above mentioned.

The Members of this Council shall be chosen by Us, and they shall be convened, at the requisition of The Chancellor of The Order, at least once in every year previously to the return of The Anniversary of The Order,—and, at other times, whenever a Meeting of them is considered necessary, in order to discuss the affairs of The Order. Moreover they may propose for Our selection, such persons as may be deserving of the grant of The Order.

11. A List of all those, upon whom The Order, or the Medal of Merit, has been conferred, shall be prepared,—which List shall contain, in addition to the name and time of Admission, also the Services of each Member, and be deposited among the Archives of The Order.

12. THE INSIGNIA of THE ORDER shall be returned to The Chancellor, after the death of the Members.

THESE STATUTES shall be for ever maintained by Us and Our Successors, and a copy of them shall be presented to each newly-admitted Member, for his guidance and observance.

The first distribution of THE ORDER was made on the 23d of December, 1815.

KNIGHT.

JAMES COLQUHOUN, Esquire, Consul General of Saxony, and Diplomatic Agent and Consul General of THE HANSE TOWNS in London, conferred upon him by His Majesty THE KING of SAXONY and His Royal Highness THE PRINCE CO-REGENT, " in acknowledgement of his peculiar and approved zeal in their Service, and for the interest of their Subjects," in having, as their Plenipotentiary, concluded a Treaty of Commerce and Navigation with the Republick of *Mexico*.—Royal Licence, 1st of March, 1836.

But I cannot allow this favourable opportunity to pass by, without further recording the amiable and exalted qualities of this excellent Man,—a MAN, who, it will be seen, has gained the endearing applause of *a whole Senate* for NATIONAL BENEFITS, " *in imitation of his Father's example*."—

On the 1st of October, 1828, THE SENATE of The Free Hanse Town of BREMEN, in consideration and acknowledgement of the many distinguished Services of JAMES COLQUHOUN, Esq., Diplomatic Agent and Consul General of The Hanse Towns in London, and Master of The Steel Yard, as well by his zeal in the fulfilment of the general duties of his Office, as by his particular care and successful management of several important Negotiations in which his endeavours to promote the Interest of THE HANSE TOWNS, their Commerce and Navigation have been so meritoriously conspicuous,—DECREED to invest him with the Honour of The Burgership *(Freedom)* of The City and Republick of BREMEN.

On the 18th of September, 1834, The Senate and Burgesses of HAMBURGH presented Mr. COLQUHOUN with the Privileges of an Honorary Citizen of The Free Hanseatic City of HAMBURGH, by Patent, written in the purest and most classical German, in a splendid Book, richly bound in Crimson velvet, and consisting of two Vellum leaves,—

On the outside of the first leaf are the Arms of HAMBURGH embossed in solid gold,—and on the outside of the second leaf, are those of the COLQUHOUN Family to correspond,—

On the first page is a representation of the General Assembly of The Burgomasters, Senators, and Citizens, with the actual Portraits of those then present. The four Burgomasters are seated in the centre,—The Senators are standing on their left,—and The Citizens in front of them. The Prothonotary of The Senate is placed between The Burgomasters and the Burgesses. The Costume is that of the Middle Ages. The period selected is, when the Senior Alderman is in the act of handing the Resolution, by which The Citizens approve of the Honorary Freedom granted to Mr. COLQUHOUN by The Burgomasters and Senate, to The Prothonotary.

This Picture is exquisitely executed by Mr. JULIUS MELDE, a Hamburgh Artist.

Over and below this Representation are Arabesques, allegorically expressing *Order* and *Liberty*, by children in peaceful attitudes and in excited action.

On the opposite page is THE FREEDOM, of which the following is a translation,—

" By virtue of a Resolution of The Senate and Burgesses adopted this day and in acknowledgement of the many meritorious Services, which JAMES COLQUHOUN, Esquire, Master of The Hanseatic Steel-Yard, and our Chargé d'Affaires and Consul General in Great Britain has, *in imitation of his Father's example*, rendered to this City during a period of Twenty years, by his prudent and unremitted exertions for her benefit, and more especially for the maintenance of her relations with the friendly Government of GREAT BRITAIN,—WE do, by the present Publick Document, award him the rights of an Honorary Citizen of The Free Hanseatic City of HAMBURGH.

Done in Hamburgh, on the 18th of September, 1834.

The Senate of The Free Hanseatic
City of Hamburgh,
for the same,
BARTELS, *J. U. Dr.*."

Over The Freedom is a Vignette, containing the Allegory of The Eleven Sources of *The Elbe*, represented by Eleven Nymphs, behind whom is the *Riesengebirge (Mountain of Giants)*, from which *The Elbe* has it's Source, represented by the head of a Giant.

Below the Grant of Freedom is a drawing, representing *Britannia* and *Hammonia* (the Tutelar Genius of *Hamburgh*) joining hands, as emblematical of the friendly relations of the two Countries, and surrounded by Tritons, Sea Nymphs, and Dolphins.

The Arabesques are executed with extreme elegance, by Mr. Erwin Speeter, a Hamburgh Artist.

The Book is incased in a beautiful Rosewood box, 21 inches long by 16 inches broad.

It may be hoped, that Mr. Colquhoun will consent to have this elegant specimen of Art engraved, to perpetuate the occasion of the Compliment, and to preserve the Portraits of the Honourable Donors.

On the 10th of October, 1835, The Senate of The Free Hanseatic City of Lubeck further conferred the quality of an Honorary Citizen of Lubeck upon Mr. Colquhoun, as a mark of the sincerest acknowledgement of the meritorious Services rendered by him, during many years, with efficacy and zeal, (especially for the interest of *Lubeck*), in all matters connected with his Official Appointments.

These Patents have not been conferred on more than three or four individuals,—that of Hamburgh is *unique*, and applies to Mr. Colquhoun only,—who, it is believed, is at present the only Honorary Citizen of the three ancient Republicks. They may be deemed "*The Order of Civil Merit for The Republicks.*"

Mr. Colquhoun has also been presented with The Hanseatic Medal, having the appropriate Motto, "*God was with us.*" It was given to those Officers, who distinguished themselves in The Hanseatic Legion, and was conferred on some British Officers. In his case it may be deemed a Civil Decoration, as occurs in regard to The Order of The Bath, The Guelphic, and others.

Mr. Colquhoun had likewise the honour to receive The Saxon Medal of Merit from the late King of Saxony, "as a mark of His Majesty's favour."

To these well deserved tokens of respect, Mr. Colquhoun has the highly gratifying reflection, that, in the course of his Official career, he has received *Seventeen* different Votes of Thanks from The Legislatures of The Islands, for which he has acted, and still continues to act, as Agent, — accompanied in many cases with votes of Pieces of Plate.

Medals.

The Gold and Silver Medal, founded by Frederick Augustus, and distributed for the first time, on the 17th of March, 1796, and destined to recompense the Military Merits of Privates and Subalterns, belongs as a *Fifth Class* to The Royal Military Order of Saint Henry,—the Regulations of which are embodied in the Statutes of that Order.

These Medals are granted in time of war, on the recommendation of The General in Chief of the Army,—and the distribution of them, which is made as publick as possible, is also announced to the Army. They display on one side the likeness of The Founder, with the inscription already mentioned,—And, on the Reverse, within a wreath adorned with warlike instruments, the words, "*Well deserved of The Country.*" They are worn at the second button-hole, with the riband of The Order of Saint Henry, but less wide by one-third.

The Gold Medal is of the intrinsic value of 12 ducats. A gratification of 100 thalers, for the Gold Medal, and one of 25 thalers, for the Silver Medal, are granted to the widow, children or heirs of the individual who has borne those Decorations. An *Exploit* more or less brilliant, never the *Rank* of the person, gives a claim to the attainment of either of those Medals.

The Civil Medal,—representing on one side the Founder, with the inscription, "*Frederick Augustus, King of Saxony, the 7th of June,* 1815," and on the other, a wreath of Oak, with the Motto, "*To Merit and Fidelity,*"—is the *Fourth Class* of The Order of Civil Merit. It is to be worn, attached to the same riband, at the button-hole, and is granted in *Gold* (8 ducats in value), or in *Silver.*

The Medal of Merit, with the inscription, "*Bene Merentibus,*" which is given in gold and silver; is not intended to be worn publickly,—but it is at the option of every one to wear it, attached to any riband they please.

There are, besides these, several kinds of MEDALS of high value,—but which are not designed as DECORATIONS,—as for instance, the MEDALS indicating the number of years of Service,—those, granted for saving the life of a person, instituted by a Decree of the 18th of May, 1831,—and the Decorations granted to the Chapters of *Meissen, Wurzen, Budissen* and of *Joachim-Stein.*

THE KINGDOM OF THE TWO SICILIES.

THE ORDER OF SAINT JANUARIUS.

THIS ORDER was founded by CHARLES, King of NAPLES and SICILY, afterwards King CHARLES the Third, of SPAIN, on the 6th of July, 1738, on occasion of his Marriage with Princess AMELIA, daughter of AUGUSTUS the third, King of POLAND.

THE KING of THE TWO SICILIES is Grand Master of THE ORDER, and names the Knights, who form only one Class,—and, according to the original intention, they were to be limited to *Sixty*, but their number is now indefinite.

On the Usurpation of The Kingdom of Naples by the *French*, in 1806, THE ORDER was abolished in that Country,—but still flourished in *Sicily*, and, since the year 1814, in both of these Countries.

THE BADGE is a gold Cross, of eight points, enamelled white, and golden Lilies in the four angles. On the Obverse is represented a half-length of the celebrated Patron of *Naples*, and the Patron of The Order, the Holy Martyr *St. Januarius*, in Episcopal attire,—in his left hand he holds an open book, and, in the right, a Bishop's crozier,—on the reverse, in the centre, is figured a gold book, and two phials half filled with the blood of Martyrs, and below is the MOTTO, "*In sanguine Fœdus.*"

THE KNIGHTS wear THE BADGE, pendent to a broad, rich, poppy coloured, watered riband, sashwise over the right shoulder,—and, on the left breast, a Silver STAR, of eight points, with golden Lilies in the angles,—in the middle, is a four pointed Silver Cross, on the border of which is the MOTTO, "*In sanguine Fœdus.*"

On occasions of Ceremony the Knights appear in a DRESS, which consists of a purple coloured Mantle, strewed with golden Lilies, lined with white taffety, and adorned with ermine. The Coat and Vest are of cloth of silver,—and on the head is worn a black Hat, with a white feather. In this Dress, THE BADGE is worn, suspended from a gold COLLAR, round the neck, which is composed alternately of mitres, croziers, and the letter B.

All persons upon whom THIS ORDER is conferred, must prove the *Nobility* of their descent for four Centuries, and they are addressed by the Title of, "*Excellency.*"

KNIGHTS.

Field Marshal His Grace The Duke of WELLINGTON, K. G., &c.,—*See*, THE TWO SICILIES, ORDER of ST. FERDINAND and of MERIT.

GEORGE MARTIN, ESQ., Vice-Admiral of the Blue Squadron of His Majesty's Fleet, and Commanding Officer of His Majesty's Ships and Vessels on the Coast of *Sicily*, G. C. B., "as a testimony of the sense which His Sicilian Majesty entertains of his Services."—Royal Licence, 6th of July, 1811.

Sir JOHN ACTON, Bart., many years Prime Minister to the King of Naples.—This gentleman commanded *The Tuscan Gallies* in the expedition of Comte O'REILY, against *The Dey of Algiers*.—Died, 12th of August, 1811.

Sir JOHN STUART, *Count of Maida*, in Calabria, Lieut. General in The

Army, K. C. B., Member of The Imperial Ottoman Order of The Crescent, of the first degree,—and late Commander of the Forces in Sicily and The Mediterranean, to accept and wear the Insignia of a Knight of The Royal Sicilian Order of *Saint Januarius*, which His Majesty FERDINAND the Fourth, King of The Two Sicilies, was pleased to confer upon him in the year 1810, as the highest personal mark of His Royal favour, "in testimony of the sense which His Majesty entertained of his great meritorious services in the command of The British Forces, during the memorable and brilliant events which led to the final security of The Kingdom of Sicily against the formidable attacks of a powerful enemy."—Royal Licence, 16th of December, 1814. —Died, 1st of April, 1815.

Sir CHARLES PHILLIPS, Knt., Major General of His Majesty's Forces, "as a testimony of the high sense which His Majesty The King of The Two Sicilies entertains of the distinguished Services rendered by him in the defence of *Sicily*, and in the expedition to the Coast of *Naples*, in 1815." — Royal Licence, 7th of March, 1817.

Sir JOHN MURRAY, Bart., G. C. H., Lieut. General of His Majesty's Forces, and Colonel of the 3d West India Regt. of Foot, "as a testimony of the approbation of His Majesty The King of The Two Sicilies of the Services rendered by him while commanding His Sicilian Majesty's troops in *Sicily*, and particularly for his distinguished conduct at the Battle of *Castalla*, in Spain, where he commanded an armed force, of which a Corps of His Sicilian Majesty's troops formed a part."—Royal Licence, 2d of August, 1817.

ALEXANDER MACKENZIE, Esq., a Lieut. General in The Army, and G. C. H., "in testimony of the approbation of His Majesty, FERDINAND the Fourth, King of The Two Sicilies, of his distinguished services in *Sicily*, during the years 1808, 1809, and 1810, and particularly of his conduct during the period that he was enabled with the Corps under his command to take possession of *Lower Calabria*, when the Division of *The French* army, under the orders of General POULONEUX, was obliged to evacuate the Fortress of *Scylla*, and the Military positions which they occupied in that Country."—Royal Licence, 23d of May, 1818.

Rear-Admiral Sir DAVID MILNE, K.C.B., to accept and wear The Insignia of The Royal Sicilian Order of ST. JANUARIUS, and also the Cross of a Knight Commander of The Royal Military Order of WILLIAM of The Netherlands, with which their Majesties The King of The Two Sicilies and The King of The Netherlands have been pleased to honour him, " in testimony of the high sense which those Sovereigns entertain of the able conduct and signal valour manifested by him as Second in Command at the memorable Attack upon the Town and Shipping of *Algiers*, on the 27th of August, 1816."—Royal Licence, 3d of October, 1820.

JAMES OCONCHAR, Baron FORBES, a General in The Army, and Colonel of the 21st Regt. of Foot (*or*, Royal North British Fusileers), " for his distinguished conduct, while serving in the Army in *Sicily*."—Royal Licence, 1st of June, 1830.

The Order of Constantine.

The History of The Order of Constantine, *or*, of The Constantine Order of St. George, has already been related under the head of The Grand Duchy of Parma,—where it is stated to have been transferred, in 1734, from *Parma* to *Sicily*,—and to have also flourished in *Parma*, since the year 1816.

It may now be only necessary further to remark, that this Order, with the other Sicilian Orders, was abolished, in 1806, by the Ephemeral King, Joseph Buonaparte, in *Naples*,—but that it still flourished in *Sicily*, and, on the expulsion of *The French* from *Naples*, in 1814, was re-instated in all it's ancient rights.

The reigning King of The Two Sicilies is Grand Master of The Order.

The Knights consist of three Classes, — *viz.*, Grand Crosses, Knights, and Serving Brothers,—the two latter Classes are again divided into several branches.

The Badge is precisely the same, as that which is described under the head of Parma. The Grand Crosses and Knights wear it round the neck, and embroidered on the dress. The former are only distinguished by having *The St. George* appended to the lower point of the Cross.

To attain to the first Class, the conditions required are, ancient, real *Nobility*, the Catholick religion, a certain degree of Wealth, and an Age of at least 16 years,—

To apply themselves to the practice of Virtue, to follow the Grand Master in war, to maintain two Soldiers at their own expense, to appear constantly with their Sword, *to play no Games of chance,* and *to pursue no Civic occupation,* are obligations imposed upon the Upper Classes.

The Collar is the same as that of the Knights of Parma. The Dress of Ceremony is very splendid.

Knights.

Captain William D'Arley, of His Majesty's Marine Forces, to accept the Rank of Knight of The Royal and Military Order of Constantine, conferred upon him by Ferdinand the Fourth, King of The Two Sicilies.—Royal Licence, 7th of December, 1801.

Sir John Carr, Knt., to accept the Rank and wear the Insignia of a Knight of The Royal and Military Order of Constantine, conferred upon him by His Majesty Ferdinand the Fourth, King of The Two Sicilies, " as a testimony of his Royal regard and esteem."—Royal Licence, 15th of September, 1810.

THE ROYAL SICILIAN MILITARY ORDER OF SAINT FERDINAND AND OF MERIT.

When THE ENGLISH took possession of the City of *Naples*, on the expulsion of the *French*, on the 26th of June, 1799, King FERDINAND the Fourth, who had retired to *Sicily*, was triumphantly brought back, on the 10th of July following,—when arriving in the Bay of *Naples*, he immediately hoisted his Standard on board *The Foudroyant*, Lord NELSON's Ship, where he remained with all his Ministers.

On the 1st of August, 1799, Lord NELSON wrote to Mr. NEPEAN, "I most sincerely congratulate their Lordships on the entire liberation of the Kingdom of Naples *from the French* ***Robbers***, for by no other name can they be called, for their conduct in this Kingdom."

To commemorate this remarkable event in the most distinguished and permanent manner, His Majesty, FERDINAND the Fourth, founded, on the 1st of April, 1800, "THE ORDER of SAINT FERDINAND and of MERIT."

THE STATUTES of foundation expressly declare, that THE ORDER is established, with a view to give a publick proof of gratitude to God, and the beloved Protector *St. Ferdinand*, King of CASTILE,—who lived in the 13th Century, was Cousin German of *St. Louis*, and a man of equal eminence,—for the complete re-conquest of the Kingdom, to reward those Subjects who, on this occasion, had displayed extraordinary proofs of fidelity and devotion to the Royal Person and Monarchy,—and, in future, to excite and inspire the Subjects, and especially *the Nobility* of The Two SICILIES, to the feeling of Honour and of true Glory.

On the *second* conquest of *Italy* and *Naples* by the *French*, in 1806, THIS ORDER, together with the other ROYAL ORDERS in the Kingdom of *Naples*, was abolished,—it, nevertheless, flourished with it's brethren in *Sicily*, whither the Neapolitan Court had retired,—and, since the year 1814, it has shone forth again with splendour in THE UNITED KINGDOMS.

At first it consisted only of two Classes,—*viz.*, GRAND CROSSES, and COMMANDERS,—but, on the 25th of July, 1810, His Majesty added to it, KNIGHTS, as a third Class.

The reigning King is always Grand Master and Chief of THE ORDER. He alone appoints the Members, of whom, according to the Statutes, the first Class, including the Royal Family, is limited to *Twenty-four*. For the other two Classes no definite number is fixed.

THE BADGE is a Star, of six branches of gold rays, and six intermediate silver Lilies, surmounted with a Royal crown. On the Obverse, on a gold ground, is the image of *St. Ferdinand*, in Royal attire and mantle, with the Crown upon his head, a drawn Sword in his right hand, and a Crown of Laurel in his left,—around, in a dark blue circle, in golden letters, is the MOTTO, "*Fidei et Merito.*" On the Reverse, the words, "*Ferd.* IV. *inst. anno* 1800," occupy the golden space of the circle.

This Badge, which is extremely elegant, and is the same for all Classes, but differs in size with every Class, is worn by The Grand Crosses, pendent to a broad dark blue riband, with a red border, being the colours of The Royal House of BOURBON, sashwise over the right shoulder,—and also, on the left breast, a STAR, exactly like the obverse of THE BADGE.

THE COLLAR consists of a Castle, the Letter F., and a Crown, resting upon military trophies, and between each of them is a *fleur de lis.* With this COLLAR, the Knights are bound to adorn their COATS of ARMS. They receive the Title of "*Excellency,*"—have the same *Entrée*, as the Chamberlains on duty,—take their place, on occasions of Ceremony, close to the steps of the Throne, on the right hand, and *cover* themselves in the presence of The King, in all circumstances in which the same privilege is granted to Spanish Grandees of the first Class. A General, who, as Commander in a battle, has gained a complete victory, shall "*de jure*" receive the Grand Cross.

THE COMMANDERS wear THE BADGE, round the neck, without the STAR on the breast. On occasions of Ceremony their place is next to the Grand Crosses, and they take rank before their Colleagues, both Civil and Military, who do not possess THE ORDER. They also receive *Pensions*, according to the pleasure of The King. He who has so defended a fortified place, as to have compelled the enemy to raise the siege,—or, who has conquered a fortress occupied by the enemy, has "*de jure*" claims to the Cross of Commander, with the assignment of a yearly income from the Revenues.

THE THIRD CLASS are decorated with a small Cross, which is worn, suspended from the button-hole of the cape of the coat, with a riband, one third part narrower in breadth than that of The Commanders.

THE OFFICERS of THE ORDER, consist of a CHANCELLOR, MASTER of THE CEREMONIES and a SECRETARY. These wear the Cross, on a narrower riband, round the neck, and the STAR on the breast, not with a *gold* ground, but entirely embroidered with *silver*.

Previous to Lord NELSON's leaving Sicily, in June 1800, The King invested him with THE ORDER of ST. FERDINAND and of MERIT, of which His Lordship was the *first* Knight, after The Royal Family.

His Majesty also, on the 25th of July, 1810, instituted the honourable distinction of a MEDAL of THE ORDER, in *gold*,—and another, in *silver*,—as a reward of Merit,—to be worn in like manner from the button-hole, and by a riband of the same breadth as that of The Knights,—and to which, according to circumstances and rank, a certain monthly *Pension* was to be added.

GRAND CROSSES.

The Right Honble. HORATIO, Viscount and Baron NELSON, *of the Nile*, K. B., and Vice-Admiral of The Blue,—and also Duke of BRONTE, *in Sicily*.

Sir WILLIAM SIDNEY SMITH, Knight, Commander and Grand Cross of The Royal Swedish Military Order of The Sword, and Rear-Admiral of The Blue Squadron of His Majesty's Fleet, " as a gracious mark of His Sicilian Majesty's favour."—Royal Licence, 25th of August, 1807.

The Right Honble. EDWARD, Viscount and Baron EXMOUTH.—*See*, SPAIN, ORDER of CHARLES the Third.

The Right Honble. Lord BURGHERSH, His Majesty's Envoy Extraordinary and Minister Plenipotentiary at The Court of Tuscany, Colonel in The Army, and Aid-de-Camp to His Royal Highness The Prince Regent, " in testimony of the approbation of His Majesty THE KING of THE TWO SICILIES of the services rendered by him in the field of Battle, during the Military operations which led to *The Restoration* of His Sicilian Majesty to the Throne of *Naples*."—Royal Licence, 27th of July, 1815.

ROBERT MACFARLANE, Esq., Lieut. General of the Forces, Colonel of the 97th Regt. of Foot, and G.C.H., " in testimony of the high sense which FERDINAND the Fourth, King of The Two Sicilies, entertains of the distinguished services rendered by him throughout the different operations against the Enemy on the Coasts of *Italy*, more particularly in command of the Sicilian Troops at the assault of the entrenchments and capture of *Genoa* in the year 1814, and in command of the Allied British and Sicilian Army in the expedition to the coast of *Naples*, in 1815."—Royal Licence, 20th of January, 1817.

Field Marshal His Grace The Duke of WELLINGTON, K. G., *&c.*, to accept and wear The Insignia of a Knight Grand Cross of The Royal Sicilian Military Order of SAINT FERDINAND and of MERIT, and also of a Knight of The Royal Sicilian Order of SAINT JANUARIUS, conferred upon His Grace by His Majesty THE KING of THE TWO SICILIES, " in testimony of the high sense which that Sovereign entertains of his distinguished merits and brilliant achievements."—Royal Licence, 16th of July, 1817.

Lieut. General Viscount BERESFORD, G.C.B.

Sir THOMAS FRANCIS FREMANTLE, G.C.B., G.C.H., Knight Commander of The Imperial Military Order of MARIA THERESA of Austria, and Rear Admiral of The Red Squadron of His Majesty's Fleet, " as a mark of the approbation of His Majesty FERDINAND the Fourth, King of The Two Sicilies, of his active and distinguished services before the Enemy while commanding a Squadron of His Majesty's Ships in *The Adriatic*, during the late war."—Royal Licence, 2d of January, 1819.

Rear Admiral Sir CHARLES VINICOMBE PENROSE, K.C.B., and Knight Grand Cross of The Most Distinguished Order of *St. Michael* and *St George*, " in testimony of the approbation of His Majesty FERDINAND the Fourth, King of The Two Sicilies, of his services against the Enemy, and more particularly for his meritorious conduct when commanding the Combined British and Sicilian forces directed against the Fortress of *Gaeta*, in the year 1815."—Royal Licence, 27th of August, 1819.

COMMANDERS.

Sir THOMAS TROUBRIDGE, Bart., Captain in The Royal Navy,—Royal Licence, 7th January, 1801.—*Re-conquest* of *Naples*, in 1799.—Lost at sea, in 1805.

Captain ALEXANDER JOHN BALL, of The Royal Navy.—Royal Licence, 7th January, 1801.—*Re-conquest* of *Naples*, in 1799. Created a Baronet, in 1801. Died, on the 20th of October, 1809.

Captain SAMUEL HOOD, of The Royal Navy, G. C. B.—Royal Licence, 7th January, 1801.—*Re-conquest* of *Naples* from the French, in 1799. Created a Baronet, in 1809. Died, 13th of June, 1815.

Captain BENJAMIN HALLOWELL, of The Royal Navy.—Royal Licence, 7th January, 1801. Services in the *Re-conquest* of *Naples*, in 1799. In 1828, he took the Surname of CAREW,—Was Vice-Admiral of the Red, and K. C. B.

Captain Thomas Louis, of The Royal Navy.—Royal Licence, 19th of April, 1802.

Robert Hall, Esq., a Post Captain in The Royal Navy, and a Brigadier General in the Service of His Majesty Ferdinand the Fourth, King of The Two Sicilies, "for the great courage and intrepidity displayed by him in several Attacks on the Enemy's *Flotilla* on the Coast of *Calabria*."—Royal Licence, 11th of March, 1813.—Knighted, in 1816.—Died, 7th of February, 1818, Commissioner of the Navy in Canada, aged 39.

Alexander Bryce, Esq., a Colonel in the Army, and Lieut. Colonel of The Royal Engineers, "in testimony of the high sense which His Sicilian Majesty entertains of his distinguished services, whilst employed with the British Army in the defence of The Kingdom of *Sicily*."—Royal Licence, 24th of August, 1814.

Lieutenant Colonel Charles Ashe à Court, late Adjutant-General of His Majesty's Forces serving in Sicily, and Major in The Greek Light Infantry Regiment,—Royal Licence, dated 5th of March, 1816, to accept and wear The Insignia of a Commander of The Royal Sicilian Military Order of St. Ferdinand and of Merit,—and the Insignia of The Royal Piedmontese Order of St. Maurice and St. Lazare,—which their Majesties The King of The Two Sicilies, and The King of Sardinia had respectively conferred upon him, "for the distinguished Services rendered by him at the period of the landing of the French Division in *Sicily*, and at the reduction of *Genoa*."

William Charles Fahie, Esq., Post Captain in The Royal Navy, Colonel of the Portsmouth Division of His Majesty's Royal Marine Forces, and C. B., "for the distinguished Services rendered by him, while commanding The British Squadron (acting in co-operation with the Austrian Forces) at the reduction of *Gaeta*."—Royal Licence, 11th of March, 1816. Afterwards Rear-Admiral of The Red, and K. C. B.

James Campbell, Esq., *(of Inverness)*, Lieut. General in The Army, "in testimony of the approbation of His Majesty The King of The Two Sicilies of the highly distinguished services rendered by him in the Field, as well in the repulse of an Attack of the

Enemy's troops near *Messina*, on the 18th of September, 1810, by the forces under his command, as during the other Military operations of the late War in the Kingdoms of *Naples* and *Sicily*."—Royal Licence, 5th of July, 1816. A Baronet, and G. C. H. Died, in 1819.

JOHN LEMOINE, Esq., Colonel in The Royal Regiment of Artillery, and C. B., "in testimony of the high sense which His Majesty FERDINAND the Fourth, King of The Two Sicilies, entertains of his signal services in the Field, while commanding the Royal Artillery in the Neapolitan Dominions during a period of nine years, particularly at the memorable Battle of *Maida* in the year 1806, and the repulse of an Attack of the Enemy's troops near *Messina*, on the 18th of September, 1810." — Royal Licence, 6th of September, 1816.

JOHN TOUP NICOLAS, Esq., a Post Captain in The Royal Navy, and C. B., "in testimony of the approbation of His Majesty FERDINAND the Fourth, King of THE TWO SICILIES, of the distinguished intrepidity manifested by him, while commanding His Majesty's Sloop, "*Pilot*," in effecting the capture and destruction of several of the Enemy's Vessels in various Actions on the East and West Coasts of *Calabria*, in the years 1810, 1811, and 1812"—Royal Licence, 27th of September, 1816.

ROBERT TRAVERS, Esq., Colonel in The Army, Lieut. Colonel of the 10th (*or*, North Lincolnshire) Regt. of Foot, and C.B., "as a testimony of the high sense which His Majesty The King of The Two Sicilies entertains of the distinguished gallantry displayed by him in the Field in different affairs with the Enemy, while commanding the advanced guard of The Anglo-Sicilian Army on the Coasts of *Italy*."—Royal Licence, 20th of November, 1816. Afterwards Major-General, and Knighted.

The Honble. FREDERICK WILLIAM AYLMER, Post Captain in The Royal Navy, and C.B., "in testimony of the sense which His Majesty The King of The Two Sicilies entertains of the distinguished gallantry displayed by him as Captain of His Majesty's Ship "*Severn*," at the memorable attack upon the Town and Shipping of *Algiers*, on the 27th of August, 1816."—Royal Licence, 7th of March, 1817.

William Gosset, Esq., Lieut. Colonel in The Corps of Royal Engineers, and C.B., " in testimony of the high sense which His Majesty The King of The Two Sicilies entertains of the distinguished intrepidity and able conduct displayed by him in the memorable attack upon the Town and Shipping of *Algiers*, on the 27th of August, 1816."—Royal Licence, 7th of March, 1817.

John Coode, Esq., Post Captain in the Royal Navy, and C.B., " as a mark of the approbation of His Majesty The King of The Two Sicilies, of the highly distinguished gallantry and able conduct displayed by him as Captain of His Majesty's Ship, " *The Albion*," in the memorable attack upon The Town and Shipping of *Algiers*, on the 27th of August, 1816, upon which occasion he was severely wounded."—Royal Licence, 6th of November, 1818.

Lieut. Colonel Richard Church, C.B., and a Lieut. General in The Sicilian Service, to accept and wear The Insignia of a Commander of The Royal Sicilian Military Order of St. Ferdinand and of Merit, and also The Grand Cross of The Royal Neapolitan Military Order of St. George of The Reunion, with which His Sicilian Majesty has been pleased to honour him, " in testimony of His Royal approbation of the distinguished services rendered by him during the Military operations, which led to *the Restoration* of that Sovereign to the Throne of *Naples*."—Royal Licence, 4th of October, 1820.

Lieutenant Colonel Gideon Gorrequer, Knight of The Royal Hanoverian Guelphic Order, " for his distinguished conduct in face of the Enemy, during the occupation and defence of *Sicily*, and at the taking of *Genoa* in 1814, on which occasion The Sicilian Troops were employed in conjunction with those of England, under the command of Lieut. General Lord William Bentinck." —Royal Licence, 1st of January, 1830.

KNIGHTS.

THOMAS STAINES, Esq., a Post Captain in The Royal Navy, and Commander of His Majesty's Ship "*Cyane*," "for the great skill and intrepidity displayed by him, in an engagement between *The Cyane*, under his Command, and the *French* Frigate *Ceres*, and several of the Enemy's Gun-boats, near the Mole of *Naples*, on the 27th of June, 1809."—Royal Licence, 21st of November, 1809.—Knighted, on the 6th of December, 1809, and K.C.B.

Sir JAHLEEL BRENTON, Bart.

WILLIAM HENRY SMYTH, Esq., Commander (*now*, Post Captain), in The Royal Navy,—" which His Majesty the King of THE TWO SICILIES has conferred upon him, as a testimony of the high sense which that Sovereign entertains of his Services, and of the Valour displayed by him against the Enemy *off Messina*, during the period in which he served in THE COMBINED FLEET."—Royal Licence, 16th of March, 1816.

JOHN HILTON, a Lieutenant in the Navy, and late Senior Lieutenant of His Majesty's Sloop, "*The Bustard*," "for the great courage and intrepidity displayed by him in various Actions with the Enemy's Vessels near *Messina*."—Royal Licence, 23d of September, 1811.

GEORGE CHARLES HOSTE, Esq., a Captain in the Corps of Royal Engineers, "for the great courage and intrepidity displayed by him in the action fought by "*The Spartan*" Frigate, against a Squadron of the Enemy in the Bay of *Naples*, on the 3d of May, 1810."—Royal Licence, 27th of November, 1811.

GEORGE WICKENS WILLES, Esq., a Commander in The Royal Navy, "for the great courage and intrepidity displayed by him as First Lieutenant of His Majesty's Ship "*Spartan*," in the Action fought by that Frigate, against a Squadron of the Enemy in the Bay of *Naples*, on the 3d of May 1810."—Royal Licence, 26th of June, 1812.

THOMAS READE, Esq., a Major in the Army, Captain in the 27th (*or*, Inniskilling) Regt. of Foot, and Assistant Quarter-Master General of the Forces in Sicily, "for his distinguished conduct, on divers occasions, whilst in the command of a *Flotilla* of Gun-boats, in the service of the British Army, in and near The Straits of *Messina*, during the attempted Invasion of the Kingdom of Sicily by the French Forces, in the year 1810."—Royal Licence, 30th of January, 1813.—Afterwards Lieut. Colonel, Deputy Adjutant-General at *St. Helena*, and C.B.—Knighted, 27th of November, 1815.

Lieutenant RICHARD MOORMAN, (*now*, Commander,) of The Royal Navy, "for the great courage and intrepidity displayed by him in several Attacks against the Enemy, near *Messina*."—Royal Licence, 4th of March, 1813.

FRANCIS LE HUNTE, Esq., Captain of The Royal Navy, "in testimony of the high sense which His Majesty FERDINAND the Fourth, King of The Two Sicilies, entertains of the valour and intrepidity displayed by him (then Lieutenant), in action with the Enemy at *Pietra Nera*, on the Coast of *Calabria*, on the 15th of February, 1813."—Royal Licence, 14th of November, 1814.

Lieutenant JOHN TURNER FLIN, of The Royal Navy, "as a testimony of the high sense which His Majesty FERDINAND the Fourth, King of The Two Sicilies, entertains of his distinguished intrepidity in Action with the Enemy, off *The Coast of Sicily*."—Royal Licence, 5th of March, 1816.

WILLIAM PLUNKETT DE BATHE, Esq., Major in the Army, and Captain in the 85th, (*or*, The Duke of York's Own) Regt. of Light Infantry, (*or*, Bucks Volunteers), "in testimony of the approbation of His Majesty FERDINAND the Fourth, King of The Two Sicilies, of the distinguished intrepidity displayed by him, while serving in *Sicily* and *Calabria*, as Aid-de-Camp to the late Lieut. General Sir JOHN STUART, *Comte of Maida*."—Royal Licence, 10th of July, 1818.

Captain RICHARD CUST, of the 31st (*or*, Huntingdonshire) Regt. of Foot, "as a mark of the approbation of His Majesty The King of The Two Sicilies of the distinguished intrepidity displayed by him in several Actions with the Enemy, particularly in the memorable and glorious Victory achieved over the *French* forces on the Plains of *Maida*, on the 4th of July, 1806, (upon which occasion he served as a Major of Brigade to the Forces,) and, subsequently, at the siege and capture of *Ischia* and *Procida*."—Royal Licence, 31st of October, 1818.

Lieutenant WILLIAM COLIN CLARKE, of The Rifle Brigade, "in testimony of the approbation of His Majesty FERDINAND the Fourth, King of The Two Sicilies, of the distinguished Services rendered by him, while serving as Aid-de-Camp to Lieut. General Sir ROBERT MACFARLANE, K.C.B., at the assault of the Entrenchments and capture of *Genoa*, in the year 1814,—and also, in the expedition to the Coast of *Naples*."—Royal Licence, 4th of February, 1819.

RICHARD H. FLEMING, Esq., Lieutenant (*now*, Commander) in The Royal Navy, "in testimony of his distinguished conduct and gallant services, while commanding His Neapolitan Majesty's Sloop "*Gieuseppe*," at *Gaeta*.—And, also, for having personally delivered the Dispatches announcing His Majesty's RESTORATION to His Kingdom.

The Royal Neapolitan Military Order of Saint George of The Reunion.

When, on the conquest of The Kingdom of Naples, by the *French*, in 1806, Joseph Buonaparte ascended that Throne, he abolished, together with all that could remind him of the previous Dynasty, the three hitherto existing Royal Orders of Knighthood, which have just been described,—they, however, flourished under the protection of their Grand Master in *Sicily*, where neither his own arm, nor that of his powerful brother, Napoleon, were able to reach them.

The new King was soon convinced, how necessary and useful in the hands of a Sovereign is such a means of conferring external Distinction and Honour, and felt how indispensable it was in particular to him, as an *Usurper*, to procure friends and partizans, in order to establish his existence firmly upon the Throne.

He, therefore, on the 28th of February, 1808, founded an Order, which he intended as a reward to those who had rendered him service, or who had traiterously laboured for his benefit in the Conquest of the Country, and which he arrogantly named, " The Order of The *Two* Sicilies," although he was master of only *one* of these Kingdoms.

The three Classes, into which THE ORDER was divided, were styled,—

DIGNITARIES,—60.,—COMMANDERS, 100.,—and Knights, 600.

Every one took, upon his Admission, an Oath, to devote his life and goods to the defence and glory of the Crown and the State,—and enjoyed *Revenues*, which were derived from those of *the suppressed Orders.*

The successor of JOSEPH BUONAPARTE on the Throne of *Naples,* his Brother-in-law JOACHIM MURAT, preserved THE ORDER with some unimportant alterations,—and it retained it's original constitution, until the year 1815, when King FERDINAND the Fourth returned from *Sicily,* and again became SOVEREIGN of BOTH KINGDOMS.

This memorable event, so honourable to BRITISH valour, is thus communicated by WILLIAM A'COURT, Esq., His Majesty's Envoy at the Court of *Naples,* to Viscount CASTLEREAGH, in a Dispatch, dated the 17th of June, 1815,—

" I have the satisfaction to inform your Lordship, that HIS SICILIAN MAJESTY made this day His Publick entry into His Capital, after an absence of nine years. The crowd that thronged the road all the way from *Portici* was immense, and nothing could exceed the enthusiasm of the people on the appearance of their legitimate Monarch. It was impossible to mistake the Publick feeling upon this occasion. The theatrical processions of MURAT drew crowds, as I am told, of curious Spectators, but curiosity was not the inducement here,—in every countenance might be read the honest expression of heartfelt joy, at the return of a beloved and native Sovereign.

" His Majesty was received, on His arrival at THE PALACE, by all the Principal Nobility of the Country, the great majority of whom appeared to partake of the enthusiasm which had been previously demonstrated by the lower Classes. In fact, never was National joy so unequivocally and so universally displayed."

And Lord BURGHERSH, in a dispatch of the same date, adds,—

"Having received the commands of His Majesty, King FERDINAND the Fourth, to attend him from *Portici*, in His entry into His Capital, I had this day the honour of being present with his Majesty, and of witnessing the enthusiasm with which He was received by his people. The King entered *Naples* at the head of His own troops, together with the Austrians and British, who defiled before him on his arrival at His Palace.

"The constant attachment the Neapolitan people are known to have ever borne their legitimate Sovereign, makes it unnecessary to detail to your Lordship their joy at His return. His Majesty re-assumes the Government of His Country, beloved and respected by all Classes of His Subjects."

THE ORDER now seemed likely to share the fate, which had befallen the ancient NEAPOLITAN ORDERS, and to be suppressed, with a view to annihilate the remembrance of a Family, which could not be altogether agreeable, especially to the reigning House.

Many considerations, however, urged the propriety of an ORDER being established,—Whereupon, FERDINAND the Fourth, King of THE Two SICILIES, by a Decree, dated the 1st of January, 1819, founded "THE ROYAL NEAPOLITAN MILITARY ORDER of ST. GEORGE of the Reunion," as a reward for Valour and Military Merit. One of the principal causes of this foundation being, THE NEW UNION of the Two KINGDOMS into ONE.

THE KING is Grand Master, and The Duke of CALABRIA, Grand Constable of THE ORDER. There is also a Grand Marshal, who is named by The King.

The ORDER comprises six Classes,—*viz.*, GRAND COLLARS,—COMMANDERS,—KNIGHTS *by right*,—KNIGHTS *by favour*,—GOLD MEDALS,—and SILVER MEDALS.

The rank of GRAND COLLAR was abolished by a Decree, of the 28th of September, 1829, and the title of GRAND CROSS was instituted in it's place.

The rank of GRAND CROSS, and KNIGHT *by right*, is regarded as a mark of eminent valour, and is conferred on distinguished Generals and Veteran officers.

THE GOLD MEDAL is destined for Under-Officers and Soldiers, distinguished for their bravery.

The Rank of KNIGHT *by favour*, and THE SILVER MEDAL, form a separate distinction, which is given for brave actions, and also for *Forty years* of service, comprising, however, two Campaigns. Under-Officers and Soldiers only obtain THE SILVER MEDAL.

The Badge is a gold Cross, of four points, enamelled red, in the angles of which are two swords crossed, of gold, placed upon a crown of laurel,—in the centre, upon a white ground, is an equestrian figure of *St. George*, slaying the Dragon, of gold,—and surrounded with a blue circle, on which is the MOTTO, "*In hoc signo vinces*,"—pendent to the lower point of the Cross is also the effigy of St. George and the Dragon, in gold,—on the reverse, is the effigy of *St. George*, and around it, the word "*Virtuti*."

It is worn by THE GRAND CROSSES, suspended from the neck by a light blue, watered riband, bordered with yellow. THE STAR presents the obverse of the Cross, the angles being occupied by the crossed swords, and silver rays.

Commanders wear the same Decoration from the neck, but of smaller dimensions, and without the *St. George* attached.

KNIGHTS *by right* wear the same Decoration, but yet smaller, from the button-hole.

The Decoration of THE KNIGHTS *by favour* is the same as the preceding, but without the crown of laurel,—and, on the reverse, it bears the word "*Merito*," instead of "*Virtuti*."

GRAND CROSS.

Lieut. COLONEL RICHARD CHURCH,—*See*, THE TWO SICILIES, Order of St. FERDINAND and of MERIT.

THE ROYAL ORDER OF FRANCIS THE FIRST.

THIS ORDER was founded by FRANCIS the First, KING of THE TWO SICILIES, on the 28th of September, 1829, and is destined as a reward for CIVIL MERIT of every description. MILITARY men may also aspire to THIS ORDER, when they have displayed any signal merit.

THE SOVEREIGN is always Grand Master of THE ORDER, which consists of five Classes, —*viz.*, GRAND CROSSES,—COMMANDERS,—KNIGHTS,—GOLD MEDALS,—and SILVER MEDALS.

THE BADGE is a gold Cross, of four broad points, enamelled white, the angles being filled with gold Lilies,—in the centre, on a field of gold, is the cypher of the Royal Founder, F. 1., surmounted by a crown, and surrounded with a wreath of Laurel,—and, on a second circle, upon a blue ground, is the MOTTO, "*Optime merito de Rege,*" in letters of gold, the whole being surmounted by a Royal crown. It is worn by the Grand Crosses, suspended from the neck, by a deep red riband, edged with blue. THE STAR, which is a Silver Cross, of four points, presents the obverse of THE BADGE, and is worn on the left breast.

The Cross of COMMANDER is a little smaller, and is worn from the neck.

The Cross of a KNIGHT is yet smaller, and is worn from the left button-hole.

The GOLD and SILVER MEDALS are also worn from the button-hole.

Medals.

On the 1st of November, 1814, The King, Ferdinand the Fourth, instituted a Medal, by which faithful services and devotion to the defence of the Country are rewarded,—and with which similar merits in all classes, up to the very highest, are recompensed.

In the year 1815, all those who had shown themselves particularly active in the downfall of Joachim Murat were likewise distinguished by an appropriate Medal.

A third Medal was struck, on the 9th of September, 1816, with which all Military persons of every rank were recompensed, who had distinguished themselves by Fidelity and Attachment to The King. It is of *bronze*, surrounded by a radiant Star. One side exhibits the bust of the King,—and, on the other, are the words, "*Fidelity, Attachment.*" It is worn in the button-hole, pendent to a riband of *Bourbon* red.

THE KINGDOM OF SPAIN.

THE ORDER OF THE GOLDEN FLEECE.

THE ORDER of THE GOLDEN FLEECE has ever been ranked among The Most Illustrious and Distinguished Orders of Knighthood in Europe.

THIS ORDER being fully described in The Orders of Knighthood in THE EMPIRE of AUSTRIA, it may not be necessary to say any thing further upon it here, than to observe, that the number of Knights of THIS BRANCH of THE ORDER, of which THE KING of SPAIN is SOVEREIGN, is not limited, though it seldom exceeds Seventy or Eighty.

In all other respects the STATUTES and COLLARS of both BRANCHES are exactly the same. *See*, AUSTRIA.

THIS ORDER was the only one, which was not annulled, like all the other Spanish Orders, by JOSEPH BUONAPARTE, in 1808,—but was expressly confirmed anew by him, in September, 1809.

GRAND CROSS.

ARTHUR, Duke of WELLINGTON, K.G.

The Order of Charles the Third.

This Order was instituted by Charles the Third, first Duke of Parma, then King of The Two Sicilies, and, lastly, King of Spain, on the 19th of September, 1771, in commemoration of the birth of his son Charles Clement. He named it after himself, and devoted it to "*The Pure Conception of The Virgin Mary.*"

The reasons which induced Charles the Fourth to revive it, are not exactly known,—but this revival took place, on the 12th of June, 1804.

It was abolished, as already stated, with the other Spanish Orders, in 1808, by Joseph Buonaparte, then King of Spain, and was only re-instated in it's original form, in 1814.

It is a universal Order of Merit for the Nobility, both *Native* and *Foreign*. It's Members constitute two Classes,—

Grand Crosses, who, beside the King and the Royal Family, are limited to 60.

Knights, 200, who enjoy a *Pension* of 375 *gulden* each.

Those who do *not* receive *Pensions*, are styled, "Supernumeraries," and form a Class by themselves.

To belong to the first Class, it is requisite to be of the age of 25 years, and to be of the Military Profession. It is only the Princes of The Royal Family, other Sovereigns and Princes, who are exempted from these obligations.

Of the second Class, the Two Hundred Pensioners comprise many Members, both Civil and Military,—but all are obliged to adduce proofs of their Nobility of descent.

THE OFFICERS of THE ORDER are, a Grand Chancellor, whose post is always filled by *The Patriarch of The Indies*,—a Secretary, a Master of The Ceremonies, and a Treasurer.

THE BADGE of THE ORDER is a Star, of eight points, enamelled white, and edged with gold, over the two upper points is a wreath of laurel, chased in gold, and golden fleur de lis in the angles,—on the centre of the Cross, is the image of The Virgin Mary, enamelled in proper colours, vestments white and gold.—On the Reverse, the letters C C., with the number III. in the centre, and this MOTTO, "*Virtuti et Merito.*" It is worn by The Grand Crosses, pendent to a broad watered riband, sashwise over the right shoulder,—the riband is striped, blue and white, two-fifths being of the one colour, and three-fifths of the other. They wear also a STAR of silver, of the figure of The Badge of The Order, embroidered on the left side of the outer garment.

The Ecclesiastical Prelates wear the Badge, round the neck,—as do also the Lay Officers of The Order.

The Knights Pensioners, and Supernumeraries, wear it of a smaller size, pendent to a narrower riband at the button-hole, and without the Star.

It is stated, that the STAR should be worn on the left breast, only by the Knights of *the appointed number* of THE ORDER of CHARLES the THIRD, to distinguish them from those that have received only *Supernumerary Crosses*, there being a *Pension* attached to the former. And, as all the Crosses that have been granted (up to the year 1817) to British Officers serving in Spain, are only *Supernumerary*, of course they are not entitled to wear *any* Star whatsoever, nor any thing more than the Cross *at the button-hole*. THE STAR belonging to the Class of Knights GRAND CROSSES, is much larger, and quite different in the centre. This Dignity has been obtained by no one in England, except their Royal Highnesses THE PRINCE REGENT and The Duke of YORK, and Lord EXMOUTH,—therefore, no other persons can wear the Star of this description. The Knights, either of the *appointed number*, or *Supernumerary*, can only wear the Cross suspended at the button-hole, and by no means, round the neck,—no one being entitled to wear it in this manner, excepting *three* persons in all Spain, who are the Ministers of The Order, *viz.*, The Secretary, the Master of Ceremonies, and the Treasurer. The Royal *Crown* over the Cross has been suppressed many years ago,—and, in it's stead, a wreath of *Laurel* has been placed, of course, the former must not be worn, even by the Knights Grand Crosses.

It appears, that these remarks had been occasioned by The Spanish Ambassador having observed, in 1817, with displeasure, several abuses that had been introduced by some of the British Officers,—but, as he had no authority in England to enable him to repress them, he had refrained from making any complaint, and had only hinted to them the impropriety of their proceedings in this respect.

As a Dress of Ceremony, both Classes wear a wide light blue Mantle, sprinkled with Stars, and the Badge round the neck, pendent to a gold Collar, the forty-one links of which consist of towers, a bundle of military trophies, and the number 3., encircled with a wreath of laurel and palm, and united by lions.

In the list of The Grand Crosses is the name of *St. Ignatius de Loyola*, who, in 1817, was invested with this Lay distinction. To honour Saints in this manner, or to appoint them Commanders in Chief on particular expeditions, is an ancient custom originating in the wars against *The Moors*. The revival of it in this instance, shows the favour in which *The Jesuits* were then holden. The horrors of *The Inquisition* are happily abolished.

Grand Crosses.

The Right Honble. Edward, Viscount and Baron Exmouth, Admiral of The Blue Squadron of The Fleet, and G. C. B., to accept and wear the Insignia of the following Royal Orders of Knighthood, *viz.*, The Insignia of Knight Grand Cross of The Royal and Distinguished Spanish Order of CHARLES THE THIRD, — of Knight Grand Cross of The Royal Sicilian Order of ST. FERDINAND and of MERIT, — of Knight of The Royal Sardinian Supreme Order of THE ANNUNCIATION, — of Knight Grand Cross of The Royal Sardinian Order of ST. LAZARUS and ST. MAURICE,—and of Knight Grand Cross of The Royal Military Order of WILLIAM of THE NETHERLANDS, with which their Majesties the King of SPAIN, the King of THE TWO SICILIES, the King of SARDINIA, and the King of THE NETHERLANDS, have respectively been pleased to honour him, "in testimony of the high sense entertained by those Sovereigns of the judgement and valour manifested by His Lordship in conducting the operations of The Fleet under his command against the City of *Algiers*, on the 27th of August, 1816, the result of which has been so beneficial to the cause of Humanity."—Royal Licence, 26th of April, 1817.

And His Royal Highness The Prince Regent taking into consideration the eminent Services of The Right Honble. EDWARD Viscount and Baron EXMOUTH, rendered to His Majesty and the Realm upon divers important occasions, during a life constantly and zealously devoted to his Country, but, especially, the signal ability, energy, and valour manifested by him in the conduct of the operations of The Fleet under his command against the City of *Algiers*, and the renunciation by it's Government of the barbarous practice of Christian Slavery, were the glorious and happy results, His Royal Highness was desirous, as an additional testimony of the just sense which he entertained of such highly distinguished conduct and merits, that the same should be commemorated by certain honourable ARMORIAL AUGMENTATIONS allusive thereto, to be borne by him and by his Descendants.— Royal Licence, 28th of February, 1817.

Sir STAPLETON COTTON, Bart., *afterwards* Viscount COMBERMERE.— *Peninsular War.*

JOHN HAY, Esq., commonly called Lord JOHN HAY, Captain in The

Royal Navy, and C. B., the Commander of The British Naval Forces on the North Coast of Spain, "in testimony of Her Catholick Majesty's Royal approbation of the important Services rendered by His Lordship at the Siege of *Bilboa*."— Royal Licence, 29th of March, 1837.

Colonel DE LACY EVANS, Lieut. General in the Service of Her Catholick Majesty, and Knight Grand Cross of The National and Military Order of St. Ferdinand, "in testimony of Her Majesty The Queen Regent's Royal approbation of his important Services, especially at the Assault of *Irun* and the Capture of *Fontarabia*." —Royal Licence, 25th of July, 1837.

KNIGHTS.

CHARLES MENZIES, Esq., Lieutenant, *now* Lieut. Colonel, Commanding The Royal Marine Artillery, K.H., and K.T.S., "for gallant and distinguished conduct," which is most honourably certified by the late brave Captain Sir WILLIAM HOWE MULCASTER, of The Royal Navy, dated at *Toulouse*, on the 23d of October, 1827,—

"These are to certify, that Captain CHARLES MENZIES commanded The Royal Marines which were attached to the boats under my command, detached from His Majesty's Ship "*Minerva*," the 20th of June, 1806. The officers and crews of which, most ably assisted by The Royal Marines, stormed *The Fort of Finisterre*, spiked the guns, and cut *Five* Spanish Luggers from under it's protection,—and I take this opportunity of acknowledging how much I was indebted to Captain MENZIES for his zeal and intrepid Spirit, from which I foresaw the decided and gallant support I received from him, and which greatly contributed to the success of the enterprise,—an enterprise, which The Earl of ST. VINCENT, then Commander-in-Chief of The Channel Fleet, characterized in his Dispatches to The Admiralty, as a very neat exploit.

"I further certify, that His Majesty's Ship "*Minerva*" laying at anchor off *Oporto*, the 10th of July, 1806, Captain MENZIES reported to me, that he could disguise *The Barge* in such a manner, as to exactly represent *The Guarda* Piratical Privateers, —and having given orders, for his suggestions to be carried into effect, and perceiving the appearance of The Barge *so completely altered*, as to give every prospect of success, I manned and armed her, and taking with me Captain MENZIES and a detachment of Royal Marines, proceeded in search of the enemy. On the following day,—being upwards of Fifty miles from where the *Minerva* lay at anchor,—we fell in with *The Buena Dicha*, Spanish Privateer of *Guarda*,—which vessel had been successful against *The Oporto Wine Company*,—and, by the help of our disguise, got within pistol shot before we were discovered, when a sharp contest ensued, and she was finally carried by boarding.

" Captain MENZIES' conduct upon this, and the former occasion, convinced me that for zeal, courage, and ability he yields not to any Officer in The Corps to which he belongs.

" The Earl of ST. VINCENT also favourably noticed the capture of *The Buena Dicha*, which was published by order of The Admiralty.

" Captain MENZIES was ever one of the first to volunteer his services, and ever ready to execute what he had volunteered."

For the former of these daring exploits Colonel MENZIES received an HONORARY SWORD of the value of £50., from THE COMMITTEE of THE PATRIOTIC FUND. He has already bequeathed to fame, his *right* arm,—notwithstanding which disaster, he heroically exults, " that he has still *one* left, at the service of his Country."

JOHN DOWNIE, Esq., Assistant-Commissary-General of His Majesty's Forces serving in The Peninsula, Brigadier-General in the Spanish Service, and Colonel of The Royal Legion of *Estremadura*, " for the signal courage and intrepidity displayed by him in several Actions with the Enemy, in *Spain*."—Royal Licence, 29th April, 1813.—Knighted, 19th of May, 1813.

ROBERT MENDS, Esq., Captain in The Royal Navy, " in testimony of the high sense which His Catholick Majesty entertains of the Military services rendered by him, whilst commanding a

British Squadron on the Coasts of *Galicia* and *Cantabria*."—Royal Licence, 16th of May, 1815.—Knighted, in 1815. Died, in 1824.

Sir Loftus William Otway, Knt., Colonel in The Army, and C.B., " in testimony of the high sense His Catholick Majesty Ferdinand the Seventh, King of Spain, entertains of his meritorious Services against the Enemy, during the glorious Campaigns of The British Army in *The Peninsula*."—Royal Licence, 8th of May, 1816.

Charles Morgan Carrol, Esq., Captain in His Majesty's 73d (Highland) Regt. of Foot, late of the 87th (*or*, Prince of Wales's Own Irish) Regt. of Foot, and Colonel in The Spanish Service, " for his meritorious and gallant conduct, in command of a Company of the 87th Regt. in the defence of the breach of *Tarifa*, on the 31st of December, 1811."—Royal Licence, 20th of May, 1816.

Emanuel Thomas Poe, Esq., Captain in His Majesty's 96th Regt. of Foot, and a Lieut. Colonel in The Spanish Service, " for his signal intrepidity and meritorious Services in several Actions with the Enemy in *The Peninsula*."—Royal Licence, 20th of May, 1816.

Sir William Parker Carrol, Knt., Lieut. Colonel in The Army, and a Major in The Spanish Service, " for his highly distinguished intrepidity and meritorious Services at the memorable Battle of *Tamames*, and in several Actions with the Enemy, during *The Peninsular War*."—Royal Licence, 20th of May, 1816.

Edmond Temple, Esq., a Captain of Cavalry in The Spanish Service, " in testimony of the high sense which His Catholick Majesty Ferdinand the Seventh, King of Spain, entertains of his meritorious and gallant conduct against the Enemy, during *The Peninsular War*,—more particularly in the Actions fought at *Arroyo del Puerco*, *Caceres*, *Arroyo de Molinos*, and at the passage of *The Bidassoa*."—Royal Licence, 6th of July, 1816.

Patrick Campbell, Esq. Major in The Royal Artillery, and Brigadier-General in the Service of His Catholick Majesty, late Colonel of The Spanish Regiment of Light Infantry, called "*Caçadores of Majorca*," "as a signal testimony of the approbation of His Catholick Majesty FERDINAND the Seventh, King of Spain, of the distinguished services rendered by him in several actions with the Enemy, during *The Peninsular War*,—particularly at *Medellin*, *Talavera de la Reyna*, *Castalla*, *Tarragona*, and *Barcelona*."—Royal Licence, 6th of August, 1816.

Sir HOWARD DOUGLAS, Bart., Colonel in The Army, Inspector-General of The Royal Military College, and Commandant of the Senior Department of that Institution, "in testimony of the approbation of His Catholick Majesty FERDINAND the Seventh, King of Spain, of the signal gallantry and able conduct displayed by him in the Field, while employed with the Spanish forces in *The Peninsula*."—Royal Licence, 6th of September, 1816.

THOMAS FELLOWES, Esq., a Post Captain in The Royal Navy, and C.B., "in testimony of the approbation of His Catholick Majesty FERDINAND the Seventh of the distinguished intrepidity displayed by him, while in command of the English Gun-boats in numerous Actions with the Enemy's Flotilla and Batteries, during the ever memorable defence of *Cadiz*."—Royal Licence, 22d of February, 1822.

EDWARD BRACE, Esq., Captain in The Royal Navy, and C.B., to accept and wear the Cross of The Royal and Distinguished Order of CHARLES the Third of Spain, with which His Catholick Majesty had been pleased to honour him, "in testimony of the high sense that Sovereign entertains of the able conduct and distinguished intrepidity displayed by him at the defence of *Cadiz*, in the year 1811,—And also, to accept and wear The Insignia, of the Third Class, of The Royal Order of WILHELM of THE NETHERLANDS, which His Majesty The King of THE NETHERLANDS had been pleased to confer upon him, "in testimony of His Royal approbation of the skill, valour, and perseverance manifested by him, while commanding His Majesty's Ship *The Impregnable*, at the memorable Battle of *Algiers*, on the 27th of August, 1816."—Royal Licence, 6th of March, 1822.

Sir ROBERT STEELE, Knt., Lieut. Colonel in the 3d Regt. of Swiss

Guards in The Service of Spain, and Knight of The Royal and Distinguished Spanish Order of CHARLES the Third, to accept and wear the Insignia granted to *The Pensioned Knights* of the said Royal and Distinguished Order, which Decoration His Catholick Majesty, FERDINAND the Seventh, has been pleased to confer upon him, " as a further testimony of the high sense which that Sovereign entertains of the meritorious services and gallant conduct displayed by him, in Action with the Enemy during the war in *The Peninsula.*"—Royal Licence, 19th of October, 1819. This brave officer has also received the Decoration of " THE LYS," from the King of France,—as well as CROSSES, for the several Battles in the Campaign of 1813, from The King of Spain.

Sir ROBERT HUGH KENNEDY, Knt., Commissary-General of the Forces, " for his distinguished conduct during *The Peninsular War.*"—Royal Licence, 23d of June, 1831.

WILLIAM WYLDE, Esq., Lieut. Colonel in The Royal Artillery, and British Commissioner at the Head-Quarters of The Spanish Army, to accept and wear the Cross of The Royal and Distinguished Order of CHARLES the Third,—the Cross, of The Second Class, of The National and Military Order of SAN FERNANDO,—and the Insignia of a Knight Commander of The Royal Order of ISABELLA the Catholick,—conferred upon him by THE QUEEN REGENT of SPAIN, " in testimony of Her Royal approbation of his services in the operations undertaken for raising the Siege of *Bilboa,* and also in the Action before *St. Sebastian,* on the 1st of October, 1836."—Royal Licence, 24th of March, 1838.

Lieut. Colonel CHARLES CHICHESTER, Brigadier-General in the Service of Her Catholick Majesty, and Knight of the Third Class of The National and Military Order of SAN FERNANDO, to accept and wear The Cross of The Royal and distinguished Order of CHARLES the Third, and the Insignia of a Knight Commander of The Royal Order of ISABELLA the Catholick, " in testimony of the approbation of The Queen Regent of Spain, of his distinguished services at the Siege and Capture of *Irun*, in May, 1838, and in the Action of the 1st of October, 1836."—Royal Licence, 3d of April, 1838.

Major Sir JOHN GASPARD LE MARCHANT, of the 20th (*or*, East Devonshire) Regt. of Foot, Knight Commander of The Military Order of SAINT FERDINAND, and Brigadier-General in the service of Her Catholick Majesty, "in testimony of The Queen Regent's Royal approbation of his services in the Action of the 1st of October, 1836."—Royal Licence, 4th of May, 1838.

SUPERNUMERARIES.

CHARLES FELIX SMITH, Esq. Lieut. Colonel in The Army, and Captain in The Corps of Royal Engineers, "in consideration of the Military services rendered by him in *The Peninsula*, but more particularly for his skill and highly distinguished merit in the defence of *Tarifa*, in December, 1811."—Royal Licence, 21st of September, 1814.—Knighted, 10th of November, 1814.

JOHN CLARKE, Esq., Colonel of Cavalry of The Spanish Armies, "in testimony of the high sense which His Catholick Majesty entertains of the military services rendered by him in the Field."—Royal Licence, 14th of April, 1815.

BURGES CAMAC, Esq., Lieut. Colonel in The Army, and Captain in The First Regiment of Life Guards, "in testimony of the high sense which His Catholick Majesty entertains of the signal intrepidity displayed by him in several Actions with the Enemy, during the recent arduous Campaigns in *The Peninsula*."—Royal Licence, 4th of August, 1815.

HUGH GOUGH, Esq., Lieut. Colonel of His Majesty's 87th (*or*, Prince of Wales's Own Irish) Regt. of Foot, "in testimony of the high sense which His Catholick Majesty entertains of the signal intrepidity displayed by him in several Actions with the Enemy, during the recent arduous Campaigns in *The Peninsula*."—Royal Licence, 28th of August, 1815.—Knighted, 4th of December, 1815.

Lieut. Colonel CHARLES DOWNIE, of The Spanish Service, "in testimony of the high sense which His Catholick Majesty entertains of the signal services rendered by him in the field of Battle."—Royal Licence, 23d of September, 1815.

PHILIP KEATING ROCHE, Esq., Lieut. Colonel in The British Army, and Lieut. General in The Spanish Service, to accept and wear The Supernumerary Cross of The Royal and Distinguished Order of CHARLES the Third,—and The Order of SAINT FERDINAND, of the Third Class,—"in testimony of His Catholick Majesty's approbation of the distinguished Military Services rendered by him in the Field, during *The Peninsular War*."—Royal Licence, 15th of March, 1816.—Knighted, 14th of May, 1816.

CHARLES SILVERTOP, Esq., late Captain in the 14th Regt. of Light Dragoons, and now a Colonel in The Spanish Service, "as a signal testimony of the approbation of His Catholick Majesty FERDINAND the Seventh, King of Spain, of the distinguished services rendered by him in several Actions with the Enemy, during *The Peninsular War*,—particularly in the memorable Battle of *Barrosa*, and at *Usagre*."—Royal Licence, 29th of July, 1816.

JAMES JONES, Esq., a Major in The Army, Captain on Half-pay of the 15th Regt. of Light Dragoons, and Lieut. Colonel in the Service of His Catholick Majesty, "in testimony of the high sense which His Majesty FERDINAND the Seventh, King of Spain, entertains of his distinguished and gallant conduct in several Actions with the Enemy, during *The Peninsular War*,—particularly in the memorable Battle of *Barrosa*, and at *Castalla*."—Royal Licence, 6th of September, 1816.

ARCHIBALD MACLAINE, Esq., Lieut. Colonel in The Army, and C. B., "in testimony of the high sense which His Catholick Majesty FERDINAND the Seventh, King of Spain, entertains of the highly distinguished intrepidity displayed by him in the arduous defence of Fort *Matagorda*, in 1810,—in the memorable Battle of *Barrosa*, in 1811,—and of the capture of *Seville*, in 1812."—Royal Licence, 12th of October, 1816.

Lieutenant ROBERT STEELE, of The Corps of Royal Marines, and Lieut.

Colonel of the Regiment of Valencia in the Spanish Service, "in testimony of the high sense which His Catholick Majesty FERDINAND the Seventh, King of Spain, entertains of the meritorious services and gallant conduct displayed by him in Action with the Enemy during *The Peninsular War*, more especially in the Attack of the Enemy's entrenchments at the passage of the River *Bidassoa*, and other Actions in *The Pyrenees*, during the year 1813."—Royal Licence, 3d of December, 1816.—Knighted 16th of February, 1817.—*See*, KNIGHTS.

ARTHUR GOODALL WAVELL, Esq., Colonel in The Spanish Service, to accept and wear The Supernumerary Cross of The Royal and Distinguished Order of CHARLES the Third,—and the Cross of The Royal and Distinguished Military Order of SAN FERNANDO, First Class, "in testimony of the sense which His Majesty FERDINAND the Seventh, King of Spain, entertains of the distinguished intrepidity displayed by him in Action with the Enemy, more especially on the night of the 16th of June, 1811, during the Siege of *Tarragona*, in an assault against the Trenches occupied by the Enemy."—Royal Licence, 16th of February, 1818.

CHARLES MACLEAN, Esq., M. D., "for his meritorious conduct, while in the Service of His Catholick Majesty FERDINAND the Seventh, King of Spain."—Royal Licence, 29th of July, 1823.

Sir JOHN BROWNE, Knt., Colonel in The Army, and Lieut. Colonel of the 13th Regt. of Light Dragoons, and Knight of The Royal Portuguese Military Order of THE TOWER and SWORD, "in testimony of the high sense which His Catholick Majesty, FERDINAND the Seventh, entertains of the distinguished Military services rendered by him in the Field, during *The Peninsular War*."—Royal Licence, 12th of July, 1824.

EDWARD JOHNSON, Esq., M.D., "in consideration of the services rendered by him to The Spanish Nation during the war with France in the year 1811, and especially on account of the number of dangerous wounds which he received, while executing a Commission intrusted to him by The Spanish Government, during the operations which preceded the Battle of *Chichana*."

—Royal Licence, 12th of August, 1835.—Knighted, on the 18th of July, 1838.

THOMAS MAITLAND, Esq., Captain in The Royal Navy, "in testimony of The Queen Regent's approbation of the important services, which he rendered to the cause of HER CATHOLICK MAJESTY." —Royal Licence, 27th of June, 1837.

ALEXANDER DUNDAS YOUNG ARBUTHNOT, Esq., a Captain in The Royal Navy, "in testimony of Her Majesty The Queen Regent's approbation of the services rendered by him to the cause of HER CATHOLICK MAJESTY."—Royal Licence, 17th of July, 1837.

Sir CHARLES WILLIAM DOYLE, Knt.,—*See*, HANOVER.

THE NATIONAL MILITARY ORDER OF SAN FERNANDO.

THE GENERAL and EXTRAORDINARY CORTES being convinced, that nothing more conduces to excite that noble Military ardour which produces distinguished Actions, than the establishment of a regular Order, or System in the rewards by which two desirable ends may be obtained, that is,

1st. That only distinguished Merit shall be rewarded,—and
2d. That favour may never occupy the place of Justice,—

and considering at the same time, that to obtain this object, it is necessary to prevent in future the granting of *Brevet* Military Rank, and to limit all Promotion to the effective employment, as also to put an end to other abuses in the distribution of Rewards and Distinctions, which have crept into the Service to the great discredit of good order, and of the very Rewards so distributed, by a Decree, dated, at *Cadiz*, the 31st of August, 1811, ordained, as follow,—

A New Military Order, called THE NATIONAL ORDER of SAN FERNANDO shall be created,—

THE CROSSES of The Order shall be of *Silver* and of *Gold*,—one order of the Gold Crosses shall be *crowned*, with a crown of *Laurel*. There shall be besides GRAND CROSSES, to which shall appertain besides the *Laurel* crown, a band of wide riband hanging from the shoulder and crossing the breast from right to left, and an embroidered

Silver Cross on the left breast. The Riband shall be, for all the Crosses, *red*, with a narrow edging of *orange* colour. The Cross will consist of four equal arms, which will unite in a common circular centre, in which will be *enamelled* in the Gold Crosses, and *engraved* in the Silver, the image of *San Fernando*. In the circle, on the obverse side, will be the following Motto, " *Al merito militar*," (*To Military Merit*),—and in the circle, on the reverse side, " *La Patria*" (*The Country*),—

Any individual of the Army from the Soldier to the General, who may perform any one of the distinguished Actions hereinafter pointed out, shall be rewarded with this Order.

The King, or he who, in his absence, may receive the operative power, will grant these Crosses by a Diploma, or Title signed by his hand, and sealed with his Seal of State, specifying in the Diploma the action for which the Cross was granted,—

The Soldiers, Corporals, and Serjeants who may obtain this Order, will receive the *Silver* Cross, *gratis*. The Cross of the Officers and Cadets will be of *gold*, and paid for by themselves,—

The *Crowned* Cross can be obtained by no one, under the Rank of General of Division, which is equal to the Rank of Lieutenant General in The British Service.

Lieutenant General Whittingham, upon whom the *Crowned* Cross of San Fernando was conferred, on the 13th of December, 1814, for his signal merit, " in the action of *Castalla*, on the 13th of April, 1813, in which he commanded the Division of *Majorca*," thus describes it,—

The edges of the arms, the eight balls at the points of the angles, the edges of the rings round the centre, the King *San Fernando*, the two globes, the crown of the globes, and the Mottos, are all in gold,—

The Motto, on the front of the Cross, is, " *Al merito militar*,"—

The Motto, on the reverse of the Cross, is, " *La Patria*,"—

The colour of the ground on which the Mottos are inscribed, is a dark sky blue,—

The colour of the ground on which are the figures of The King *San Fernando*, and the two crowned globes, is a light sky blue,—

The Crown of Laurel is green,—

The arms of the Cross are of white enamel.

It may not be improper to direct attention to the style of The Diplomas, at different periods of this horrid Civil War, which has so deplorably devastated this fine Country. Prior to the year 1836, they were described, as " The *Royal* Spanish Military Order of Saint Ferdinand,"—but, soon after *the renewal* of The Constitution of 1812, in 1836, they resumed the *democratical* style of " The *National* and MILITARY ORDER of SAINT FERDINAND."

GRAND CROSSES.

Lieutenant General The Right Honble. WILLIAM CARR BERESFORD, Lord BERESFORD, G.C.B., &c., " in testimony of the high sense which His Catholick Majesty entertains of the signal services rendered by him in the Field, as a Captain-General of The Spanish Armies, during *The Peninsular War*."—Royal Licence, 24th of July, 1815.

Sir SAMUEL FORD WHITTINGHAM, Knt., Colonel in The Army, and Aide-de-Camp to His Royal Highness The Prince Regent, also Lieut. General of The Spanish Armies, and Knight of The Royal Spanish Military Order of San Fernando, " as a signal testimony of His Catholick Majesty's approbation of the distinguished services rendered by him in the Field of Battle, during *The Peninsular War*."—Royal Licence, 3d of November, 1815.

ARTHUR GOODALL WAVELL, Esq.,—*See*, SPAIN, ORDER of CHARLES the Third.

Sir STAPLETON COTTON, Bart., afterwards Viscount COMBERMERE.—*Peninsular War.*

Captain RICHARD R. BASSET.

The Right Honble. JAMES, Earl of FIFE, Major-General in the Service of FERDINAND the Seventh, King of Spain, " in testimony of the sense which His Catholick Majesty entertains of the highly distinguished intrepidity displayed by His Lordship, during *The Peninsular War,*—and more especially, his active and zealous services during the memorable siege of *Cadiz,* where, assisting in defence of the Castle of *Matagorda* against the Enemy, he was severely wounded."—Royal Licence, 27th of February, 1818.

DUNCAN MC DOUGALL, Esq., Brigadier-General in the service of The Queen of Spain, (the *first* and *third* Classes,) " in approbation of his conduct, while in the actual service of *Spain.*"—Royal Licence, 7th of December, 1836.

JOHN KINLOCH, Brigadier-General in The Spanish Service, " in testimony of The Queen Regent's approbation of his services, especially at the Battle of *Arlaban,* on the 16th and 17th of January, 1836."—Royal Licence, 15th of February, 1837.

Colonel DE LACY EVANS, Lieut. General in the Service of Her Catholick Majesty, to accept and wear the Insignia of The Grand Cross, together with the Crosses of the Third and Fifth Classes, of the Spanish Order of ST. FERDINAND, " in testimony of The Queen Regent's approbation of his services at the battle of *Arlaban,* on the 16th and 17th of January, 1836, and before *St. Sebastian,* on the 5th of May following."—Royal Licence, 4th of March, 1837.

Lieutenant Colonel CHARLES CHICHESTER, a Brigadier in the service of The Queen of Spain, (the *first* and *third* Classes) " in testimony of The Queen Regent's Royal approbation of his conduct in the Actions of *Arlaban,* on the 16th and 17th of January, 1836, and of *Ayeta,* on the 5th of May, in the same year." — Royal Licence, 1st of May, 1837.

Captain WILLIAM FERGUSON BEATSON, of the 54th Regt. of Native Infantry, in the Service of The East India Company, on the Bengal Establishment, and late a Lieut. Colonel in the service of The Queen of Spain, "in testimony of Her Majesty The Queen Regent's approbation of his conduot in the Actions on the 28th of May and the 6th of July, 1836, while attached to the British Auxiliary Legion."—Royal Licence, 11th of September, 1837.

Lieutenant ROBERT CANNAN, of the 40th Regt. of Native Infantry, in the service of The East India Company, on the Madras Establishment, and late a Lieut. Colonel in the service of The Queen of Spain, "in testimony of The Queen Regent's Royal approbation of his conduct in the Action of the 5th of May, 1836." Royal Licence, 9th of October, 1837.

Captain HICKMAN ROSE KIRBY, of the 4th Regt. of Native Infantry, in the Service of The East India Company, on the Madras Establishment, and Colonel of the late British Auxiliary Legion in Spain, "on account of the important services rendered by him in the Action of the 6th of June, 1836."—Royal Licence, 12th of March, 1838.

Lieut. Colonel CLAUDIUS SHAW, of The Royal Artillery, British Auxiliary Legion, The Royal and Military Order of SAN FERNANDO, "for meritorious conduct generally, and particularly on the 28th of May, 1836."—*See*, THE SECOND CLASS.

Captain EAST APTHORP, of the 2d Regt. of Madras Native Infantry, and late a Lieut. Colonel in The British Auxiliary Legion, "in testimony of Her Catholick Majesty's Royal approbation of his services in her cause."—Royal Licence, 17th of July, 1837.

Lieutenant EDWARD WILLIAM CROFTON, of The Royal Artillery, "in testimony of Her Catholick Majesty's Royal approbation of his services during the Action of *Lucena*, on the 19th of November, 1837."—Royal Licence, 8th of November, 1838.

Lieut. EDWARD WALTER CROFTON, of the Royal Artillery, "in testimony of Her Catholick Majesty's Royal approbation of his services during the Action of *Lucena*, on the 19th of November, 1837."—Royal Licence, 8th of November, 1838.

Lieutenant GEORGE DROUGHT WARBURTON, of The Royal Artillery, "in testimony of Her Catholick Majesty's Royal approbation of his services in the various field actions which took place from the siege of *Bilboa*, in June 1835, to the 4th of May, 1837."—Royal Licence, 26th of December, 1838.

Charles William Thompson, an Ensign in The 81st Regt. of Foot, and late a Captain in The Spanish Service, to accept and wear the Cross, of The First Class, of The National and Military Order of *St. Ferdinand*, "in testimony of Her Catholick Majesty's Royal approbation of his services in the Action before *San Sebastian*, on the 5th of May, 1836."—Royal Licence, 11th of March, 1839.

Second Class.

Major Edward Brackenbury, "for his meritorious conduct, during the War of *Independence* in Spain."—Royal Licence, 19th of December, 1828.

Captain William Considine, Colonel in the Service of The Queen of Spain, "in approbation of his conduct, while in the actual Service of Spain."—Royal Licence, 7th of December, 1836.

Captain William Ellis, of the 45th Regt. of Bengal Native Infantry, and late a Lieut. Colonel in the Spanish Service, "in testimony of Her Catholick Majesty's approbation of his conduct in the Action of *Ayeta*, on the 5th of May, 1836."—Royal Licence, 17th of April, 1837.

Graves-Chamney Swan, Esq., Captain in The Army, and late a Colonel in The Spanish Service, "in testimony of the Royal approbation of The Queen Regent of Spain of his conduct at the Action of *Ayeta*, on the 5th of May, 1836."—Royal Licence, 22d of February, 1838.

William Wylde, Esq.,—*See*, Spain, Order of Charles the Third.

William-Frederick Lapidge, Esq., a Captain in The Royal Navy, and Knight Commander of The Royal Order of Isabella the Catholick, to accept and wear the Cross of The National and Military Order of San Fernando, "on account of the Services which he has afforded, on different occasions, on the Coast of Spain."—Royal Licence, 3d of April, 1838.

Colonel Claudius Shaw, of The Royal Artillery, British Auxiliary Legion, The National and Military Order of San Fernando, of the Second Class, "for the assault and occupation of *Pueblo* and the capture of *Yrun*, on the 16th and 17th of May, 1837."

Colonel SHAW is also, by permission from The Queen of Spain, entitled to wear a MEDAL, for the capture of *Yrun*, which was granted to the whole of THE LEGION serving there,—to the Officers, a *gold* Medal,—and to the Non-commissioned Officers and Privates, one of *silver*.

THIRD CLASS.

Colonel Sir SAMUEL FORD WHITTINGHAM, Aide-de-Camp to His Royal Highness The Prince Regent, and Lieutenant General in The Spanish Armies, to accept and wear the Insignia of The Royal Military Order of Merit of San Fernando of Spain, with which His Majesty FERDINAND the Seventh hath been pleased to honour him, "in approbation of the recommendation of His Supreme Council of War, according to the Institutes of the said Order, and in testimony of the high sense which that Sovereign entertains of the Military services rendered by him, before the Enemy in the Field."—Royal Licence, 9th of May, 1815.

PHILIP KEATING ROCHE, Esq.,—*See*, SPAIN, ORDER of CHARLES the Third.

Sir JAHLEEL BRENTON, Bart., K. C. B., Rear Admiral of The Red, and Lieut. Governor of Greenwich Hospital.

JOSHUA CROSSE, Esq., Captain in the 36th (*or*, Herefordshire) Regt. of Infantry, "in testimony of the sense which His Catholick Majesty entertains of the intrepidity displayed by him in several Actions with the Enemy in *The Peninsula*."—Royal Licence, 30th of July, 1819.

RICHARD CHARLES STEELE, Esq., Captain in The Royal Marine Artillery, in testimony of Her Catholick Majesty's approbation of his distinguished Services with the intrepid Marine Battalion, which saved the Artillery and the whole Spanish Army at *Ernani*.

Major JOHN GASPARD LE MARCHANT, Brigadier-General in the Service of The Queen of Spain, "in approbation of his conduct, while in the actual service of Spain."—Royal Licence, 7th of December, 1836. *See*, THE ORDER of CHARLES the Third.

Lieutenant CHARLES SHAW, to accept and wear the Insignia of The Spanish Order of SAINT FERDINAND, of the Third class,—and the Insignia of a Knight Commander of The Portuguese Order of THE TOWER and SWORD,—"in approbation of his conduct in the Action of *Ayeta*, on the 5th of May, 1836, while in Her Catholick Majesty's actual service,—and also, for his conduct in many Actions, and especially that of the 25th of July, 1833, while in Her Most Faithful Majesty's actual service."—Royal Licence, 21st of February, 1837.

Lieutenant Colonel CHARLES CHICHESTER,—*See*, SPAIN, ORDER of CHARLES the Third.

SWYNFEN THOMAS CARNEGIE, Esq., (commonly called, The Honble. SWYNFEN THOMAS CARNEGIE) "in testimony of The Queen Regent of Spain's approbation of his services in the various Actions which took place from the Siege of *Bilboa*, in June, 1835, to the 4th of May, 1837."—Royal Licence, 14th of September, 1838.

Second Lieutenant DUNCAN BLANCKLEY SHAW, of The Royal Artillery, of The British Auxiliary Legion, The National and Military Order of SAINT FERDINAND, "for meritorious conduct generally, and for distinguished gallantry during the operations at *Hernani* and *Yrun*, from the 14th to the 17th of May, 1837,"

The Order of Saint James, *of Compostella.*

The origin and history of this Order is the same, as that of The Order of Civil Merit of Saint James, in Portugal. *See,* Portugal.

The Badge of The Order is a red Sword, formed like a Cross, in the manner of the hilt of the antient Sword, and is worn upon the breast, or on an oval gold Medal.

This, like The Military Orders of *Calatrava* and *Alcantara,* is highly esteemed in Spain, and conferred only upon persons of Noble Families.

The Military Order of Calatrava.

The Institution of The Knights of Calatrava was somewhat more romantick in it's origin, than that of The Order of St. James, of *Compostella.* The town of *Calatrava,* from it's situation on the frontiers of the Moorish territory of *Andalusia,* where it commanded the Passes into *Castile,* became of vital importance to the latter Kingdom. The defence of it had accordingly been intrusted to the valiant Order of The Templars, who, unable to keep their ground against the pertinacious assaults of *The Moslém,* abandoned it, at the expiration of eight years, as untenable. This occurred about the year 1158, and The Castilian Monarch, Sancho *the Beloved,* as the last resort, offered it to whatever good Knights would undertake it's defence.

The emprise was eagerly sought by a Monk (Don Raymond) of a distant Convent in *Navarre,* who had once been a Soldier, and whose military ardour seems to have been exalted, instead of being extinguished, in the solitude of the Cloister. The Monk, supported by his Conventual brethren, and a throng of Cavaliers and more humble followers, who sought Redemption under the Banner of The Church, was enabled to make good his word. From the Confederation of these Knights and Ecclesiastics sprung The Military Fraternity of Calatrava, which received the confirmation of The Pontiff, Alexander the Third, in 1164. The Rules which it adopted were those of *St. Benedict,* and it's discipline was in the highest degree austere.—Prescott.

It flourished several centuries, and became very powerful. Their number, strength, influence, and possessions, were, at length, so considerable as to render them objects of the jealousy of the Crown, to which their revenues and the office of Grand Master were inalienably annexed by Pope INNOCENT the Eighth, in 1489.

The Badge of The Order is a gold Cross, *fleury*, enamelled red, upon a golden lozenge shaped medal, which is worn, pendent to a broad red riband at the breast.

The Military Order of Alcantara.

This celebrated Military Order derives it's origin from The Order of St. Julian, *or,* Of the Pear-Tree, which was instituted in the Kingdom of *Leon*, about the beginning of the Twelfth century,—from that time until the City of *Alcantara* was taken from the Moors, it continued to be known by the name of "The Order of St. Julian,"—but Alphonso, the ninth, King of Leon, after taking the city, made Martin Fernandez *de* Quintana, Grand Master of *Calatrava*, Governor of it,—and, in a few years afterwards, The Order of Calatrava gave the City and Castle over to The Order of St. Julian, on condition that the same should be holden and remain under the subjection of The Grand Master of *Calatrava*, and his Successors.

Alcantara being thus made the Chief Seat of The Order of St. Julian, the Knights laid aside the old device of *a Pear-tree*, by which they were distinguished, as well as the appellation by which they were known,—and they assumed the name of Knights of "The Order of Alcantara." They remained subject to The Order of Calatrava until the year 1411, when their Grand Master, being authorized by Pope Lucius the Second, threw off all obedience to the Superior Order, and became independent.

After this event, it continued in high estimation, and acquired vast possessions, until the year 1495, when the Grand Master of it, being made a Cardinal and an Archbishop, resigned his office, which Pope Alexander the Sixth conferred on King Ferdinand of *Arragon*, and his Queen ISABELLA of *Castile*, and annexed it inalienably to the possession of The Spanish Crown. Since that time the Kings of Spain have enjoyed the revenues of The Grand Master and of the Commanderies belonging to THIS ORDER, of which they are called, "*Perpetual Administrators.*"

It is never conferred, but upon persons of the most Illustrious and Antient Families.

THE BADGE of THE ORDER is a gold Cross, *fleury*, enamelled green, upon a golden lozenge shaped medal, which is worn, pendent to a broad green riband on the breast.

The Military Order of Our Lady of Montésat.

This Order was founded by James the Second, King of *Arragon* and *Valencia*, and confirmed by Pope John the Twenty-second, in 1317. The Order of The Templars having been abolished in 1311, the King solicited from The Pope the cession of all The Revenues which The Templars had possessed in his Kingdom, with a view to create a New Order of Knighthood. This request was refused by Clement the Fifth, but was granted by his Successor. The Order was then founded, on the 22d of July, 1319, at the Palace of *Barcelona* by The King, in presence of The Bishop of the City, and many Knights and other persons.

This New Order was established in the City of *Montésat*, endowed with the possessions of The Templars, and subjected to the jurisdiction of The Order of Calatrava, whence it's first Knights were taken. In 1399, it was further enriched with the possessions of St. George, *of Alfama*, which had been founded about the year 1201, in the Diocese of *Tortosa*.

The Order of Montésat had a particular Grand Master until the year 1587, when the Grand Mastership was united to the Crown of Spain.

The Badge of The Order is a red Cross, upon a lozenge shaped shield of gold,—and is worn, pendent to a red, watered riband.

On days of Ceremony the Knights wear a White Mantle, on which is The Badge of The Order.

THE ROYAL AND MILITARY ORDER OF SAINT HERMENÉGILDE.

THIS ORDER was founded by King FERDINAND the Seventh, on the 28th of November, 1814, and destined, by a Decree of the 10th of July, 1815, as a recompense for Officers of the Royal Armies and the Navy, who distinguish themselves in the Military Service.

The King is Grand Master. Captains General are by right Grand Crosses of THIS ORDER,—as are also Generals, who have served as Officers, during *Forty years*,—these form a first Class, with the title of "*Excellency*."

The second Class is composed of Officers from the rank of Brigadier downwards, who have served *Forty years*, as Officers.

The third Class is conferred upon those, who have served *Twenty-five years*, and holden the rank of Officers for *ten years*.

The King has reserved to himself the right to decree a diminution in the period of Service in particular cases,—as for the winning of a Battle, an important Siege, or other brilliant Exploit.

Those individuals who have been members of THE ORDER *ten* years, and still continue in the service, enjoy the following Pensions,—*viz.*, Grand Crosses, 10,000 reals,—the Second Class, 4800,—and the Third Class, 2400.

THE BADGE of THE ORDER is a Cross, of four points, enamelled white, and edged with gold,—in the centre of the obverse, upon a deep blue ground, is an Equestrian figure, in gold, surrounded with a blue circle, on which is inscribed in letters of gold, "*Militar Premio a la Constancia.*" It is worn, pendent to a wide red and white, watered riband, passed as a Scarf from right to left.

THE STAR consists of eight points, of gold, with rays of Silver in the angles, and with beads on each point,—the centre is the same as the obverse of the Badge, with the addition of a green wreath of laurel.

The second Class wear the Star and the Cross, without the Scarf,— And the third Class attach the Cross to the button-hole.

SAINT HERMENÉGILDE received the Crown of Martyrdom on Easter Eve, the 13th of April, 586, in the Tower of *Seville*.

GRAND CROSS.

Lieut. General The Right Honble. WILLIAM CARR BERESFORD, Lord BERESFORD, G.C.B., Knight Grand Cross of The Royal Portuguese Military Order of THE TOWER and SWORD, and of The Royal Spanish Military Order of SAN FERNANDO, *&c.*, to accept and wear the Insignia of a Knight Grand Cross of The Spanish Order of SAINT HERMENÉGILDE, " in testimony of the high sense which His Majesty FERDINAND the Seventh, King of Spain, entertains of the signal Services rendered by His Lordship in the Field, as Captain General of The Spanish Armies, during *The Peninsular War*."—Royal Licence, 17th of May, 1817.

The Royal Order of Saint Isabella, *The Catholick.*

This Order was instituted in the year 1815, by the late King Ferdinand the Seventh, in commemoration of Queen Isabella, of Spain, under whose remarkable reign the discovery of *America* was accomplished. Religious zeal was a leading feature in her character,—and the desire of propagating the Christian faith in parts of the World where it was yet unknown, was the chief motive of the encouragement she gave to the projects of Columbus, which eventually added so much to the Spanish Monarchy. Her merits towards the Church were rewarded by the title of, "*The Catholick.*" She died with the great regret of her Subjects, in November, 1504, in the 54th year of her age.

In addition to this motive, the then commencing troubles in *Spanish America* gave occasion for it's establishment, for it was *promised beforehand* to all who should distinguish themselves in the struggle of the Spaniards against The Revolters in The New World. Hence, it is also called, "The Royal American Order of Isabella, The Catholick." It is not, however, destined for *America* alone, as meritorious *Natives* of *Spain,* as well as *Foreigners* are decorated with it.

The Badge of The Order is a Star, of eight points, enamelled red, bordered with gold, and with golden beads at the extremities of each of the points. The centre is composed of a circular Medallion, in the middle whereof, on a gold ground, are two Pillars, round which is entwined a scroll, with the words, "*Ne plus ultra*," also two regal globes of blue, with golden crowns, the whole surrounded with a circle of white enamel, on which are the words, "*A la lealtad acrisolada.*" In each of the four angles, are rays of gold, and the whole is surmounted by a wreath green laurel.

THE STAR is the same as THE BADGE, excepting that it has no golden beads,—that the device has the additional words, "*Por Isabella Catolica*,"—is surrounded with a green border of laurel leaves,—has the cypher, F. R., at the upper part, and is surmounted by a regal crown.

COMMANDERS.

WILLIAM WYLDE, Esq.,—*See*, SPAIN, ORDER of CHARLES the Third.

Lieut. Colonel CHARLES CHICHESTER,—*See*, SPAIN, ORDER of CHARLES the Third.

WILLIAM FREDERICK LAPIDGE, Esq., — *See*, SPAIN, ORDER of SAN FERNANDO.

KNIGHTS.

AUGUSTUS ALGERNON VILLIERS, Esq., Mate, late of Her Majesty's Ship, *Ringdove*, "in testimony of the Royal approbation of The Queen Regent of Spain of his services at the siege of *Bilboa*." —Royal Licence, 3d of April, 1838.

Captain FREDERICK APTHORP, of the 20th Regt. of Native Infantry, in the Service of The East India Company on the Bombay Establishment, and late a Lieutenant Colonel in the British Auxiliary Legion, "in testimony of Her Catholick Majesty's Royal approbation of his services in the Actions of the 10th and 15th of March, 1837, on the heights of *San Marcos* and of *Oriamendi*."—Royal Licence, 16th of October, 1838.

AUGUSTUS HEARN GILBERT, Gentleman, Purser in the Royal Navy, "in testimony of Her Catholick Majesty's approbation of his services in the operations undertaken for raising the siege of *Bilboa*."—Royal Licence, 19th of November, 1838.

THE ORDER OF MARIA LOUISA.

THIS FEMALE ORDER was instituted, in April, 1792, by MARIA LOUISA THERESA, Princess of Parma, and mother of the late King of Spain, as a distinction for Ladies of High and Noble Rank.—Being *abolished* with the rest of the Spanish Orders by King JOSEPH BUONAPARTE, in 1808, it was *renewed* by the late Queen MARIA ISABELLA FRANCESA, a Princess of Portugal, in the year 1816.

THE BADGE of THE ORDER is a Star, of eight points, enamelled white, edged with gold, and having a narrow border of a violet colour around it. In the centre, on a white enamelled oval Medallion, and surrounded by a violet coloured circle, is the figure of *St. Ferdinand*,—in the angles of the cross are Castles and golden Lions alternately, connected together by a golden chain, and on the superior point is a golden wreath of laurel. On the reverse, is the cypher of MARIA LOUISA, with the inscription, "*Rl. Ordn. Dla Reina Maria Louisa.*"

THIS BADGE is worn, pendent to a broad violet coloured and white riband, passed over the right shoulder.

Crosses, and Medals of Merit.

King Ferdinand the Seventh, in order to reward the conduct of his Subjects, during his absence from the Kingdom, created, by various Ordinances, a great number of Decorations of Honour, which were bestowed on single Towns, separate Divisions of the Army, or for single Occurrences.

The great variety of Ribands thus introduced, occasioned the appearance of a special Royal Ordinance, in the year 1817, as to the manner in which they should be worn.

Medal of Merit of Bailen.—This Medal was granted by The Supreme Junta of Seville, in the name of The King, to the Army of Andalusia, present at the affair at *Bailen*, on the 19th of July, 1808, under the command of General Don Francisco Xavier *de* Castaños, at which the French, under the conduct of General Dupont, surrendered at discretion.

Cross of Merit of The North,—instituted, in the name of The King, by a Decree of the 23d of March, 1809, for all those individuals, who had formed a part of the Troops under the command of Captain General The Marquess *de la* Romana. The inscription is, "*My Country is my Polar Star.*"

Cross of Merit of Gerona,—was founded, in the name of The King, by a Decree of The Regency, of the 14th of September, 1810, for all those who had taken part in the defence of *Gerona*, in 1809. The Medallion of the Cross bears on the Reverse, the inscription, "*The Country, for Valour and Constancy.*"

Cross of Merit of Talavera *de la* Reyna.—This Cross was conferred by The Council of The Regency of Spain, the 8th of December, 1819, on the Generals and Officers, who were present at the battle of *Talavera de la Reyna.*

Cross of Merit for Fidelity at Valençay,—founded by The King, by a Decree of the 25th of August, 1814, and conferred on Spaniards who accompanied him into a Foreign Country, and remained with him until his return. The Medallion of The Cross bears on the Reverse, the inscription, "*Valençay*, 1808," in the centre of which is a Chain of gold, beneath which is engraven, "*Fides.*"

Cross of Merit of Saragoza,—instituted, on the 30th of August, 1814, by The King, for all his Generals, Chiefs and Officers, who had distinguished themselves at the Siege of *Saragoza*. This Cross is of *gold*,—but it is cast in *copper* for the Private Soldiers. The Medallion of the Cross bears a representation of The Virgin,—and, on the Reverse, the inscription, "*The King to the Defenders of Saragoza.*"

Cross of Merit of St. Marcial,—This Cross was instituted by a Royal Decree, of the 24th of October, 1814, for all Military men who had been present at the Battle on the Heights near *St. Marcial*, on the *Bidassoa*. It is of *gold*, for the Generals and Officers,—and, of *copper*, for the Privates.

Medal of Distinction for Prisoners of War,—instituted by a Royal Decree, of the 6th of November, 1814, and conferred on Military men, confined in the Prisons of France. It is of *gold*, for the Officers, —and, of *silver*, for the Privates. By a Royal Decree, of the 26th of June, 1815, this Medal was conferred on persons in The Civil Service, in similar confinement. It was worn on a yellow riband, with a green border, attached to the button-hole,—having on one side, a Tower encircled by a chain,—and, on the other, the inscription, "*Suffering for our Country.*"

Cross of Merit of The Escurial,—instituted, by The King, on the 5th of December, 1814, as a reward for those of his Subjects, who, by their Fidelity and Attachment to his person, on the occasion of his Arrest at The Castle of *St. Laurent*, by order of The Prince of Peace, Don Manuel *de* Godoy, exposed themselves to exile and persecution. The Medallion of the Cross represents a Palm branch and a Gridiron, and the Reverse bears the inscription, "*The King to recompense Innocence.*"

Cross of Merit of Ciudad Rodrigo,—instituted, by The King, on the 6th of December, 1814, for the Defenders of *Ciudad Rodrigo*, the 10th of July, 1810,—that for the Officers, in *gold*,—and, in *silver*, for the Privates. This Cross is bestowed on those only, who were present at the Sorties. All the others wear the riband alone. On the Reverse is inscribed, "*Valour, proved at Ciudad Rodrigo.*"

Cross of Merit conferred on The Army of Reserve of Andalusia,—instituted, by The King, on the 28th of December, 1814, as a reward for the distinguished Merit of The Warriors of The Army of Reserve of Andalusia, during the Campaign of 1813,—and particularly for the capture of *Pancorbo*, and the battles of *Sorauren*, the *Pyrenees*, and *Nivelle*, the 13th of October. The Cross is of *gold*, and enamelled, for the Generals and Officers,—and, of *copper*, for the Privates. The Medallion of the Cross bears the portrait of the King, with the inscription, "*The King to The Army of Reserve of Andalusia,*"—and, in the centre of the Reverse, is the year 1813, with the inscription, "*Pancorbo, Pirineos, Nivelle.*"

Cross of Merit of Toulouse,—instituted, by The King, on the 30th of January, 1815, and conferred on the Generals and Officers of the Fourth Army, who distinguished themselves at the Battle of *Toulouse*, on the 10th of April, 1814. The Cross bears the inscription, "*Valour and Discipline,*"—and the Reverse, "*Battle of Toulouse*, April 10, 1814."

Cross of Merit of Chiclana,—conferred on the Troops of the Fourth Army, who were present at the Battle of *Chiclana*, on the 5th of March, 1811. Instituted, by The King, on the 13th of February, 1815.

Cross of Merit of Albufera,—instituted, by The King, on the 1st of March, 1815, for the glorious Battle of *Albufera*, on the 16th of May, 1811, gained by the Generals Don Xavier *de* Castaños and Don Joaquim Blake,—and intended for all the Generals, Officers, and Soldiers of the Army of *Estremadura*, who had distinguished themselves in the Conflict.

Cross of Merit of Seville,—instituted, by The King, on the 17th of March, 1815, in memory of the re-capture of Seville, on the 27th of August, 1812. It is cast in *gold*, for the Officers,—and, in *bronze*, for the Soldiers. In the centre of the Reverse, are the words, "27th of August, 1812,"—and around, "*The King to the Re-captors of Seville.*"

Cross of Merit of The First Army,—instituted, by The King, on the 31st of March, 1815, in memory of the brave Soldiers of The First Army, who distinguished themselves,—principally in *Catalonia*,—during the War of Independence. The Cross bears beneath the Crown, the Arms of *Catalonia*, four red rays,—and, around, the inscription, "*Defence of my King and of Royalty.*"

Cross of Merit of The Second Army,—instituted, by The King, on the 31st of March, 1815, in memory of the brave Soldiers of The Second Army, who distinguished themselves,—principally in The Kingdom of *Murcia*,—during the War of Independence. The letter L., in the centre of the Cross, signifies, "*Loyalty*,"—and round the Medallion of the Cross, is the inscription, "*Recompense of Military Virtue.*"

Cross of Merit of The Third Army,—instituted, by The King, on the 31st of March, 1815, in memory of the brave Soldiers of The Third Army, commanded by The Duke d'Albuquerque in the Isle of Leon, particularly for the defence of *Cadiz*. In the centre of the Cross, are the two Pillars of *Hercules*, in gold, with the inscription, "*Conquerors from The Straits to The Pyrenees.*"

Cross of Merit of Vitoria,—instituted, by The King, on the 2d of April, 1815, in memory of the glorious Battle of *Vitoria*, the 21st of June, 1813, gained by The Divisions of The Fourth Corps of The Army, commanded by The Captain General The Duke of Ciudad Rodrigo and Field Marshal Don Francisco Tomas *de* Longa. In the centre of the Cross is, "*Irurac-Vat,*"—and, on the Reverse, "*Recompense for the Battle of Vitoria.*"

Cross of Merit of St. Laurent *del* Puntal,—instituted, by The King, on the 10th of April, 1815, as a recompense for services rendered by The Officers of a Batallion of Infantry, at the defence of the Castle of *St. Laurent del Puntal*, in the environs of Cadiz, during the War of Independence. The inscription on the Cross is, "Valour proved by the Artillerymen of *St. Laurent del Puntal,*"—and, on the Reverse, "*By King* Ferdinand VII., *in the year* 1814." The Under-Officers and Privates wear the Medallion of the Cross, of a larger size, on the left arm.

Cross of Merit of Astorga,—instituted, by The King, on the 10th of April, 1815, and conferred on the Troops for the glorious Defence of *Astorga*, a small Fortress commanded by Field Marshal José Maria *de* Santocildes, against a Corps of The French Army under General Junot. The Cross for the Officers is, of *gold*,—that for the Privates, of *silver*. The inscription is, "*We gained this Honour at Astorga by our Valour.*"

Cross of Merit of Valls,—instituted, by The King, on the 27th of April, 1815, in honour of the Battle of *Valls* in Catalonia, the 25th of February, 1809, under the command of General Theodore Reding. The Medallion of the Cross comprises the Arms of Catalonia, and bears the inscription, "*The King, for efforts of Valour.*" The Reverse of the Medallion of the Cross is red, and contains the name, "*Valls,*" with the inscription, "On the 25th of February, 1809."

Cross of Merit of Ordal,—instituted, by The King, on the 1st of May, 1815, and conferred on the Troops of a Column of Grenadiers of The Third Division of The Second Corps of The Army, for the bloody Conflict of *Ordal* in Catalonia, on the night of the 12th and 13th of September, 1813. The inscription on the Cross is, "*Battle of Ordal, 13th of September,* 1813,"—and on the Reverse of the Medallion of the Cross, "*King, Country or Death.*"

Medal of Distinction of Tarancon,—instituted by Royal Decree, of the 12th of June, 1815, to recompense the Valour of the Vanguard of the Centre of the Army, commanded by General Don François Xavier *de* Venegas, and which was attacked by a Corps of the enemy of 900 Dragoons. The Medal bears in the centre, the inscription, "*Invincible Infantry,*"—and, on the border, "*At Tarancon,* 25th *of December,* 1808." It is of *gold,* for the Officers,—of *silver,* for the Under-officers,—and of *copper,* for the Privates.

Cross of Merit of Tarragona,—instituted, by The King, on the 14th of May, 1815, and conferred on the Troops, for the Siege of the Fortress of *Tarragona,* which lasted two months, under the command of General Don Juan Senen *de* Contreras. The inscription on the Cross is, "*Death rather than surrender.*"

Cross of Merit of Alcannitz,—instituted, by The King, on the 14th of May, 1815, for the Members of the small Corps of the Army of Arragon, under the command of Don Joaquim Blake, which distinguished itself at the Battle of *Alcannitz,* on the 23d of May, 1809, against Marshal Suchet.

Cross of Merit of The Army of The Left,—instituted by three Royal Decrees, of the 14th of May, the 12th of June, and the 25th of June, 1815, as a reward to the Generals, Officers and Soldiers of The Army of *Galicia,* or, *The Army of The Left,* or *Sixth Army,* who distinguished themselves at the sanguinary Conflicts of *Rioseco, Sornaza, Gueces, Espinosa* in the Mountains, *&c.*

CROSS of MERIT of THE SEVENTH ARMY,—instituted, by The King, on the 19th of May, 1815, for the Troops of The Seventh Army, commanded by General DON GABRIEL *de* MENDIZABEL, during the War of Independence in *Old Castile*, in *The Asturias*, in *Arragon*, and *Navarre*, and in *The Basque Provinces.*

CROSS of MERIT for the CONFLICT near THE BRIDGE of ALCOLEA,—instituted, by The King, on the 3d of June, 1815, for the Troops who had distinguished themselves, during the Conflict near *The Bridge of Alcolea*, the 7th of June, 1808, against a French Corps commanded by General DUPONT. The Cross for the Officers is, of *gold*,—and, of *silver*, for the Soldiers. The inscription on the Reverse is, "*Liberty of Spain*, 7th *June*."

CROSS of MERIT of TARIFFA,—instituted, by The King, on the 4th of June, 1815, for the Troops of The Fourth Corps of The Army, and for The Navy, as a reward for the Defence of the City of *Tariffa*, in December 1811. The Cross is of *gold*, for the Officers,—and, of *silver*, for all other persons.

CROSS of MERIT for CIVIL PRISONERS, — instituted, by Royal Decree, of the 4th of June, 1815, for persons, not Military, who, from attachment to The King and the good Cause, refused to recognise the unlawful Government of NAPOLEON, and were carried away to the Prisons of France. The inscription on the Cross is, "*For Exile, for King and Country*."—and, on the Reverse, "F. VII."

CROSS of MERIT of PAMPELUNA *and* BAYONNE,—instituted, by The King, on the 4th of June, 1815, in honour of the Generals, Officers and Soldiers, who carried on the Sieges of *Pampeluna* and *Bayonne*, in 1813 and 1814, under the command of General ESPAÑA. The inscription on the Cross is, "*For Valour and Discipline*,"—and, on the Reverse, "*At Pampeluna and Bayonne, in the years* 1813 *and* 1814."

CROSS of MERIT for THE TROOPS of THE ASTURIAS,—instituted by a Royal Decree, of the 4th of June, 1815, and distributed among the individuals forming part of The Corps of The Army, which had protected The Province of *Asturia*, against the Attack of The French troops under Marshal NEY, and the Generals KELLERMAN and BONNET. In the centre of the Cross are the Arms of Asturia, a white cross on a field *azure*, with the inscription, "*The Asturians have never been conquered*,"—and, on the Reverse, "*For the Asturian Army*, 1808."

CROSS of MERIT of ALBUQUERQUE,—instituted by a Royal Decree, of the 5th of June, 1815, as a Mark of Distinction for the individuals belonging to The Corps of the Army of Estremadura, which took part in the retreat to *The Isle* of *Leon*,—which step had become absolutely necessary, in order to save the government established in the name of King FERDINAND, and which was executed with much skill by The Commander in Chief, THE DUKE of ALBUQUERQUE. The obverse represents the Pillars of Hercules, with the sea and vessels, and the inscription, "*He saved the sinking Ship*." On the Reverse, is an Eye upon a Medallion of gold, with the inscription, "*To the Duke d'Albuquerque and his Army*."

MEDAL of MERIT of THE COUNT CASA ROXAS,—granted by a Decree of The Spanish Regency, on the 10th of June, 1810, to the eldest son of Count CASA ROXAS, DON JOSÉ ROXAS, to recompense his eminent Patriotism and disinterested Sacrifices.

CROSS of MERIT of LERIN,—instituted by a Decree of The Regency, of the 23d of July, 1811, in the name of King FERDINAND, and conferred on individuals forming part of The Batallion of Chasseurs of Cadiz, who courageously defended the City of *Lerin*, on the 25th, 26th, and 27th of September, 1808. The Reverse bears, on a green Medallion, the inscription, " The 25, 26, and 27 of September, 1808, *Lerin*." The Officers only wear this Cross. The Soldiers wear an Escutcheon with the same inscription.

CROSS of MERIT of UTIEL,—instituted by a Royal Decree, of the 20th of June, 1815, for those who fought at the Battle of *Utiel*, on the 25th of August, 1812.

Cross of Merit *de la Fugà de Portugal*,—instituted by a Royal Decree, of the 22d of June, 1815, and conferred as a Mark of Distinction on those individuals of The Spanish Corps, who, being in Portugal, in 1808, returned voluntarily on the breaking out of the War of Independence to defend the rights of His Catholick Majesty, and the Cause of their Country. On the Reverse are the words, "*Portugal, year* 1808."

The bravery of the Spanish Troops in the Capture of Carthagena, in 1814,—and the great difficulties, toils, and hardships to which they were then exposed, were, in the year 1816, recompensed by a Medal,—which displays on one side, "*Attachment and Fidelity to their King, Ferdinand* VII.,"—and, on the other, the words, "*Conquerors of Carthagena in India.*"

An appropriate Decoration was also conferred on those *Societies* of *Ladies*, by Ferdinand the Seventh, who took care of the wounded, during the war.

In 1817, the Inhabitants of Madrid, who had shown themselves so determined against the Troops of Napoleon, in the three first days of December, 1808, during the Siege of The Capital, received a Decoration, which differs from all other decorations of a similar kind, inasmuch as it presents entirely the form of a *Badge*. It is a gold Cross, enamelled white and green, with eight rays cut through at the ends. In the middle of the obverse, is an arm, with a sword in a silver field, encircled with the inscription, "*Courage and Fidelity of Madrid*,"—and, on the reverse, are, "*The first days of December*, 1808." The pretensions of the applicant are very severely examined, and the following Parties are excluded from Admission,—*viz.*, All the purchasers of The National Estates of the first, second, or third hands,—All those who mediately or immediately have had a share in the buying or selling the said Estates,—All who received places or marks of Honour from the Usurping Government, — All who acted as Officers in the Civic Guard or Cavalry Guard of Honour,—and all who officiated as Præfects, Sub-Præfects, Members of The Criminal Court of Justice, Officers of Police, *&c.*, under The Usurper.

Medal of Merit of Tamames,—instituted by Royal Decree, of the

2d of July, 1815, to reward those who had fought on the 18th of October, 1809, at the battle of *Tamames.*

Medal of Merit of Medina *del Campo*,—instituted by Royal Decree, of the 2d of July, 1815, for all those who had taken part in The Battle of *Medina del Campo*, on the 23d of November, 1809.

Cross of Merit for The Couriers of The Cabinet,—instituted by Royal Decree, of the 9th of July, 1813, for the Royal Spanish Couriers of The Cabinet, who had been in active service with the Armies, during the War of Independence. It bears the inscription, "*Courage and Valour for his King and Country*,"—and, on the reverse, are the words, "*The Couriers of the Cabinet.*"

Bracelet of Distinction of The Patriotick Junta of Ladies,—This Bracelet decorated with the cypher of King Ferdinand the Seventh, was conferred on those Ladies who composed The Patriotick Society of Ladies at Cadiz, during the War of Independence. It was instituted by a Royal Decree, of the 27th of July, 1815, and bears the inscription, "*To the Patriotick Junta of Ladies of Cadiz.*"

Cross of Merit of The Victims of the 2d of May,—instituted by a Royal Decree, of the 27th of October, 1815, and conferred on all the children, widows, and nearest relations of those persons who perished at *Madrid*, on the 2d of May, 1808, in the Revolt which on this day broke out against the French. To this distinction are also added pecuniary and other advantages.

Cross of Merit for the Navy,—This Cross, which serves as a mark of Distinction for the Chiefs, Officers, and other individuals of The Royal Spanish Navy, who, by their services in any part of the World, may have contributed to the successful issue of the War of Independence, was founded by a Royal Decree, of the 22d of September, 1816. The Officers wear it, in *gold* and *enamel*,—the others, in *silver*. The Medallion on the reverse is of gold, and bears the letter F. VII., with the inscription, "*For the valour of the Mariners.*"

Cross of Merit of Cartagena *de Indias*,—This Cross for Merit at *Carthagena* in America was instituted by a Royal Ordinance, of the 1st of April, 1816, for all those individuals of The Spanish Army and Navy, who, under the command of Lieut. General Don Pablo Morillo, Captain General of The Provinces of *Venezuela*, took part in the siege and capture of *Carthagena*. It is of *gold* and *enamel*, for the Generals and Officers,—and of *gilt metal*, without enamel, for all Ranks from that of Serjeant downwards. It bears the inscription, "*Perseverance and Fidelity for his King* Ferdinand VII." — and, on the reverse, on a white medallion, are the words, "*Conquerors of Cartagena de Indias.*"

Cross of Merit of Menjibar,—instituted by a Royal Decree, of the 18th of April, 1816, for the troops of The first Division of The Army of Andalusia who had fought under the command of General Reding at the battle of *Menjibar*, on the 16th of July, 1808. The Medallion, on the reverse, is white, and bears the cypher, "F. VII.," surrounded by the words, "*Menjibar*, 16th *July*, 1808."

Cross of Merit of Bubierca,—instituted by a Royal Decree, of the 30th of May, 1816, and conferred on the Generals, Officers, and Soldiers, who fought at the battle of *Bubierca*, on the 29th of November, 1808.

Cross of Merit of Aranjuez,—instituted by a Royal Decree, of the 30th of May, 1816, and conferred on individuals of the Spanish Corps, who took part in the affair of *Aranjuez*, on the 5th of August, 1809.

Cross of Merit of Almonacid,—instituted by a Royal Decree, of the 30th of May, 1816, and granted to all those individuals of The Spanish Corps, who fought at the battle of *Almonacid*, on the 11th of August, 1809.

Decoration of Honour of the Honourable Council of The Mesta, —This Decoration was instituted by a Royal Decree, of the 31st of

May, 1816, and granted to all the Members of *The Mesta*, who were present at the Re-unions of this Society which took place, on the 26th of April and the 3d of May, 1816, over which his Majesty presided in person. *The Mesta* is the Privileged Corporation of the Possessors, &c. of migratory Flocks of Sheep. It bears the inscription, "FERDINAND VII. *to the Honourable Council of The Mesta.*"

CROSS of MERIT of SAN JORGE.—The Cross of St. George was conferred by the Supreme Junta of The Kingdom, on the 15th of January, 1810, on The Members of The Provincial Junta of Catalonia, as a recompence for their zeal and patriotism. It's institution was confirmed by a Decree of His Catholick Majesty, on the 12th of March, 1815.

CROSS of MERIT for THE DIVISION of MAJORCA,—This Cross was confirmed by a Decree, of the 27th of June, 1816, upon the Troops of all the Armies who, during the war of Independence, formed the Division of *Majorca*, under the command of Lieut. General WHITTINGHAM. The four letters upon the cross, A. L. D. M., signify, "*To the Division of Majorca.*" The Medallion, of the reverse, is white, and bears the words, "*Valour and Discipline.*"

CROSS of MERIT of CASTALLA,—conferred by a Royal Decree, of the 27th of June, 1816, upon all those individuals of the Division of *Majorca*, who fought at the battle of *Castalla* against the French troops, commanded by Marshal SUCHET.

MEDAL of MERIT of LUCENA,—granted by a Royal Decree, of the 23d of October, 1816, to four individuals, Don FERNANDO RAMIREZ *de* LUQUE, Don ANTONIO ORTIZ REPIRO, Don FRANCISCO POLO VALENZUELA, and Don FRANCISCO ASIS *de la* CARRERA, as a reward for the services rendered by them, at the risk of their lives, on the 25th of September, 1810. It bears the inscription, "*Hill of Capitulation,*"—and, on the reverse, upon a Medallion of gold, the words, "*Loyalty rewarded by* FERDINAND VII. *Lucena,* 25th *September,* 1810."

CROSS of MERIT of VILLA-FRANCA *del Vierzo,*—for those individuals of the Corps of the Army in *Galicia*, commanded by General Don Nicholas Mahy, who took an active part in the recapture of *Villa-Franca del Vierzo*, on the 19th of March, 1809, and in the battles of *Lugo*, on the 18th and 19th of March, 1809. It was granted by a Royal Decree, of the 13th of March, 1817, and is of *gold*, for the Generals and Officers,—and, of *bronze*, for persons of inferior Rank. The Medallion, of the obverse, displays the Arms of *Lugo*, a Chalice with a Host,—that of the reverse, contains the Arms of *Villa-Franca*, a golden Lion, on a mountain, *vert*, upon a field, *azure*,—there are likewise the words, "*Capture of Villa-Franca del Vierzo*, 19th *March*, 1809."

Medal of Merit for Villar *de Ciervos*,—instituted by a Royal Decree, of the 16th of October, 1823, and conferred upon 27 inhabitants of *Villar de Ciervos*, who distinguished themselves, on the 27th of August, 1823, by a bold attack upon The Constitutional band of Don Alonzo Martin, Lieut. Colonel of the Regiment *Algarbe*, (brother of the famous *Empecinado*). The obverse displays the bust of King Ferdinand VII., with the inscription, "To the valorous Defenders of their King Ferdinand VII." On the reverse are the words, "*The Royalists of Villar de Ciervos in the year* 1823."

Escutcheon of Fidelity,—King Ferdinand the Seventh,—in order to give a proof of the satisfaction with which he regarded the courageous resolution of those persons who, moved by their Fidelity, had quitted their Country and exposed themselves to dangers of every kind, in favour of the legitimate rights of Sovereignty, and for the defence of the State and Religion, from 1820 to 1823,—instituted, on the 14th of December, 1823, the Escutcheon of Fidelity, and authorised the Captains General of Provinces to cause the necessary *Brevets* to be delivered to all persons, as well Civil as Military, who, by their courage, fidelity, perseverance, good conduct, and attachment to the person of the King, had rendered themselves worthy of this mark of Distinction.

This Decoration is of *silk*, for the Under-officers,—and of *woollen*, for the Soldiers. It is worn on the left breast of the coat, and bears the inscription, "*The King, for Fidelity.*"

The Captains General not having distributed this Decoration according to the Royal instruction, the nominations made by them, were annulled by a Decree of the 14th of January, 1824, and will not be henceforward valid, unless accompanied by the Royal sanction.

Cross for Military Fidelity,—founded by the 12th article of the Royal Decree, of the 9th of August, 1824, and intended as a reward for the intrepid Royalist troops who, during the first epoch, from the 7th of March, 1820, to the 30th of June, 1822,--and the second, from the 1st of July 1822 to the end of February 1823,—defended the Sovereignty of the King, with courage, submission, and exemplary conduct, without regard to the dangers to which they exposed themselves. A third epoch embraces from the 1st of March to the 1st of May, 1823. The Cross, for the first epoch, bears on the obverse the inscription, "*The King to Military Fidelity*," and, on the reverse, the Royal Arms, with the words, "Ferdinand VII. *to those who defended Religion and the Throne in a heroick and eminent manner.*"

The Cross, for the second epoch, has no laurel branches,—it bears, on the obverse, the Motto, "*The King to Military Fidelity*,"—and, on the reverse, the Royal Arms, with the inscription, "Ferdinand VII. *to the Defenders of Religion and the Throne.*"

Those who took part in the two epoohs, bear the Cross suspended from the neck,—those who were concerned but in one, wear it at the button-hole.

The Decorations are of *gold*, for the Superior Officers,—of *silver*, for the Subaltern Officers,—and of *copper*, for all other individuals.

As the object of this Institution, and that of The Escutcheon of Fidelity, are the same,—*viz.*, a proof of Fidelity towards the King, it has been ordered, that no person, however distinguished his merit may be, shall obtain and wear the two Decorations at the same time.

The Queen, Maria Christina, of Spain, during the malady of her Royal Husband, founded, as Regent, on the 10th of October, 1832, at *St. Ildefonso*, a Military Institution, named, "Military Constancy." It is composed of four Classes,—the first of which confers a monthly Supplementary pay of 4 reals, after 10 years' service,—the second, of

10 reals, after 15 years' service,—the third, of 20 reals, after 20 years' service,—and the fourth, of 30 reals, after 25 years' service.

Cross of Maria Isabella Louisa,—The Journal of The Court of Madrid, of the 20th of June, 1833, contains a Royal Decree, by which a particular Cross is instituted for the Army, by land and sea, in commemoration of the solemn Oath taken to *The Infanta*, eldest daughter of The King. The Decoration of this Order, which will take it's name from *The Infanta*, will consist of a Cross, suspended from a blue riband, —96 Crosses will be distributed amongst each Batallion, and a similar number amongst each regiment of Cavalry. The form of the Cross is not yet known.

THE

KINGDOM OF SWEDEN AND NORWAY.

Preliminary Observations.

Gustavus Adolphus, by an Ordinance, dated, at *Stockholm*, the 26th of November, 1798, confirmed the Four following Orders of Knighthood,—*viz.*,

The Order of The Seraphim,
The Order of The Sword,
The Order of The Polar Star,
The Order of Wasa,—

And, at the same time, he promulgated The Statutes of the several Orders, with some alterations, in which form they are at present in force.

THE ORDER OF THE SERAPHIM.

The most distinguished and most ancient of THE SWEDISH ORDERS is, THE ORDER of THE SERAPHIM, also styled that of "*The Blue Riband.*" King MAGNUS of Sweden, who lived about the year 1280, is supposed to have been it's Founder. Proofs of this opinion, however, are wanting,—but the founding of a chivalrous Decoration, suggested by that of *The Knights of Malta*, and destined for the same object, appears to have been undertaken by him. His nephew MAGNUS ERICKSON, whom the more prevalent opinion makes THE FOUNDER of THE ORDER, probably because he introduced alterations in it, created, in 1386, on his Coronation many Knights of The Order,—and it would seem, that, in earlier times, these creations took place only on similar occasions,—for, at the Coronation of JOHN the Second, in 1497, Fifty Knights were then created.

The existence of THE ORDER dates with certainty as far back as 1336. It has, however, since undergone so many alterations, both as regards it's DECORATIONS and it's STATUTES, having in fact under King CHARLES the ninth, on the introduction of *Lutheranism* into Sweden, been totally abolished, that it's original intention has little or no similarity with it's objects at the present time.

These alterations were effected by King FREDERICK the first, who renewed THIS ORDER, as well as those of THE SWORD and of THE POLAR STAR, on the 17th of April, 1748, on completing the Seventy-second year of his age. He replaced it in it's ancient rights, and promulgated special STATUTES for it's government, which, in the principal points are still in force.

In pursuance of these Regulations, THE ORDER can only be conferred upon Sovereigns, Princely and other high Personages, as well as the highest Officers of The State, whose Rank must at least be that of Lieutenant General.

The King and the Princes of the Blood are born Knights of THE ORDER. With the exception of them, the number of The Knights, —who form only one Class,—is limited to *Twenty-four* NATIVES, and *Eight* FOREIGNERS. The King, however, has the Privilege of presenting the *Insignia* of THE ORDER to *Foreign* Princes.

The Grand-Mastership is inseparable from The Crown.

No SWEDE can receive THE ORDER, unless he is already a COMMANDER of one of the two other Swedish Orders, THE SWORD *or* THE POLAR STAR. On receiving THE ORDER of THE SERAPHIM, however, he becomes, if he is not so previously, *Commander* of the other ORDER which he possesses.

KNIGHTS of THIS ORDER may not accept any *Foreign Order* of Knighthood, under the penalty of forfeiting his right to THE ORDER of THE SERAPHIM.

The Investiture takes place at Stockholm in *The Ritterholm Church*, or, in The Church of *The Royal Palace*.

The Knights are entitled to wear *gilt Spurs*.

A Fee of 800 dollars, together with a dollar for each year of the age of The Knight, is paid on admission to the Treasury of THE ORDER, from which the expenses of THE ORDER are defrayed. This sum, however, is subject to variation by the 23d section of The Statutes.

THE KING addresses the Knights, both verbally and in writing, "*Sir.*"

The 28th of April, the birth-day of King FREDERICK the First, is the grand Anniversary of THE ORDER.

On the death of a Knight, the decorations are returned to The Treasurer of THE ORDER.

In the year 1784, King GUSTAVUS appointed a BISHOP of this ORDER, who should have the Superintendence of the *Ecclesiastics*, who have the care of THE HOSPITALS for THE SICK and for FOUNDLINGS, which are placed under the care of THE ORDER.

THE BADGE of THE ORDER is a gold Star, of eight points, enamelled white, the centre blue, with the Arms of Sweden, and the initial letters, I. H. S., (*Jesus Hominum Salvator*), over the H., a Cross, the arms inclosed with four Seraph's heads, with expanded wings, and in the arms under the lowest crown, the Passion nails. The Seraph's heads are between the double points of the Star, and over the upper points is the regal Crown of Sweden. The obverse of the Star displays the initials F. R. S. (*Fridericus Rex Sueciæ*), the Restorer of The Order.

It is worn, on ordinary occasions, pendent to a sky-blue, watered riband, scarfwise, over the right shoulder,—the Knights having also a STAR, of eight points, embroidered in silver on the left side of their outer garment. But, on extraordinary occasions and days of Ceremony, they wear The Badge pendent to THE COLLAR of THE ORDER, which is composed of eleven golden heads of Seraphs, with wings expanded, and eleven blue Patriarchal Crosses, enamelled on gold, and all joined together with chains of gold.

By the 15th section of The Statutes every Knight must constantly wear upon his Clothes, and around his Coat of Arms, THE BADGE of THE ORDER, under the penalty of forfeiting his Knighthood.

Of the OFFICERS of THE ORDER, the Treasurer, the Secretary, and the Master of the Ceremonies, as Grand Officers, wear indeed the STAR on the left breast.—but wear the BADGE, *not over* the shoulder, but attached to a riband two fingers wide, *round the neck*. The others, as *Inferior* Officers, wear in the button-hole a smaller Cross, appended to a blue riband two fingers wide.

On solemn occasions, or when a CHAPTER is holden, the Knights wear a DRESS of Ceremony. On such days The King dines at the same table with the Knights, who constantly *keep their Hats on.*

THE ARMS of the Knights are exhibited, engraven on copper-plates, in the Choir of *The Ritterholm Church* at Stockholm, where they have their Seats. The MOTTO of the Knight is added, as is his Name, together with the day and year of his Reception. His Coat of Arms is also entered in the Genealogical Archives of THE ORDER, for which the Knight pays 8 dollars to their funds. This sum, however, is subject to variations, according to the 17th section of The Statutes.

The Order of The Sword.

The Order of The Sword, which is also styled "*The Order of The Yellow Riband*," was first instituted by Gustavus Vasa, in the year 1522, — but, having soon declined, and fallen into oblivion, it remained *dormant,* until Frederick the First, the great Restorer of The Swedish Orders, revived it on his 73d birth-day, the 17th of April, 1748, as a Military Order, destined for those, who "*have drawn the Sword with honour in the service of their Father-land.*" The Princes of the Blood are, however, born Knights.

King Adolphus Frederick, his Successor, confirmed this revival three years afterwards, having first made some trivial alterations.

Up to the reign of Gustavus the Third, it consisted only of three Classes,—but, in 1773, His Majesty added another, that of Commanders with the Grand Cross, the present first Class. The Class of Knights with The Grand Cross can be attained only in time of War, and the Recipient must, in quality of Commander in Chief, have obtained a Victory, either by Sea or Land,—Even the King himself, according to the Statutes, can only receive this Cross by the decision of the Army. If, however, the Army under his immediate command gain a Victory, and ascribe this honour to him, he is bound to accept it. This was the case with GUSTAVUS the third, on whom it was conferred, in 1789, by his victorious troops.

THE KING, however, is always MASTER of THE ORDER,—And Princes of The Royal House, on the male side, are born COMMANDERS of THE ORDER.

The four Classes, of which THE ORDER consists, are called,—*viz.*,

1. COMMANDERS, *with* THE GRAND CROSS,—to which The Knights of *The Order of The Seraphim*, who are of the Military Class, naturally belong.
2. COMMANDERS.
3. KNIGHTS, *with* THE GRAND CROSS.
4. KNIGHTS.

THE COMMANDERS *with The Grand Cross* are limited to *Twenty-four*, —COMMANDERS are limited to the same number, independent of The Knights of THE ORDER of THE SERAPHIM, who are also COMMANDERS of THIS ORDER. The Master, however, has the power to extend the number beyond *Twenty-four*, by the admission of *Foreigners*. The number of Knights is unlimited.

No one dares apply for THE ORDER, under the penalty of being declared for ever unworthy of it. The Admission rests solely with The King. He makes his choice known to The Chapter, and the Appointment takes place, unless satisfactory reasons are adduced against it. He, who in time of peace had served with zeal and approbation for *Twenty years*, may become a Knight. A year's service in time of *War* is reckoned equal to several years in time of *Peace*. In order to attain the First or Second Class, he must already possess one of the other two Classes, and the Rank of Colonel.

Those who have been wounded in war, or have captured trophies of Victory from the Enemy, may become Knights without regard to length of service.

THE CELEBRATION of THE ORDER of THE SWORD is not fixed for any certain day, and THE CHAPTER assemble whenever The King wishes. In this Assembly every Commander has a place and a voice, and is consulted in all that relates to the affairs of THE ORDER. The Commanders sit according to Seniority, but The Grand Crosses always take the Chief places.

The Members of THIS ORDER are bound to wear the *Insignia* upon their Dress, and around their Coat of Arms, during life, or to forfeit their right to the Distinction. Persons also who accept a *Foreign Order*, without the permission of The King, forfeit their Knighthood.

The Admission of a Knight takes place without any Religious service in The King's Chamber. The Knights of *The Seraphim*, the Commanders of the other *Swedish Orders*, and the Grand Crosses and Commanders of THE ORDER of THE SWORD must, however, be present. The OATH of a Knight is to defend *The Evangelical Lutheran Belief*, at the risk of life and fortune,—to serve the King and the State with fidelity,—and to meet courageously the Enemies of the Kingdom.

If a *Foreigner* is appointed Grand Cross or Knight, the *Insignia* of THE ORDER are sent to him, and he is bound to furnish a list of the services which he has rendered, which is placed in the Archives of THE ORDER.

Commanders and Knights shall pay such sum to The Treasury as now is or shall be decided upon. The same payment is imposed upon *Foreigners*, which, however, shall be paid to *The Lazaretto* at Stockholm. In addition, Commanders shall pay to *The Charta Sigillata* a fee of twelve rix-dollars twenty-four shillings,—and Knights two rix-dollars and twenty-four shillings, and two rix-dollars to The Chancery.

REVENUES are assigned to THE ORDER, which are expended in *Pensions*, that are determined by THE CHAPTER.

On the decease of a Knight, the Decoration is to be returned to The Treasurer of THE ORDER.

THE BADGE is a gold *St. Andrew's Cross*, of eight points, enamelled white, and edged with gold,—in the centre, which is enamelled blue, is a Sword, erect, white hilt, and pommel of gold, between the thre Swedish crowns,—between the four points of the Cross are four golden crowns, and towards the ends of the points are six swords, of gold, *in chief*, two in *saltier*, the scabards enamelled blue,—*in base*, two in *saltier* also,—on the dexter and sinister sides, one *in pale*, with belts environing the others,—over the two in chief is the Regal crown of Sweden, the cap enamelled blue. The Reverse of the Badge is enamelled blue, thereon a Sword *in pale*, white, hilt and pommel of gold, enfiled with a wreath of laurel, and, over the wreath, this MOTTO, "*Pro Patria.*" It is worn by the first Class, pendent to a broad, watered yellow riband, with a blue border, sashwise over the right shoulder,—and at the same time, a STAR, embroidered in silver on the left breast, and on the mantle, which represents the Obverse of the shield, but without the Swords which surround it. On solemn occasions it is suspended from the neck, by a COLLAR, of gold, consisting of eleven Swords, their belts and guards joined by chains,—on the guards are enamelled Helmets, blue.

The Cross of the Second Class is somewhat smaller, but is worn in the same manner, only, instead of the STAR, a small Sword erect, embroidered in silver or massive silver, is worn on the left breast. He who advances from the third to the second Class by way of distinguishing this circumstance, wears two such small swords laid across, on the left breast, embroidered in silver.

The third Class, and the Knights with the Grand Cross, wear the Badge appended to a smaller riband, round the neck. Their Cross differs, however, from that of the first two Classes, as Swords are only placed between the two upper points. Whoever, belonging to this Class, receives subsequently *The Order of The Seraphim*, or becomes Commander of the collective Swedish Orders, wears under the Star of The Seraphim, the small erect Sword.

The Knights wear the Cross, like the third Class, but smaller, pendent to a still narrower riband in the button-hole.

THE HABIT appointed for Ceremonial occasions, and which resembles the National Costume, is sky blue, with white borders.

THE OFFICERS of THE ORDER are the same, as those of all the other Swedish Orders. The two HERALDS wear, attached to a Scarf of blue and yellow silk, an oval Shield, in the middle of which are the three Ducal coronets, with the erect swords, in a semicircle, with the Motto, "*Pro Patria.*" Their Official Dress is similar to that of the two first Classes, except that, instead of the Mantle, they have only a Mantlet, which is adorned with gold ties, and fringes. On the breast and on the back is a Sword erect, embroidered in gold, and, on the shoulders, they bear the MOTTO, "*Pro Patria.*"

GRAND CROSS.

Sir WILLIAM SIDNEY SMITH, Knt.—*See*, THE TWO SICILIES, ORDER of ST. FERDINAND and of MERIT.

Invested with the *Insignia* of Commander and Grand Cross by His late Majesty GEORGE the Third, at *St. James's Palace.*

To express the sense of GUSTAVUS, King of Sweden, of the share our brave Englishman had in the defeat of the Russian fleet in the Gulf of *Viborg*, he Knighted him on the spot with The Grand Cross of THE NOBLE ORDER of THE SWORD. Not satisfied with this Honour, he asked Sir SIDNEY, what he should give him, as a remembrancer of *Schvenko-Sund?* The Knight replied, "Your Majesty's picture," —*No, my brave friend,*" returned the King, "*you shall have your own!*"— and he the next day sent him a fine Portrait of CHARLES *the Twelfth.* The compliment was worthy of the Prince, and of the British Hero.—PORTER's Travels.

WILLIAM HENRY WEBLEY PARRY, Esq., (*late* WEBLEY), Captain in The Royal Navy, and C. B., "in testimony of the sense which His Majesty GUSTAVUS ADOLPHUS the Fourth, King of Sweden, entertains of his gallant conduct with His Majesty's Ship *The Centaur*, under his command, in the Engagement with the Russian Fleet in *The Gulf of Finland*, on the 26th of August, 1808."—Royal Licence, 19th of January, 1809.

Sir JAMES SAUMAREZ, Bart., K.B., and Vice-Admiral of The Red Squadron of His Majesty's Fleet, "for the high sense which His Majesty The King of Sweden entertains of his great merit and services."—Royal Licence, 4th of May, 1813.

Invested with the *Insignia* of THE ORDER by His Royal Highness THE PRINCE REGENT, at *Carlton House*, the 24th of June, 1813.

ARTHUR, Marquess of WELLINGTON,—*See*, AUSTRIA, Grand Cross of MARIA THERESA.

The Honble. Sir CHARLES WILLIAM STEWART,—*See*, RUSSIA, ORDER of ST. GEORGE, of the Fourth Class.

COMMANDERS.

The Right Honble. Lord MACLEOD,—His Majesty, GEORGE the Third, was pleased to comply with the request of THE KING of SWEDEN, to invest His Lordship with the Ensigns of THE ORDER,—the Ceremony being performed, on the 9th of December, 1778, at *St. James's*.

Captain THOMAS BYAM MARTIN, of The Royal Navy,—

I should act with great injustice to this brave Officer, (*now*, Admiral Sir T. B. MARTIN, and G.C.B.), if I omitted to perpetuate his fine, spirited Letter to Vice-Admiral Sir JAMES SAUMAREZ, which breathes the very soul of a true-hearted BRITON, —and is a LESSON, that should ever be made to thrill through the perfidious breasts of every foe of ENGLAND,—

"H. M. S. *Implacable*,
off *Percola Point*,
8th of July, 1809.

"Sir,

The position taken by *The Russian Flotilla* under *Percola Point*, seemed so much *like a defiance*, that I considered something was necessary to be done, *in order to impress these Strangers with that sense of respect and fear*, which His Majesty's *other Enemies are accustomed to show to The British Flag*, — I, therefore, determined to gratify the anxious wish of Lieutenant JOSEPH HAWKEY to lead the boats of the Ships of The *Implacable*, *Bellerophon*, *Melpomene*, and *Prometheus*, which were assembled by 9 o'Clock last night, and proceeded with an irresistible zeal and intrepidity towards the Enemy, who had the advantage of local knowledge, to take a position of extraordinary strength within two rocks, serving as a cover to their wings, and from whence they could pour a destructive fire of grape upon our boats, which, notwithstanding, advanced with perfect coolness, and never fired a gun '*till actually touching the Enemy*, when they boarded Sword in hand, *and carried all before them*.

" I believe a more brilliant achievement does not grace the records of our Naval History,—each Officer was impatient to be the leader in the attack, and each man zealous to emulate their noble example, and the most complete success has been the consequence of such determined bravery,—of *eight* Gun-boats, each mounting a thirty-two, and twenty four pounder, and forty-six men, *six* have been brought out, *one* sunk, —and the whole of the Ships and Vessels (*Twelve* in number,) under their protection, laden with Powder and Provisions for the Russian army, brought out, *and a large armed Ship* taken and burnt,—I have deeply to lament the loss of many men killed and wounded, and especially that most valuable officer, Lieutenant HAWKEY, who, after taking one Gun-boat, was killed by a grape-shot, in the act of boarding the second. No praise from my pen can do adequate justice to this lamented young man,—as an Officer, he was active, correct, and zealous to the highest degree,—the leader in every kind of enterprise, and regardless of danger,—he delighted in whatever could tend to promote the Glory of his Country,—his last words were, 'HUZZA! PUSH ON! ENGLAND FOR EVER!' "

THOMAS BERTIE, Esq., Rear Admiral of The Red Squadron of His Majesty's Fleet, " in testimony of the sense which His Majesty The King of SWEDEN entertains of his Merits and Services."—He had also the honour of Knighthood conferred upon him.—Royal Licence, 24th of June, 1813. Died, 13th of June, 1825.

Major-General Sir JAMES LYON, K.C.B., G.C.H., and Knight of The Royal Swedish Military Order of THE SWORD, to accept and wear the Insignia of a Commander of the said Royal Swedish Military Order of The Sword, " in testimony of the high sense which His Majesty The King of SWEDEN and NORWAY entertains of the distinguished services rendered by him to the common cause in the field of Battle, in the year 1813."—Royal Licence, 19th of May, 1817.

Sir JOHN ROSS, Knt.,—*See*, RUSSIA, ORDER of ST. ANNE, of The Second Class.

KNIGHTS.

HENRY CLEMENTS THOMPSON, Esq., a Commander in The Royal Navy, "for his services in the engagement with the Russian Fleet in *The Gulf of Finland*, on the 26th of August, 1808.—Royal Licence, 7th January, 1809.

HENRY FREDERICK COOKE, Esq.,—*See*, RUSSIA, ORDER of ST. GEORGE.

JAMES LYON, Esq., Lieut. Colonel of the 97th (*or*, Queen's Own) Regt. of Foot, and a Colonel in the Army, serving as a Major General on the Continent of Europe, "in testimony of the high sense which The King of Sweden entertains of his great merit and services."—Royal Licence, 1st of June, 1814.

ARTHUR FARQUHAR, Esq., a Post Captain in The Royal Navy, Commander of His Majesty's Ship *Desirée*, and Senior officer of His Majesty's Vessels employed in the rivers *Elbe*, *Weser*, and *Ems*, in co-operation with the Allied Troops, "in testimony of the high sense which The King of Sweden entertains of the distinguished gallantry and conduct displayed by him at the siege and reduction of the Fortress of *Glückstadt*."—Royal Licence, 31st of May, 1814.—Now Rear-Admiral of The Blue, C.B., and K.C.H. Knighted.

THOMAS SYDENHAM, Esq.,—*See* RUSSIA, ORDER of ST. GEORGE, of the Fourth class.

KENNETH MAC-KENZIE, Esq., Post Captain in The Royal Navy, "in testimony of the high sense which The King of Sweden entertains of the services rendered by him, during the Blockade of *Norway*."—Royal Licence, 21st of February, 1815.

JOHN MARSHALL, Esq.,—*See*, RUSSIA, ORDER of ST. GEORGE, of the Fourth Class.

JAMES ROSE, Esq., Captain in The Royal Navy ,and Commander of His Majesty's Sloop *Hearty*, "in testimony of the high sense which The King of Sweden entertains of the services rendered by him to the common cause, during the siege and reduction of the Fortress of *Glückstadt*."—Royal Licence, 16th of February, 1815.

FRANCIS BANKS, Esq.,—*See*, RUSSIA, ORDER of ST. ANNE, of the Second Class.

ANDREW PELLAT GREEN, Esq., Captain in The Noyal Navy, K.H., and commanding The Batteries, "as a mark of the approbation of His Majesty CHARLES the Thirteenth, King of Sweden, of his distinguished services at the taking of *Glückstadt*."—Royal Licence, 27th of March, 1818.—Knighted, and K.C.H.

Admiral Sir THOMAS BAKER.

Captain GEORGE.

Lieut. Colonel HALKETT.

Colonel HAMILTON.

Captain G. REYNOLDS.

Captain THOMAS FOX STRANGWAYS.

Captain WRIGHT.

Major MEADE.

KNIGHTS of THE FOURTH CLASS.

PHILIP DUMARESQ, Esq., a Post Captain in The Royal Navy, and late Captain of His Majesty's Ship *The Victory*, "in testimcny of the high sense which His Majesty The King of Sweden entertains of his merits and services."—Royal Licence, 28th of June, 1813.

THOMAS MANSELL, Esq., a Commander in The Royal Navy, and Commander of His Majesty's Sloop *The Rose*, "in testimony of the high sense which His Majesty The King of Sweden entertains of his merits and services."—Royal Licence, 28th of June, 1813.

GEORGE JOHNSTONE HOPE, Esq., Rear Admiral of The White Squadron of His Majesty's Fleet, "in testimony of the high sense which His Majesty The King of Sweden entertains of his merits and services."—Royal Licence, 19th July, 1813.—Died, 2d May, 1818.

JOHN ROSS, Esq., Commander in The Royal Navy, and Captain of His Majesty's Sloop *Briseis*, "as a testimony of the high sense entertained by His Majesty The King of Sweden, of the great courage and intrepidity displayed by him on divers occasions."—Royal Licence, 4th December, 1813.

The Order of The Polar Star.

The Swedish historians assure us, that The Order of The Polar Star is of ancient origin,—but they do not state, when and by whom it was founded.

King Frederick the first, who appears to have been fond of splendour and decoration of this kind, and hence drew forth from their obscurity the forgotten Orders of The Seraphim and The Sword, renewed also with them the equally neglected Order of The Polar Star. This he did, on the 17th of April, 1748, his 73d birth-day,—whence the supposition has arisen, that the Order is a *new* one, entirely of his own creation. It received it's name, however, from the King, who wished, in *The Polar Star* which *never sets,* to exhibit to the Knights a Symbol, which should warn them never to allow the ancient reputation of The Swedish Name *to decline.*

The intention of The Order is to reward those Servants of The State, whether *Ecclesiastical* or *Civil,* who by their talents, integrity and fidelity, have rendered important services to their Country. It is also conferred upon *Foreigners.*

The King of Sweden is always Master of the Order.

The Princes of The Blood are born Knights,—and the *Civilians,* who possess The Order of The Seraphim, are also *Commanders* of this Order.

THE ADMISSION is the same, as in THE ORDER of THE SWORD,—and the OATH is in substance the same.

The Members are also bound to wear THE BADGE in the same manner during life,—and not to accept any *Foreign Order* of Knighthood without the permission of The King. The Decoration of THE ORDER is to be returned to the Treasurer, on the death of the possessor.

THE ORDER consists only of two Classes,—of which the First are called, COMMANDERS, and are *Twenty-four* in number, besides The Knights of THE ORDER of THE SERAPHIM, and FOREIGNERS,—the other are styled, KNIGHTS, whose number is limited to *Fifty*.

Six Ecclesiastics are chosen as *Commanders*, and *Sixteen* as *Members* of THIS ORDER.

No *Ecclesiastic* can be received as a *Commander*, who is not of the Rank of a *Bishop*, and a Member of THE ORDER.

The Members are chosen from Bishops and other high Dignitaries, who have rendered themselves worthy of the Distinction by their Genius and their Learning.

THE BADGE, which is one of the most tasteful extant, is a golden Star, of eight points, enamelled white, with beads of gold,—on the centre is a circle, enamelled blue, thereon a Polar Star of five rays, enamelled white, and encircled with this MOTTO, "*Nescit Occasum*,"—between the angles are four Ducal Coronets, of gold, and, on the two upper points, a Regal crown of Sweden.

The first Class wear THE BADGE, pendent to a broad black riband, round the neck,—hence THE ORDER is sometimes styled, "*The Black Riband*." On days of Ceremony it is worn, suspended to a COLLAR, composed of eleven double F's, back to back, surmounted by a Regal crown, with twelve Polar Stars, of five points, connected together by gold chains. They also wear a SILVER STAR, enriched with the Polar Star in the centre, on the left breast.

The second Class wear a smaller STAR, on a narrower riband, in the button-hole. It is only after possessing the second Class, that a Native can attain to the first.

The same Regulations have been established in THIS ORDER, respecting *Pensions*, as in THE ORDER of THE SWORD.

COMMANDER.

Sir EDWARD THORNTON.

KNIGHT.

Sir WILLIAM CHAMBERS, the celebrated Architect, Surveyor General of The Board of Works, F.R.S., and F.S.A.,—Died, 8th of March, 1796.

The Order of Vasa.

On an attentive perusal of the History of the Kings of Sweden of The House of Vasa, it will be observed, that the first three of the name of Gustavus effected much towards awakening Industry, and particularly in promoting Commerce, Agriculture, Mining, and Manufactures.

Gustavus the third, who ascended the Throne with the conviction, that the welfare of the Kingdom could only be promoted by imitating the example of his forefathers, conceived that he should accomplish this object by founding a peculiar mark of Distinction, by which meritorious conduct in the advancement of Agriculture, M ning, Commerce, and Manufactures should be rewarded and distinguished.

In pursuance of this noble design he instituted The Order of Vasa on the day of his Coronation, the 26th of May, 1772,—and he named it, "The Order of Wasa," because the Swedish word *Wasa* means a *Sheaf*, and is the Family name of the ancient Race, from which Gustavus the first in the year 1523 ascended the Swedish Throne, and this Race bore a *Sheaf* in their Coat of Arms. He could not have found a name which more clearly expressed his grateful remembrance of his valiant Ancestors, or formed a more expressive Symbol of that Agricultural industry, which it was intended to reward.

THE ORDER of VASA, which in the Kingdom is also called, "*The Green Riband*," consists of three Classes,—*viz.*,

COMMANDERS, *with* GRAND CROSSES, 6,—COMMANDERS, 8,—and KNIGHTS, 50, but this number is constantly exceeded.

Foreigners are not included.

It is an ORDER of MERIT for men of every Class.

THE KING, who is always MASTER of THE ORDER, appoints the Members, but has no power to do so, *before* his Coronation. If not a Member of THE ORDER, prior to his Coronation, he receives it solemnly on the Coronation day from the hands of The Bishop, who performs the Ceremony.

The same Regulations, respecting wearing *The Insignia* during life,—accepting *Foreign Orders*,—and returning The Insignia after death, are applied to THIS, as to the other SWEDISH ORDERS.

THE BADGE, which is of gold and alike on both sides, is an oval, in the middle of which is a *Sheaf* bound with a riband, and encompassed with the MOTTO, on a dark red ground of enamel, "*Gustaf den tredje, instiktare* MDCCLXXII," (*Gustavus* 3., *Founder*, 1772).

THE GRAND CROSSES, who pay 800 dollars to the Treasury of THE ORDER, on their admission, wear THE BADGE on a pale green, watered riband, sashwise over the right shoulder,—and, on the left breast, a SILVER STAR, of eight rays, with the gold crowned *Sheaf* in the centre, and four *nettle leaves* in the corners of the Star.

THE COMMANDERS wear THE BADGE of THE ORDER attached to a broad green, watered riband, passed over the right shoulder, but without The Star on the breast or mantle. They pay on their admission 400 dollars to the Treasury of THE ORDER.

THE KNIGHTS, who are exempted from any payments of this kind, wear THE BADGE round the neck, but without the Star.

The two first Classes are attainable at once, without having possessed the third,—as Knights belonging to the third class, do not usually receive the first classes.

On days of Ceremony the two first Classes wear a DRESS of THE ORDER, of green velvet, with a Mantle of the same, lined with white crape.

The Grand Crosses also wear a gold COLLAR, chased and enamelled, composed of four Wheat-sheaves, eight shields of the Arms of Sweden, each regally crowned and embellished with two *Caducei*, and as many *Cornucopiæ* reversed,—and four shields of the arms of *Holstein*, *viz*, a shield *per* fess, white and red, between three Passion nails, all white,—the whole joined with small chains of gold.

The Ceremony of Admission is the same, as in THE ORDER of THE SWORD,—and the OATH prescribed for The Knights, is essentially the same.

COMMANDER.

THOMAS, Earl of KELLIE,—" graciously conferred upon him by GUSTAVUS ADOLPHUS the Fourth, King of Sweden." — Royal Licence, 6th of July, 1808.—Died, 6th of February, 1828.

KNIGHTS.

Sir JOHN HILL, M.D.,—who, immediately after his Investiture, which was performed *by The Swedish Ambassador in London*, was received and recognised, at The Levee of GEORGE the Third, as a Knight of THIS ORDER.

Sir JOHN JAMIESON, M.D.—Knighted, 19th of May, 1813.

JOHN HAWKER ENGLISH, Esq., Member of The Royal College of Surgeons, London, and late Surgeon to The Head Quarters of His Royal Highness THE CROWN PRINCE of SWEDEN, " as a testimony of the high sense which The King of Sweden entertains of his Professional Services in the Field in presence of the Enemy, during the last Campaign."—Royal Licence, 11th of October, 1814.

VALENTINE DUKE, Esq., Surgeon in The Royal Navy, " in approbation of the Services which he rendered to The Swedish Navy in *The Baltick* in the years 1808 and 1809, while Surgeon of His Majesty's Ship *Victory*."—Royal Licence, 29th of July, 1833.

THE ORDER OF CHARLES THE THIRTEENTH.

THIS ORDER was founded, on the 27th of May, 1811, by CHARLES the Thirteenth, as an encouragement and reward for social and benevolent endeavours for the advantage of the Publick, and of all who require Assistance.

The King, who is always Master of THE ORDER, is bound to wear it, as well as The Heir Apparent and The Princes of THE HOUSE of SWEDEN, appointed to that dignity by THE KING. It is also conferred upon *Thirty* Native Swedes, being 36 years of age, appointed by The King, of whom *three* are of *The Ecclesiastical Order*.

It is believed, that THIS ORDER is a peculiar degree of *The Order of Free Masons*, which has been raised to the Rank of an ORDER of the Kingdom. It is, however, certain, that it is only conferred upon *Free Masons*, and those of High Rank, and hence must stand in intimate connection with THIS ORDER. This is confirmed also by the letter B., the meaning of which is known to *Free Masons*.

THE BADGE, surmounted with the Royal Crown, is a Cross, of four points, of ruby red, with a gold border. In the centre of the Obverse, on a white ground, are the initials of The Royal Founder, *viz.*, the number XIII., between two C's, intertwined with each other,—and, on the Reverse, in a triangle, the letter B. It is worn by all the Knights, who form only one Class, pendent to a red, watered riband, round the neck.

Knights can be installed on the 28th of January only.

The Order of Amaranta.

This Order was instituted, in the year 1645, by Christina, Queen of Sweden, the romantick daughter of Gustavus Adolphus *the Great,* in honour of a Lady of the name of Amaranta, equally celebrated for her beauty and virtue,—But, like it's Mistress, it shed it's proudest honours even in it's early blossom.

This Institution, in the reign of it's Foundress, was supported by Her Majesty, with great pomp,—but since that time, as the more *Martial* Orders have arisen, it is only considered as "*Un Ordre de Galanterie,*"—an occasion for social Meetings, Balls, and Festivals being fixed at particular periods, which are numerously attended by the Ladies and Chevaliers of The Order,—all of whom are of the first Rank and Respectability, uniting the exhilaration of Pleasure with the sweets of Innocence.

The Ensign of The Order is a Jewel of gold, composed of two A's, adorned with diamonds on both sides, and joined together by reversing one of them,—being set within a garland of Laurel,—wreathed about with white riband,—and on the sides thereof, this Motto, "*Dolce nella Memoria.*" This Jewel the Knights wear either pendent to a gold chain, or a crimson or blue riband, as they please.

Knights.

Sir Bulstrode Whitelock, Ambassador from The Lord Protector Cromwell.

Sir Robert Ker Porter,—who thus describes his Investiture.— Having been already, with all serious Rites, admitted a Knight of *Mars,*—my next honour of the sort, was to issue from a gayer Court. The Star of the beauteous Queen CHRISTINA, the guide of Love, and "*The Order of The Amaranth,*" the emblem of its never-dying brightness, was to enlist me for ever one of Beauty's Vassals. Accordingly, a little time ago, at the hour appointed, I went to *The Hall of Installation,*—and after going through the usual Ceremonies of Knighthood, taking the delightful Vow of being on all occasions the Defender of Loveliness in distress, the Sword of Consecration fell on my shoulders. The Count DE LA GARDE proclaimed me *Champion of the Fair!* —and, still on my knees, the hands of a beauteous Lady, the Representative of Queen CHRISTINA, fastened the shining BADGE upon my breast,—and I rose, another leaf in The Royal wreath of AMARANTH.—Travels, vol. 2. p. 227.

MEDALS, AND MARKS OF HONOUR.

THE MEDAL "FOR VALOUR IN TIME OF WAR."—is worn on the left breast. It is cast in *gold,* for the Officers,—and, in *silver,* for the Subalterns and Soldiers. There are three different kinds,—one, of King GUSTAVUS the third,—the second, of GUSTAVUS the fourth (ADOLPHUS),—and the third, of CHARLES the fourteenth (JOHN.)

THE MEDAL of "SVENSKSUND'—is of gold, and comprises two kinds,—one of the weight of ten ducats, the other, which is smaller, of five ducats.

THE MEDAL "OF THE SERAPHIM," *or,* "FOR THE DIRECTORS OF THE POOR,"—is cast in gold, of the weight of five ducats, and is granted to persons who have distinguished themselves by their zeal and charitable efforts in behalf of the Poor. On one side is the likeness of the King, with the inscription, "*Fridericus D.'G. Rex Sueciæ,*"—and, on the other, the words, "*Ordo Eq. Seraphim. Restauratus natali Regis* LXXIII.", surrounded by the COLLAR of THE ORDER of THE SERAPHIM, with the words, "*Proceres cum Principe nectit* 1748."

THE MEDAL "FOR THE ANNIVERSARY of 1772,"—is of gold, of the value of 8 ducats, and was struck to commemorate the change of Government, which then took place.

In consequence of the Siege and Capture of *Glückstadt* in 1813-14, by the Army of The Crown Prince of Sweden, and a Squadron under the command of Captain, now Rear Admiral, Sir ARTHUR FARQUHAR, several British Officers, for their distinguished valour and eminent services, received the GOLD MEDAL, bearing on one side THE ORDER of THE SWORD of SWEDEN, and on the other, the words "*For Tapperhet i Falt,*" (*For valour in the Field*), suspended by a blue and yellow riband. Not being an ORDER of KNIGHTHOOD, it is not registered in The College of Arms, but the Officers to whom it was presented by The King of Sweden, received permission to wear it, and they do so on State occasions. The following are the names of the Officers, on whom this Decoration was at that time conferred,—*viz.*,

JOSH. KNEESHAW, Esq., Lieutenant (*now* Commander) in The Royal Navy, and commanding His Majesty's Gun-vessel *Piercer*.

Sir GEORGE KEITH, Bart., Lieutenant in The Royal Navy, and commanding His Majesty's Ship, *The Red Breast*.

CHARLES HAULTAIN, Esq., K.H., Lieutenant (*now* Commander) in The Royal Navy, and Senior Lieutenant in *The Batteries*.

JOB HANMER, Esq., Lieutenant (*now* Captain) in The Royal Navy, and Senior Lieutenant of *The Gun Boats*.

Lieutenant (*now* Captain) STRANGWAYS, of The Royal Artillery,—who succeeded to the command of *The Rocket Brigade*, on the death of Captain BOGUE in the neighbourhood of *Leipsic*, on the 19th of October, 1813.

THE SWISS CONFEDERATION.

MEDALS.

Decrees, concerning the distribution of a Mark of Honour for Officers, Under-Officers and Privates of THE SWISS TROOPS, who returned *from France*, on the summons of THE DIET, of the 12th of June, 1815, declare,—

That the Decoration should consist of a SILVER MEDAL, displaying on one side the antient Banner of The Federates, a white Cross on a red ground, with the legend, "*Swiss Confederation*," and the date 1815, —and, on the reverse, the words, "*Fidelity and Honour*," within a crown of Oak. This Medal to be worn at the button-hole, attached to a red and white riband,—

That this Decoration shall likewise be conferred on Officers of The Swiss Regiments in *Spain*, in similar circumstances as those who returned from *France*.

The Decree, of the 24th of August, 1815, declares, that this Decoration, conferred on the four Swiss Ex-Regiments, should likewise be granted to those individuals of THE COMPANY of THE CENT-SWISSES, who, after the Revolutionary troubles of the month of March of this year, returned to *Switzerland* with their Officers, joined The King of France, and proceeded to *Ghent*.

The Decree, in order to perpetuate the memory of the Conflict of the 10th of August, 1792, declares,—That, on all the Officers, Under-Officers and Privates still living of the Ex-Regiment of The Swiss Guard, who were present at *Paris*, at the attack of the 10th of August, 1792, there should be conferred a MEDAL *of cast iron*, bearing on the obverse, the Cross of The Confederation, with the words, "*Fidelity and Honour*,"—and, on the reverse, the date, "10th *August*, 1792." The Medal to be worn at the left breast, attached to a red riband.

THE GRAND DUCHY OF TUSCANY.

The Order of St. Stephen.

The Order of St. Stephen was first instituted at *Florence,* in the year 1561, by Cosmo the First, Grand Duke of Tuscany, in commemoration of the Victory gained by him near *Marciano,* on the Festival of *St. Stephen,* the 6th of August, 1554, over the French, under the command of Marshal di Strozzi.

It was partly a Military, and partly a Religious Order, like that of *Malta,*—but, in a short time, it fell into disrepute, and was at length entirely disused.

The Grand Duke Ferdinand the First was a zealous Patron of The Order of St. Stephen,—and it was further renewed, on the 22d of September, 1817, by The Grand Duke Ferdinand the Third.

The Order now consists of four Classes,—*viz.*, Grand Crosses, Priors, Baillies, and Knights. The Knights are again divided into two Classes,—Knights *by right,* and Knights *by favour.*

Every one has pretensions to The Order, who can prove his *Nobility* by four descents on the paternal and maternal sides,—and possesses a free income of 300 *scudi* of his own property.

All the Knights are of the *Catholick* religion,—and the reigning Grand Duke is the Grand Master.

THE BADGE of THE ORDER is a gold Cross, of eight points, with rubies or enamelled red. The riband is red, and is worn by the three first Classes, round the neck,—and by the fourth, at the button-hole.

THE STAR which is of silver, is worn by all the four Classes on the left breast, differing only in it's size. THE DECORATIONS are provided at the expense of THE KNIGHTS.

KNIGHTS.

Sir JOHN FRANCIS EDWARD ACTON, Bart., an Officer in The Imperial service in Tuscany, in 1763,—and having particularly distinguished himself in 1775, before *Algiers*, was raised to the Rank of Major-General, and received several marks of favour from The Great Duke LEOPOLD, from The Emperor JOSEPH the Second, and was much honoured by The Empress MARIA THERESA, in several commissions, between these Courts and that of The Two Sicilies. He was sent to *Naples*, in August 1778, named Secretary of State and Lieut. General, with the Department of the Navy and Commerce,—in 1780, the Department of War was united, under his direction,—in 1785, he was made Counsellor of State, and Knight of The Royal Order of *St. Janivier*, and of *St. Stephen* of Tuscany,—and, in 1789, was Minister of the Foreign Affairs. All these he resigned,—but he continued, without any Ministerial influence, to be honoured with the particular confidence and regard of His Sicilian Majesty, for his great services.

Sir PARK GODDARD.

The Tuscan Order of Saint Joseph.

This Order was founded by Ferdinand the Third, Grand Duke of Austria, in 1807, at *Würtzburg*, of which he was then Grand Duke,—and, on being reinstated as Grand Duke of Tuscany, in 1814, he carried this Order with him, and made it the *Second Order* of Tuscany.

It consists of three Classes,—*viz.*, Grand Crosses, Commanders, and Knights.

It is particularly required that the Recipients shall profess *The Roman Catholick* religion.

The Badge of The Order is a gold Cross, of six points, enamelled white,—in the centre of which, upon a gold ground, is the effigy of the Saint, surrounded by the Motto, "*Ubique similis,*" and surmounted by a Crown, of gold,—on the reverse, are the letters, S. J. F. (*Sancto Josepho Ferdinandus.*) It is worn, pendent to a broad red, watered riband, edged with white, from right to left. And, on the left side, a Silver Star, bearing the obverse and the same Motto, as The Badge.

The second Class wear it smaller, and without The Star, also round the neck,—And the third Class, in the button-hole.

The Collar is only worn on solemn occasions.

Grand Cross.

The Right Honble. Lord Burghersh, Envoy Extraordinary and Minister Plenipotentiary to The Court of Tuscany, Colonel in The Army, Aid-de-Camp to The Prince Regent, C.B., G.C.H,, and of The Royal Sicilian Order of Saint Ferdinand and of Merit, and Knight of The Imperial Austrian Military Order of MARIA THERESA, to accept and wear the Insignia of a Knight Grand Cross of THE TUSCAN ORDER of SAINT JOSEPH, conferred upon His Lordship by His Imperial and Royal Highness The Grand Duke of Tuscany, "in testimony of His Imperial and Royal Highness's approbation of the Military services rendered by him against the Enemy in the field, during *The Neapolitan Campaign*, and throughout the period in which the troops of His Imperial and Royal Highness were engaged in active operations with *The Allied Army in Italy*."—Royal Licence, 3d of November, 1818.

KNIGHTS.

Captain THOMAS AUBIN, "as a mark of the approbation of His Imperial and Royal Highness The Grand Duke of Tuscany, of the Military services rendered by him, while serving as Aid-de-Camp to Colonel The Right Honble. Lord BURGHERSH, during the Campaign of *Naples*, and throughout the period in which the Troops of His Imperial and Royal Highness were engaged in active operations with *The Allied Army in the South of Italy*." —Royal Licence, 3d of November, 1818.

GEORGE AIREY, Esq., Major General in The Army, and Quarter-Master-General of The Forces serving in Ireland, "as a mark of the approbation of His Imperial and Royal Highness The Grand Duke of Tuscany, of the services rendered by him to the Tuscan Government in the defence of *Porto Ferrajo*, in October, 1801."—Royal Licence, 24th of September, 1819.—Knighted, in 1820, and K.C.H.

THE ORDER OF THE WHITE CROSS.

THIS ORDER was founded by The Grand Duke of Tuscany, FERDINAND the Third, Archduke of AUSTRIA, in 1814, for the reward of Military merit. It received it's name from the colour of it's symbol, and is also denominated, "*The Cross of Fidelity.*"

Nothing has been publickly made known, as to it's establishment.

MEDALS.

THE MILITARY MEDAL, of SILVER,—THIS MEDAL, which is conferred on Under-Officers and Privates, as a recompense for merit and services rendered to The State, was instituted in the year 1815. It bears on one side, the Portrait of The Grand Duke FERDINAND the third,—and, on the other, the inscription, "*Ai prodi e fedeli Toscani,* 1815." It is worn at the button-hole.

COPPER MEDAL, for LONG MILITARY SERVICE.—THIS MEDAL was also instituted in 1815. It bears the inscription, "*Al lungo e fedel servizio,*"—and is conferred upon Under-Officers and Privates, after 25 years of Service. A Pension of three *liri per* month is added to it. After the first ten years of service Under-Officers and Privates bear a *Chevron à la Française,* and receive a monthly pension of one *lire.* Half a *lire* and a second Chevron is conferred after 15 years of service. After 20 years, the pension is again augmented one *lire,* and they wear a third Chevron.—On the completion of 25 years service, they bear no more Chevrons, but receive in their place, the MEDAL,—which is worn at the button-hole.

THE REPUBLICK OF VENICE.

THE ORDER OF SAINT MARK.

Until the period of it's overthrow by The French, THE REPUBLICK of VENICE conferred THE ORDERS of ST. MARK, and THE GOLDEN STOLE.

THE ORDER of ST. MARK was established in the year 828, when the Body of the Saint was brought from *Alexandria*, in Egypt, to *Venice*.

The Knights who were honoured with THIS ORDER, enjoyed all the rights of Patricians, and free Citizens.

THE COLLAR of THE ORDER is a gold Chain, to which is pendent a MEDAL of gold, on one side whereof is the Emblem of *St. Mark, viz.*, a winged Lion, *sejant*, with wings elevated, holding in his sinister paw a drawn Sword, erect, and, in his right, an open book, with this MOTTO, "*Pax tibi, Marce, Evangelista meus.*" On the reverse, the portrait of the reigning Doge, with the image of *St. Mark* delivering a Standard to him.

The Order of the Golden Stole.

The Republick of Venice likewise appropriated to herself the right of creating Knights, the number of whom was not limited. Such Knights were styled, "*Cavalieri della Stola d' Oro,*" or, "*Knights of The Golden Stole,*" on account of the Decoration which they wore, and which was suspended over their left shoulder.

This consisted of a Stole, of gold tissue, richly embroidered with flowers of gold in relief. It was of the breadth of a hand, and was so worn over the shoulder, that it fell behind and before, as low as the knees.

The Senate conferred this Dignity only upon persons, who were descended from the oldest *Nobility* of The State,—or, upon those who had rendered themselves worthy of it, by their having filled with honour the post of Ambassador, on some extraordinary occasion.

THE KINGDOM OF WÜRTEMBERG.

The Order of The Crown.

His Majesty William, King of Würtemberg, having determined to unite the hitherto separately existing Orders of "*The Golden Eagle*," and "*The Civil Order of Merit*,"—by an Ordinance, dated, at *Stuttgard*, the 23d of September, 1818, declared,—

That the Union of these two Orders is to be solely applicable to the system of distribution, from henceforward,—and that it has no retrospective reference to the present existing Members of either Order, whose relative connections therewith, as well as the respective Statutes thereof, (now jointly merged into The New Order), are to continue in full force, with the exception of such alterations as are prescribed in the following Rules,—

The Order now substituted for the *two* Orders above mentioned, is to be called, "The Order of The Crown of Würtemberg."

It is to consist of three Classes, *viz.*, Grand Crosses, Commanders, and Knights.

The number of Members to be included in each separate Class is not yet settled, but, in determining the same, regard will be had to the number admitted into the Statutes of "*The Civil Order of Merit.*"

Besides The Order, thus divided into three Classes, Gold and Silver Civil Medals of Merit will be distributed.

The Badge of The Order is a gold Cross, of eight points, enamelled white, in the four angles of which appear the Leopard-Lions of The Royal Arms. In the middle of the Cross is a golden, white enamelled shield, on which is the cypher of King Frederick the First, who died in 1816, surrounded with a crimson coloured border, bearing in golden letters the Motto, "*Fearless and True.*" The reverse is similarly bordered, and has in the middle, a Royal Crown of gold, emblazoned on a white ground. Above the Cross is a crown of gold, connected by gold rings.

The Cross is worn, suspended over the right shoulder, from a crimson riband, edged with black, descending towards the hip. The Grand Crosses also wear, on the left breast, a Silver Star, of eight points, the middle of which exhibits the centre shield of the Cross, and the crown of gold, upon a white ground.

The Commanders and Knights wear the Cross, *only*, suspended from the neck of the former, and from the button-hole of the latter, by a similarly coloured, but narrower riband.

The Gold and Silver Civil Medals of Merit are worn, suspended from the button-hole, by the riband of The Order.

The distinctions conferred by The Order are to be added by the Members to their other Honorary Titles,—and The Badge is to be introduced into their Armorial Bearings.

The Order confers upon Commoners, *Personal Nobility*, as well as the privilege of attending at Court,—but no virtual or *hereditary* Rank.

Knight.

Augustus Bozzi Granville, M. D.—*See*, Bavaria, Order of St. Michael.

THE MILITARY ORDER OF SAINT CHARLES.

THIS ORDER was instituted by CHARLES, Duke of WÜRTEMBERG, on the 11th of February, 1759, being the Anniversary of his birth,—for the purpose of rewarding those Officers, who had served in his Corps of Troops, which was joined to The Imperial Army, during *The Seven Years'* War.

THIS ORDER did not undergo any change, either during the reign of this Prince, or of those of The Dukes LOUIS-EUGENE, and FREDERICK-EUGENE, who were his immediate Successors,—but, when, towards the end of the month of August, 1799, the French Troops attempted to make an incursion into The Duchy of WÜRTEMBERG, on the side of *Manheim*, His Serene Highness, the reigning Duke FREDERICK the Second, having caused some Batallions of new raised Recruits to advance, these seconded solely by The Imperial Regiment of *Albert*, Cuirassiers, were so fortunate as to give the enemy a considerable check, and repulsed the French, with the loss of some thousands of Prisoners. As, by this exploit, they had particularly merited the approbation of their Sovereign, THE DUKE resolved upon that occasion, and the more to augment their courage and bravery, *to renew The Order of Saint Charles*, naming it henceforward, "THE DUCAL ORDER of MILITARY MERIT of WÜRTEMBERG,"—considering this renewal the more necessary, as during the preceding reigns, particularly that of Duke LOUIS, the distribution of THIS ORDER had not been made, in a manner conformable to The Statutes.

On the 6th of November, 1799, almost all the Knights *of the old*, or *former* ORDER of SAINT CHARLES, obtained the one, *then renewed*,—and, on the death of the last non-acceptant, THE ORDER of SAINT CHARLES will become *extinct*.

The Ducal Order of Military Merit.

This Order was founded by Frederick the Second, reigning Duke of Würtemberg, on the 6th of November 1799, the Anniversary of his birth. It was modified a second time, on the 6th of November, 1816.

In the Preamble to The Statutes, His Serene Highness expresses himself to the following effect,—

As, during the course of the present War, Our troops, on the behalf of their Mother-Country and their natal Soil, have taken a part in the fatigues the dangers, and the glory of the field,—and *that*, for the preservation and maintenance of Publick security, and general good Order, —We find it, therefore, necessary during this actually existing state of circumstances, *to renew this Order of Knighthood*, and to reward them for their toils and sufferings, for the hardships and perils to which they have been exposed, and for the fidelity and bravery they have shown in the field, by conferring upon them the honourable Badge of The Order, founded by our Ancestors,—but, as various changes have happened, during a considerable lapse of time, Our will and desire is, that, henceforward, this Order be renewed and re-established, in manner as then prescribed.

The Order, formerly denominated "*The Military Order of Saint Charles of Würtemberg*," thereupon became, "The Order of Military Merit of Würtemberg."

The reigning Sovereign of this Illustrious House is the Grand Master, by right of inheritance. The reception and investiture of The Knights Companions depend upon him only. But, unless he be arrived at the age of majority, he cannot confer it upon any one.

The Princes of The House of Würtemberg are the Senior Knights.

It consists, moreover, of three Classes,—*viz.*,

GRAND CROSSES,—of which Class, the Senior enjoys a yearly Pension, during life, of 400 Rhenish florins.

COMMANDERS,—of which Class, the two oldest each receive a yearly Pension of 200 florins.

KNIGHTS,—of which Class, the four oldest each receive a yearly Pension of 100 florins.

A great part of the Revenues of THIS ORDER have arisen, since 1810, from the produce of *The Maltese Commanderies* in Würtemberg, which were suppressed at that time.

THE ENSIGNS of THE ORDER consists of the following Decorations,—

THE STAR is of gold and silver embroidery. It is of the Maltese form, and in the centre, is a field, *azure*, in the middle of which is embroidered the letter W.,—and, in the four principal points of the Star, are the words, embroidered in letters of gold, "*Bene Merentibus*," which is the MOTTO of The Order. The Grand Crosses alone wear this Star, upon the left side of their upper apparel.

The Cross, which is of the Maltese form, is of gold, enamelled white. In the centre, is a field *azure*, upon the middle of which is the letter W., —and, in the four principal points, is the MOTTO, as already described. It is surmounted with a Ducal Hat. It is worn, pendent on the breast by The Grand Crosses, and by the Commanders, attached to a rich yellow, watered riband, distinguished by a black border, — such being The Imperial colours, and also those of The Circle of Suabia, in which The States of His Highness are situate.

The Riband worn by The Grand Crosses, is somewhat broader than that, which is assigned to The Commanders.

The Junior Knights wear the same Cross, but it is somewhat smaller, and is not adorned with the Ducal Hat. It is worn, attached to the third and fourth button-hole of the Regimentals.

Such Officers as are invested with this honourable Badge, are obliged, as a proof of their regard and esteem for so evident a mark of their services and deserts, to wear the same constantly. In case any one neglects so to do, he shall be obliged to pay a fine of 20 Rix dollars to the Treasury Chest, established in favour of the Children of indigent or invalid Soldiers,—and should he be culpable of notorious neglect for any long space of time, he shall be degraded from his Dignity.

Extraordinary Merit, or long and faithful Services, can alone entitle any one to be received as a Knight Companion of THIS ORDER.

The affairs of THE ORDER are transacted by the oldest Knight, together with a Secretary.

KNIGHT.

HENRY FREDERICK MULLER, Esq., Colonel in The Army, late Lieut. Colonel of the 1st, *or*, Royal Scots Regt. of Foot, and now Lieut. Colonel of The Ceylon Regt., (*Riflemen*), " conferred upon him by His Majesty THE KING of WÜRTEMBERG, for his services during the late war."—Royal Licence, 1st of October, 1825.

The Royal Würtemberg Order of Frederick.

William, King of Würtemberg, animated by the wish to honour and preserve the memory of his late Father, for his great services to The Royal House, and to The State, resolved to found a New Order of Knighthood, and to give it the name of "The Royal Würtemberg Order of Frederick."

The Regulations are contained in nine articles, and are dated, at *Stuttgard*, the 1st of January, 1830.

The Order consists of one Class only, that of the Knights.

The Badge of The Order is a gold Cross, of four points, enamelled white, the angles being occupied by golden rays,—on a white circle, in the centre, is the Motto, "*To Merit,*" surrounded by a purple ring, inscribed with the Motto of the deceased Monarch, "*God and my Right.*" It is worn, pendent to a broad blue riband, from right to left.

The Star, radiating in four points, contains in the centre, the effigy of the late King Frederick, with the legend, "*Frederick, King of Würtemberg,*" on a purple circle.

MEDALS, AND OTHER MARKS OF HONOUR.

King WILLIAM, on the 23d of September, 1818, founded THE MEDAL of CIVIL MERIT, in gold and silver, to recompense services rendered to the State. It is worn at the button-hole, attached to the riband of THE ORDER of THE CROWN.

On the same day the same King established THE MEDAL of MILITARY MERIT, also in gold and silver. It is destined for Under-Officers and Soldiers, and is worn at the button-hole, attached to the riband of THE ORDER of MILITARY MERIT. Since the renewal of The Order of Military Merit, however, on the 23d of September, 1818, no distribution of MEDALS has taken place. The *gold* Medal will be conferred as low as the rank of Corporal inclusive,—that, in *silver*, as usual.

THE ANCIENT MEDALS of MILITARY MERIT were worn, attached to a yellow riband with a black border.

King FREDERICK the First established the following MEDALS, *for the Campaign* of 1814,—*viz.*,

1. THE MEDAL, in gold or in silver, for the Battle of *Brienne*.
2. THE MEDAL, in gold or in silver, for the Battle of *Fére-Champenoise*.
3. THE MEDAL, in gold or in silver, for the Battle of *Paris*.

The *gold* Medals were intended for Officers,—and those, in *silver*, for Subalterns and Soldiers, who had distinguished themselves in these Battles.

King FREDERICK the First also instituted *for the Campaign* of 1815, CROSS, in gold, and in silver,—and a SILVER MEDAL.

The *gold* Cross was destined for Superior Officers,—The Cross of *silver*, having the same form as that of gold, for Subaltern Officers,—And the MEDAL, which is of the same form as the MEDAL for MILITARY MERIT, was intended for Under-officers and Privates.

THE UNITED KINGDOM

OF

GREAT BRITAIN AND IRELAND.

The Patriotic Fund Institution.

It was my original intention to have confined my descriptions strictly to Foreign Orders of Knighthood,—but a brave Veteran,—Lieut. Colonel Charles Menzies,—having suggested the propriety of introducing those Honorary Rewards which have been bestowed by The Committee of The Patriotic Fund, I could not resist such a National impulse,—indeed, little persuasion was necessary,—for, as my heart was already warmed with the gallant exploits which I was recording, I dismissed at once all imaginary restraints,—and who would have hesitated ?—to embellish my Pages with the splendour of Rewards, consecrated and enhanced by purity and justice of Distribution, and the real Worth of the meritorious, and rendered invaluable when received at the hands of such a lofty Committee, who will ever be remembered, as the firmest Supporters of The Brit-

ISH THRONE. I confidently anticipate, that these daring acts of individual Bravery, cannot fail to be useful and interesting to Military men of all Ranks, as well as afford historical and gratifying information to the general Reader,—especially to those who have served in THE NAVY or ARMY, or who have RELATIVES in the Service.

Authentick accounts of the Actions of the Great, the Valiant, and the Loyal, have always been of paramount interest with a brave and generous People. GREAT BRITAIN has produced a Race of HEROES, who, in moments of danger and terror, have stood "*firm as the Rocks of their native Shore*," — and when half of the World has been arrayed against them, they have fought the Battles of their Country with unshaken Fortitude, and unparallelled Victories.

What Language, then, can be used sufficiently vigorous to describe their brilliant Achievements, and constitute such a resplendent and spirited Record, as shall revive the memory of the brave, and place their gallant Exploits in the brightest array before us?—I find this animating applause in the wise, the dignified, and the humane Reports of THE PATRIOTIC FUND COMMITTEE.

IN the Preliminary observations to their PROCEEDINGS, dated the 1st of March, 1804, THE COMMITTEE state, that, at a period in which such signal display has been made of BRITISH LIBERALITY and ENERGY, in which the spirit of unanimity and patriotism pervades all Ranks, and animates every Heart, they cannot but be sensible of the general interest taken by their Countrymen in an

INSTITUTION founded upon these noble principles. They feel it, therefore, their duty, to lay before the Subscribers and the Publick, the following Report of their Proceedings, hoping it will evince that they have not been unmindful of the great Trust committed to their care,—

They are anxious to inform those who are nobly foremost in fighting the battles of the Country, that the attention of THE COMMITTEE to their merit and sufferings is constant and unwearied, and their observation watchfully directed to reward the one, and to alleviate the other,—

They are also proud to notice, that on every occasion which has presented itself in the confined and partial warfare to which the Enemy has hitherto given opportunity, BRITISH vigour, enterprise, and courage, have maintained with equal merit and success, their wonted PRE-EMINENCE,—

The sense and love of Liberty, Independence, and Honour, those high characteristics of BRITONS, which have raised to such enviable distinction the Fame and Power of THESE ISLANDS, promise, under the favour of DIVINE PROVIDENCE, that future and more arduous Contest will be productive only of additional Glory, and ensure successful Issue to every effort for National security and honour,—

With Patriotick exultation THE COMMITTEE present to The Publick, the Report of a Subscription great beyond any recorded in the Annals even of BRITISH generosity, although the Establishment of this Fund is hardly yet known to many great Dependencies of The

Empire, from whose experienced Liberality and proved Attachment, considerable aid may be expected,—Under these circumstances, The SEAMAN, The SOLDIER, The VOLUNTEER, may confidently trust, that those who are dear to him while living, will, in the event of his falling in the sacred Cause, FIND FRIENDS AND GUARDIANS IN A GRATEFUL AND GENEROUS COUNTRY.

At the Meeting of The Subscribers to *Lloyd's*, holden on the 20th of July, 1803, when THE PATRIOTIC FUND was established,—

BROOK WATSON, Esquire, in the Chair,—

It was RESOLVED,

That in a conjuncture, when the vital interests of our Country, when the peculiar Blessings which, under our beloved Sovereign and happy Constitution, endear our social state, are involved in the issue of the present Contest,—when we are menaced by an Enemy, whose haughty presumption is grounded only on the present unfortunate position of *The Continental Powers*,—and when we seem to be placed, for the moment, *as the last barrier against the total subjugation of Europe* by the overbearing influence of *France*,—it behoves us to meet our situation as men,—as FREEMEN!—but, above all, as BRITONS! On this alone, with the DIVINE AID, depends our exemption from the yoke of *Gallic Despotism*,—on this alone, depends, under the same Protecting Power, whether this Empire shall remain what it has for Ages been, the strenuous Supporter of Religion and Morals, the Assertor of it's own, and the Guardian of the Liberties of Mankind, the Nurse of Industry, the Protector of the Arts and Sciences, the Example and Admiration of the World,—or, whether it shall become an obsequious Tributary, an enslaved, a plundered, and degraded Department of a *Foreign* Nation,—

RESOLVED,

That to give more effect and energy to the measures adopted by Government for the Defence of our Liberties, our Lives and Property,—to add weight to those personal exertions we are all readily disposed to contribute, it behoves us to hold out every encouragement to our Fellow-subjects, who may be in any way instrumental in repelling or annoying our implacable Foe,—and to prove to them, that we are ready to drain both our purses and our veins in THE GREAT CAUSE, which imperiously calls upon us to unite the Duties of Loyalty and Patriotism, with the strongest efforts of zealous exertions,—

RESOLVED,

THAT TO ANIMATE THE EFFORTS OF OUR DEFENDERS BY SEA AND LAND, IT IS EXPEDIENT TO RAISE, BY THE PATRIOTISM OF THE COMMUNITY AT LARGE, A SUITABLE FUND FOR THEIR COMFORT AND RELIEF,—FOR THE PURPOSE OF ASSUAGING THE ANGUISH OF THEIR WOUNDS, OR PALLIATING IN SOME DEGREE THE MORE WEIGHTY MISFORTUNE OF THE LOSS OF LIMBS,—OF ALLEVIATING THE DISTRESSES OF THE WIDOW AND ORPHAN—OF SMOOTHING THE BROW OF SORROW FOR THE FALL OF DEAREST RELATIVES, THE PROPS OF UNHAPPY INDIGENCE OR HELPLESS AGE,—AND OF GRANTING PECUNIARY REWARDS, OR HONOURABLE BADGES OF DISTINCTION, FOR SUCCESSFUL EXERTIONS OF VALOUR OR MERIT,—

RESOLVED,

That a Subscription, embracing all the objects in the foregoing Resolution, be now opened,—and to set an example to the Publick Bodies throughout The United Kingdom and its Dependencies, and to our Fellow-subjects of every Class and Denomination, that, independently of our individual Contributions, the sum of TWENTY THOUSAND POUNDS *Three per Cent. Consolidated Annuities,* part of the Funded Property of THIS SOCIETY, shall be appropriated to this purpose.

N. B. In the month of August, 1809, £5000 additional was voted.—and, in January, 1813, £10,000, more, — making altogether £35,000 *Consols.*

RESOLVED,

That all Sums, *however small,* which shall be offered by the Patriotism of the poorer Classes of our Fellow-subjects, shall be accepted, —THE CAUSE affecting equally the Liberties and Lives of Persons of every description.

On the 29th of July, 1803, at a GENERAL MEETING of Subscribers, it was RESOLVED,—

That a Committee be then formed from the first Fifty Subscribers of £. 100. and upwards,—

And, on the 2d of August, 1803, at a Meeting of THE COMMITTEE for managing The Patriotic Fund, it was RESOLVED,—

That the following Gentlemen be appointed TRUS-

TEES for the purchase and sale of Stock, or other Government Securities, for the purposes of this Institution, —and that the same be invested in their Names,—

Sir FRANCIS BARING, Bart.
JOHN JULIUS ANGERSTEIN, Esq.
THOMSON BONAR, Esq.

It will be observed, that THE HONORARY REWARDS were almost wholly confined to NAVAL OFFICERS, who had had *the first* glorious opportunity of sweeping the Ocean, and every lurking Inlet of the rancorous Enemy, —and that the distribution of these proud Trophies of their valour was *discontinued*, *before* the brave ARMY had been favoured with a similar opportunity of demonstrating their unrivalled prowess on shore,—and *before* THE MILITARY OPERATIONS of the Country had commenced upon a scale which astonished all Europe, and eventually hurled the disturber of Mankind from his boastful eminence.

At a General Meeting of The Merchants, Bankers, and Underwriters, and other Subscribers to THE PATRIOTIC FUND, holden at *Lloyd's* on the 24th of August, 1809,—

Sir FRANCIS BARING, BART., in the Chair.

The following Statement of the Receipt and Expenditure was laid before the Meeting by JOSEPH MARRYAT, Esq., on the part of THE COMMITTEE,—

Amount of Subscriptions . .	£354,677	1	6
Interest and Profit on Stock re-sold	70,155	15	1
	£ 424,832	16	7

Gratuities and Annuities to 2050 Widows, Orphans, and Relatives of Officers and Men killed in Action with the Enemy, to 7642 Officers and Men wounded or disabled, to the relief of British Prisoners of War in *France*, and various expenses, ...	331,611 12 2
	£ 93,221 4 5

It further appeared by The Report of THE COMMITTEE, that they had Lists of 3469 men killed, and 6010 men wounded, for whose cases no provision was yet made,—and that they estimated these demands upon The Fund, on the present system of expenditure, at £ 129,480.—It was, therefore,

RESOLVED UNANIMOUSLY,

That the enlarged Scale upon which THE MILITARY OPERATIONS of the Country are now carrying on, makes it necessary to reconsider the Original Resolutions under which THE COMMITTEE have hitherto acted,—and, *that in future the Donations from this Fund be appropriated exclusively to the Widows, Orphans, and Relatives who depended for support on those who fall in the Cause of their Country,—to the relief of those whose Wounds are attended with the loss of limb or disability from future service,—to the annual Allowance made towards the aged and infirm British Prisoners of War, and to the support of the Schools and Hospitals established at the different Depôts in France.*

N. B. The annual allowance was discontinued in 1812.

That an Appeal be made to the Patriotism and Liberality of the Publick, for NEW CONTRIBUTIONS to THE PATRIOTIC FUND, that THE COMMITTEE may be provided with the means of continuing the accustomed relief to these meritorious objects.

It may be proper to mention, that the Report, which was read at the second General Meeting of The Subscribers, contains a full account of the labours of THE COMMITTEE up to that period, — and that, with the *discontinuance of The Honorary Rewards,* ceased that *interesting, noble,* and *general Excitement,* which the publication of *The Proceedings of* THE COMMITTEE had heretofore produced,—and, consequently, the *subsequent* exertions of THE COMMITTEE were limited to the performance of the less ostentatious but pleasing task of administering pecuniary assistance to the Widows, Relatives, and Orphans of our brave Defenders who were *killed,*—and to those who were *disabled* in fighting the Battles of their Country.

It may, also, be further noticed, that THE COMMITTEE in the discharge of the sacred trust reposed in them, continued in their work of Benevolence for several years, *after* the Conclusion of the War,—until, at length, it was considered advisable and proper to name a definite period for the Admission of Claimants. In pursuance of a Resolution, therefore, which was in consequence passed at a General Meeting of THE COMMITTEE, holden at *Lloyd's,* on the 28th of May, 1823,

JOSEPH MARRYAT, Esq., in The Chair.

due notice was given, by Advertisement,

in the Newspapers of The United Kingdom, that all out-standing claims on THE FUND should be presented before the 20th of July, 1824,—At their Meeting, holden on the 9th of July, 1824, the period which was limited for the Admission of Claims, was extended to the 25th of February, 1825,—And, at their subsequent Meeting, on the 24th of May, of that year,

ROBERT SHEDDEN, Esq., in The Chair,

a General Statement of the Accounts was laid before them, and The Committee having taken the same into consideration,—and, also, that in the month of July next ensuing, a period of *Ten years* will have elapsed *since the termination of the War*, during which space of time sufficient opportunity will have been afforded to all those who were interested, to prefer their claims for the benefit of The Fund,—It was, therefore,

RESOLVED,

That, on the 1st of July, 1825, the operations of THE PATRIOTIC FUND shall cease and terminate, as far as relates to the late War, ending in the year 1815,—and no further Claims shall be admitted, or Donations given, except under some special and extraordinary circumstances submitted for the decision of a COMMITTEE summoned for that particular purpose.

That a printed Circular, signed by The Secretary, announcing the *closing* of THE PATRIOTIC FUND, be returned to all future Claimants,—

That all such Balance of THE PATRIOTIC FUND, as now appears to be in hand, or which hereafter may be in hand, after paying the necessary expenses of Salaries,

Postage, Rent of Office, and such other Disbursements as may occur, shall be holden in TRUST, by the three present Trustees or Treasurers, or such other as THE COMMITTEE for the time being may appoint, for the purpose of providing a FOUNDATION for a Subscription *in the event of a future War,* and thereby perpetuating the benefits of an INSTITUTION, so honourable in it's commencement to THE MERCHANTS and UNDERWRITERS at *Lloyd's,* and so zealously and unanimously supported by every Class and Description of Persons throughout THE UNITED KINGDOM,—a SUBSCRIPTION which, in it's amount, stands pre-eminent of all other Subscriptions ever known in this Country, and which most happily and effectually answered those purposes for which it was originally intended, *viz.,* to animate our Defenders both by Sea and Land, and to relieve the pains and the sorrows of those who might suffer during the sacred and arduous struggle for our Existence, as a NATION.

It may be proper here to observe, that the sum of £21,274 17*s.* 2*d.* was expended in HONORARY REWARDS, —*viz.,*

SWORDS, - -	£10,241	2	6
PLATE, - -	11,033	14	8
	£21,274	17	2

And that THE COMMITTEE recommended, that these should be *discontinued,*—as claims of *Merit alone,* ought to yield to claims of *Merit and Distress combined.*

By a General Statement of the Receipts, from the commencement of THE FUND in July 1803 to March 1826, it appears, that they amounted to the magnificent sum of SIX HUNDRED and TWENTY-NINE THOUSAND EIGHT HUNDRED and TWENTY-THREE POUNDS, FOURTEEN SHILLINGS, and ONE PENNY ! ! !

The present TRUSTEES are,—

GEORGE SHEDDEN, Esq.
GEORGE ROBERT SMITH, Esq., M. P.
WILLIAM GEORGE SHEDDEN, Esq.
WILLIAM THOMPSON, Esq., M. P., and Alderman.
ABEL SMITH, Esq., M. P.

But I hasten to inscribe upon an imperishable Tablet THE NAMES of THE HEROES, who have so nobly acquired these PROUD and GRATEFUL MEMORIALS of their VALOUR.

Honorary Swords.

	Value.
Lieutenant Charles Adams, H.M.S. *Renommée*, for his gallant conduct, in the command of one of the boats belonging to that Ship, in boarding and carrying The Spanish Schooner, *Giganta*, on the 4th of March, 1806, . .	£. 50
Lieutenant James Ayscough, H.M.S. *Centaur, off Martinique*, for his bravery, in assisting in the complete destruction of a Battery in *Petit Ance d'Arlette*, mounting 6 twenty-four pounders, on the 26th of November, 1803, . .	50
Captain Thomas Baker, H.M.S. *Phœnix*, for his skill and gallantry, in the capture of The French Frigate *La Didon*, of 44 guns and 330 men, on the 10th of August, 1805,	100
See, Captain The Honble. Alan Gardner.	
Lieutenant Barber, H.M.S. *Magicienne*, for his gallant conduct, in the capture of The Dutch Gun-boat *The Schrik*, on the 23d of March, 1804, when he was wounded, .	50
Captain Henry William Bayntun, *v.* Captain Charles Bullen.	
Lieutenant George Beatie, of The Royal Marines, H.M.S. *Blenheim, off Martinique*, for his gallant conduct in storming *Fort Dunkirk*, a battery of nine guns, on the 17th of November, 1803, ...	50
Lieutenant George Bettesworth, H.M.S. *Centaur*, for his intrepidity in the capture of *The Curieux*, French Corvette, *off Martinique*, on the 4th of February, 1804,—when he was wounded, ...	50
Austin Bissell, Esq., Commander of H.M.S. *Racoon*, for the capture of The Brig *La Petite Fille*, the Schooner *La Jeune Adele*, and the Cutter *L'Amelie*, off the coast of *Cuba*, on the 13th of October, 1803, ...	100

	Value. £.
Captain The Honble. H. BLACKWOOD, *v.* Captain CHARLES BULLEN.	
Lieutenant BLIGH, H.M.S. *Cerberus*, for most gallantly boarding and bringing out, a French Schooner, and a Sloop, near to *The Pearl Rock,* Martinique, under a most tremendous fire of cannon and musquetry from the shore,—on the 2d of January, 1807, ...	50
Lieutenant B. BLUETT, H.M.S. *Scorpion,* for gallant conduct in command of Boats, in boarding and carrying The Dutch National Brig, *Atalante,* on the 3d of April, 1804,	50
Captain WILLIAM BOLTON, H.M.S. *Fisguard,* for his determined bravery and conduct, in storming the forts, citadel, and town of *Amsterdam,* Curaçoa, on the 1st of January, 1807, ...	100
Lieutenant JAMES BOWEN, H.M.S. *La Loire,* who, with Lieut. FRANCIS TEMPLE, cut out *The Venteux,* French Gun Brig, from under the Batteries of *The Isle de Bas,* on the 27th of June, 1803, ...	50
Lieutenant JAMES BOXER, H.M.S. *Antelope,* for his gallant conduct, in the Capture of The Dutch Gun-boat *The Schrik,* on the 23d of March, 1804,—when he was wounded,	50
Lieutenant BOYD, of The Royal Marines, H.M.S. *Blenheim, off Martinique,* for his gallant conduct in storming *Fort Dunkirk,* a battery of nine guns, on the 17th of November, 1803, ...	50
Captain (*now* Rear Admiral, Sir) JAHLEEL BRENTON, (Bart.), of H.M.S. *Spartan,* for the very gallant Action fought by that Ship against a French Squadron, carrying 96 guns and 1108 men, in *The Bay of Naples,* on the 3d of May, 1810, when Captain BRENTON was severely wounded, about the middle of the Action, which lasted two hours, —and who nobly adds, " my place was most ably supplied by Mr. WILLES, my First Lieutenant, whose Merit becomes more brilliant by every opportunity he has of	

Value. £.

shewing it,—*he is without exception one of the best and most gallant Officers I ever met with,"* 100

Captain CHARLES BRISBANE, H.M.S. *Arethusa*, for his bravery and enthusiastic zeal, in the capture of the Spanish Frigate *Pomona*, of 38 guns, with a complement of 347 men, and the destruction of 12 *Gun-boats*, each carrying a twenty-four pounder, with a complement of 100 men each, and the explosion of a *Castle* mounting 16 thirty-six pounders, near *The Moro Castle*, in the Island of *Cuba*, on the 23d of August, 1806, ... 100

Captain ROBERT HUNTER BROWN, *v.* Captain NATHANIEL DANCE.

Captain CHARLES BULLEN,—

LLOYD'S,

RESOLVED,

That Swords of the value of £100 each, with appropriate Inscriptions, be presented to the surviving Captains and Commanders of His Majesty's Ships, who shared in the dangers and glory of the memorable Action with The Combined Fleets of *France* and *Spain*, *off Cape Trafalgar*, on the 21st of October, 1805,—*viz.*,

Captain CHARLES BULLEN, of The *Britannia*.
Captain ELIAB HARVEY, of The *Temeraire*.
Captain JOHN CONN, of The *Dreadnought*.
Captain CHARLES TYLER, of The *Tonnant*.
Captain GEORGE DUFF, of The *Mars*.
Captain WILLIAM P. CUMBY, of The *Bellerophon*.
Captain CHARLES JOHN M. MANSFIELD, of The *Minotaur*.
Captain ROBERT MOORSOM, of The *Revenge*.
Captain ISRAEL PELLEW, of The *Conqueror*.
Captain HENRY WILLIAM BAYNTUN, of The *Leviathan*.
Captain JOHN PILFOLD, of The *Ajax*.
Captain EDWARD CODRINGTON, of The *Orion*.
Captain Sir FRANCIS LAFOREY, Bart., of The *Spartiate*.

Captain HENRY DIGBY, of The *Africa*.
Captain WILLIAM HARGOOD, of The *Belleisle*.
Captain JAMES NICOLL MORRIS, of The *Colossus*.
Captain RICHARD KING, of The *Achille*.
Captain ROBERT REDMILL, of The *Polyphemus*.
Captain WILLIAM G. RUTHERFORD, of The *Swiftsure*.
Captain GEORGE HOPE, of The *Defence*.
Captain JOHN STOCKHAM, of The *Thunderer*.
Captain P. C. DURHAM, of The *Defiance*.
Captain RICHARD GRINDALL, of The *Prince*.
Captain Sir EDWARD BERRY, of The *Agamemnon*.
Captain The Honble. H. BLACKWOOD, of The *Euryalus*.
Captain WILLIAM PROWSE, of The *Sirius*.
Captain T. DUNDAS, of The *Naiad*.
Captain LAPENOTIERE, of The *Pickle*.
Lieut. ROBERT YOUNGE, of The *Entreprenante*.

	Value. £.
Lieutenant GEORGE CANNING, H.M.S. *Desirée*, for gallant conduct in bringing out of *Monte Christe* Roads, six sail of Schooners, under a smart fire from the batteries, on the 18th of August, 1804, ...	50
Lieutenant WILLIAM CARR, of H.M. Schooner *L'Eclair*, for his spirited conduct in engaging *The Grande Decidée*, from *Guadaloupe*, carrying 22 nine-pounders, and 230 men, on the 5th of February, 1804, ...	50
Captain WILLIAM STANLEY CLARKE, *v*. Captain NATHANIEL DANCE.—This gallant officer now discharges, with consummate zeal and ability, the important office of a DIRECTOR of The Affairs of THE EAST INDIA COMPANY, —whose honour and interest he so nobly defended.	
Lieutenant NICHOLAS BRENT CLEMENTS, H.M.S. *Mediator*, being the leading Fire-ship in the attack and destruction of four French Ships of the Line, and driving several others on shore, in *Aix Roads*, when he was wounded in this gallant enterprise,—on the 11th of April, 1809, .	50
Captain Lord COCHRANE, H.M.S. *Pallas*, for his brave and	

	Value. £.
spirited conduct, in the capture of The National Corvette *La Tapageuse*, in the river of *Bourdeaux*, and in driving on shore three other *National Corvettes*, where they were wrecked, on the 5th of April, 1806, ...	100
Captain EDWARD CODRINGTON, *v.* Captain CHARLES BULLEN.	
Lieutenant THOMAS COLE, H.M.S. *Blenheim*, for his gallant conduct, in the attack and capture of The French Privateer Schooner *L'Harmonie*, in the harbour of *Marin*, in the Bay of *St. Anne*, Martinique, on the 17th of November, 1803, ...	50
Lieutenant WILLIAM COMPSTON, H.M. Sloop *Drake*, for gallant conduct in cutting out a Schooner in the Harbour of *Trinité*, Martinique, on the 19th of February, 1804, when he was wounded in the arm, ...	50
Captain JOHN CONN, *v.* Captain CHARLES BULLEN.	
Captain WILLIAM COOMBE, to whom, when Lieutenant of H.M.S. *Galatea*, a Sword of the value of £50 was voted, on the 21st of April, 1807, for gallant conduct,—and who was killed in Action with the enemy, before the Sword could be delivered to him,—his Widow requests, that she might be allowed to receive it, fondly expressing, that it might "*be kept in the Family, as a Memorial of his Services*,"	50
Lieutenant WILLIAM COOTE, H.M.S. *Cerberus*, for most gallantly boarding and bringing out, a French Schooner, and a Sloop, near to *The Pearl Rock*, Martinique, under a most tremendous fire of cannon and musquetry from the shore,—when he received a most desperate wound in the head, which deprived him of his eye-sight,—on the 2d of January, 1807, ...	50
Lieutenant ROBERT CORNER, H.M.S. *Thisbe*, for his gallant conduct, in boarding and carrying, *with five men*, The *Véloce* Privateer, of 3 guns and *forty* men, in The Mediterranean, on the 1st of May, 1804, ...	50
Mr. J. C. CRAWFORD, Mate of H.M.S. *Pallas*, for his bravery,	

	Value.
in the capture of The National Corvette *La Tapageuse*, in the river of *Bourdeaux*, on the 5th of April, 1806, .	£. 30
Lieutenant George Crofton, H.M.S. *Cambrian*, for his gallant conduct, in boarding and carrying two ships and The Spanish Privateer Schooner *Maria*, on the 7th of July, 1805, in the River *St. Mary's*, ...	50
Captain Acheson Crozier, of The Royal Marines, H.M.S. *Centaur, off Martinique*, for his bravery in assisting in the complete destruction of a Battery in *Petit Ance d'Arlette*, mounting six twenty-four pounders, on the 26th of November, 1803,—where he was wounded, ...	50
Captain William P. Cumby, *v.* Captain Charles Bullen.	
Captain James Richard Dacres, *Jun.*, H.M.S. *Bacchante*, for his distinguished and spirited conduct, in company with H.M. Ship *Mediator*, in the capture of The French National Schooner *Dauphin*, and the destruction of the Fort at *Samana*, "that nest for Privateers," on the 25th of February, 1807, ...	100

Captain Nathaniel Dance,—

Lloyd's, 14th of August, 1804.

Resolved Unanimously,

That the circumstance of an Enemy's Fleet of Ships of War, commanded by an Admiral (Linois), in an 84 gun Ship, *being attacked, defeated,* and *chased* by a Fleet of Merchant Ships, protecting an immense property, is highly honourable to The British Naval character, and *affords a brilliant example to the present and future ages,*

Resolved,

That a Sword of £100 value, with a suitable Inscription, be presented to Captain Nathaniel Dance, of The East India Ship *Earl Camden*, who acted as Commodore on that occasion,—

Resolved,

That Swords of £50 value each, with appropriate Inscriptions, be presented to The Commanders of the other Fifteen Ships engaged in that Action,—*viz.*,

Captain John Fann Timins, of The *Royal George*.
Captain William Moffatt, of The *Ganges*.
Captain James Farquharson, of The *Alfred*.
Captain Archibald Hamilton, of The *Bombay Castle*.
Captain James Pendergrass, of The *Hope*.
Captain John Christopher Lockner, of The *Ocean*.
Captain John Kirkpatrick, of The *Henry Addington*.
Captain William Ward Farrer, of The *Cumberland*.
Captain William Stanley Clarke, of The *Wexford*.
Captain Henry Meriton, of The *Exeter*.
Captain Thomas Larkins, of The *Warren Hastings*.
Captain Robert Hunter Brown, of The *Dorsetshire*.
Captain Henry Wilson, of The *Warley*.
Captain Robert Torin, of The *Coutts*.
Captain John Wordsworth, *Jun.*, of The *Earl of Abergavenny*.

	Value.
	£.
Lieutenant William Dawson, H.M.S. *St. Fiorenzo*, in testimony of the gallant manner in which he continued the Action with The French National Frigate *La Piedmontaise*, (after the death of Captain Hardinge, who fell gloriously in the early part of the action), and finally conducted it to a successful issue,—off *Ceylon*, on the 8th of March, 1808, ...	100
Captain Henry Digby, *v.* Captain Charles Bullen.	
Captain William Henry Dillon, H.M. Sloop *Childers*, for his gallantry and able conduct, in action with a *Danish Ship of War*, of much superior force, off the Coast of *Norway*, on the 13th of March, 1808, in which he was very severely wounded, ...	100
Lieutenant William Domett, H. M.S. *Centaur*, remarkably active in annoying the Enemy during the war,—dis-	

	Value.
tinguished himself particularly in destroying a Battery of three thirty-two pounders, at *Point d'Arlette*, and a Battery of two eighteen pounders, at *Point Margot de Diamant*, both on the Island of Martinique,—and lost his life by the explosion of a Sloop, which he had re-captured from the enemy, on the 9th of February, 1804,—RESOLVED, That a Sword of £50 value, in memory of his late son, be presented to the Father, Mr. ROBERT DOMETT.	£.
Lieutenant P. I. DOUGLAS, Third of H.M.S. *Franchise*, for his gallant and meritorious conduct, in the capture of The Spanish Monarch's Brig *Raposa*, in The Bay of *Campeachy*, on the 7th of January, 1806, . . .	50
Mr. WILLIAM PITT DOWLER, Midshipman, of H.M. Sloop *Swift*, who, in command of a boat, *with six men*, boarded and carried *La Caridad Perfecta*, a Spanish Schooner, of *twelve guns, and Fifteen men*, from under the Batteries protecting the Harbour of *Truxillo*, on the 13th of August, 1805, ...	30
Lieutenant DRURY, Second of H.M.S. *Hydra*, for his daring gallantry, in the capture of three armed *Polaccas* in the harbour of *Begu*, on the Coast of *Catalonia*, and the storming of a *Fort* containing four twenty-six pounders, and under a heavy discharge of Musketry, on the 7th of August, 1807, ...	50
Captain GEORGE DUFF, *v.* Captain CHARLES BULLEN.	
Captain T. DUNDAS, *v.* Captain CHARLES BULLEN.	
Captain P. C. DURHAM, *v.* Captain CHARLES BULLEN,—and HONORARY PLATE.	
Captain E. ELPHINSTONE, H.M.S. *Greyhound*, for his brave and spirited conduct, in company with H.M. Sloop, *Harrier*, in the capture of The Dutch Republican Frigate *Pallas*, of 36 guns, the *Victoria*, a two-decked Ship of about 800 tons, and the *Batavia*, a Ship of about 500 tons,	

	Value. £.
both of them armed for the purpose of war, and richly laden with the produce of *The Moluccas*, in The Java Seas,—on the 26th of July, 1806.	100
Lieutenant, *now* Captain, MATTHIAS EVERARD, of the 2d *or* Queen's Royal Regiment of Foot, for his great gallantry, in leading on *The Forlorn Hope*, in the assault and capture of *Monte Video*, on the 3d of February, 1807. .	50
Captain ARTHUR FARQUHAR, H.M. Bomb *Acheron*, for so nobly supporting the honour of The British Flag, and successfully protecting the Convoy under his care, against *two French Frigates*, in The Mediterranean, on the 4th of February, 1805.	100
Captain JAMES FARQUHARSON, *v.* Captain NATHANIEL DANCE.	
Captain WILLIAM WARD FARRER, *v.* Captain NATHANIEL DANCE.	
WILLIAM FERRIS, Esq., Commander of H.M. Sloop *Drake*, who volunteered his services, in the attack and capture of The French Privateer Schooner *L'Harmonie*, in the harbour of *Marin*, in the bay of *St. Anne*, Martinique, and the storming of *Fort Dunkirk*, on the 17th of November, 1803.	100
Lieutenant MICHAEL FITTON, H. M. Schooner *Pitt*, for his gallantry and superior professional abilities, in the capture of The French Schooner *La Superbe*, off *Cape Maize*, on the 26th of October, 1806.	50
Lieutenant JOHN FLEMING, H.M.S. *Franchise*, for his distinguished merit and bravery, in the capture of The Spanish Monarch's Brig *Raposa*, in the Bay of *Campeachy*, on the 7th of January, 1806.	50
Lieutenant JOHN FORREST, First of H.M.S. *Emerald*, for his gallant conduct in cutting out *The Mosambique* French Schooner Privateer, of 10 eighteen-pounder carronades, from under the Batteries, near *The Pearl*, on the 13th of March, 1804.	50
Lieutenant ROBERT FOWLER, of The Royal Navy, for his able	

Value. £.

services in the memorable Action which took place in the *China Seas*, on the 15th of February, 1804, between a Division of *The East India Company's Ships and a French Squadron*, ... 50

Lieutenant Thomas Furber, H.M.S. *Blenheim*, for his gallant conduct in the attack and capture of The French Privateer Schooner *L'Harmonie*, in the harbour of *Marin*, in the bay of *St. Anne*, Martinique, on the 17th of November, 1803, ... 50

Lieutenant James Wallace Gabriel, H.M.S. *Phœbe*, in consideration of his gallant conduct, on the 14th of July, 1803, in an engagement with a French Privateer, and having been severely wounded in the Action, . . 50

Captain The Honble. Alan Gardner.

Lloyd's,

Resolved,

That Swords of the value of £100 each, with appropriate Inscriptions, be presented to the surviving Captains and Commanders of His Majesty's Ships, who shared in the engagement with a French Squadron, *off Ferrol*, on the 4th of November, 1805,—*viz.*

Captain The Honble. Alan Gardner, of The *Hero*.
Captain Richard Lee, of The *Courageux*.
Captain L. W. Halsted, of The *Namur*.
Captain Wilson Rathbone, of The *Santa Margaritta*.
Captain Thomas Baker, of The *Phœnix*.
Captain Henry Hotham, of The *Revolutionnaire*.

Lieutenant Gibson, H.M.S. *Galatea*, for his gallantry, in the capture of The French Imperial Corvette, *Le Lynx*, off the Coast of *Caraccas*, on the 21st of January, 1807, . 50

Mr. John Green, Master's Mate, of H.M.S. *Galatea*, who

	Value.
commanded a boat at the capture of The French National Corvette *Le Lynx*, as recorded in The London Gazette, of the 14th of April, 1807, ...	£. 30
Captain Richard Grindall, *v.* Captain Charles Bullen.	
Captain Archibald Hamilton, *v.* Captain Nathaniel Dance.	
Lieutenant J. M. Hanchett, H.M.S. *Antelope*, for his gallant conduct, in the capture of The Dutch Galliot *The Schrik*, on the 23d of March, 1804, ...	50
Captain George N. Hardinge, H.M.S. *Scorpion*, for his intrepidity in the capture of The Dutch National Brig *Atalante*, off *The Vlie*, mounting 16 long twelve-pounders, and 76 men, on the 31st of March, 1804, . . .	100
Captain Eliab Harvey, *v.* Captain Charles Bullen.	
Lieutenant John Haswell, First of H.M.S. *Pallas*, for his bravery and meritorious conduct, in the capture of The National Corvette *La Tapageuse*, in the river of *Bourdeaux*, on the 5th of April, 1806, ...	50
Mr. George Hawkins, Midshipman, of H.M.S. *Magicienne*, on whom the command of one of the Boats, which attacked The Dutch Gun-boat *The Schrik*, on the Coast of Holland, on the 23d of March, 1804, had accidentally devolved,—and by whose sword The Dutch Commander fell, while attempting to cut down a British Seaman,	30
Lieutenant John Hawkins, H.M.S. *Atalante*, for his spirited conduct, on the 9th of October, 1803, in *Quiberon Bay*,	50
Lieutenant Robert Hayes, of The Royal Marines, H.M.S. *Hydra*, for his determined gallantry, in the capture of three armed *Polaccas* in the harbour of *Begu*, on the coast of *Catalonia*, and the storming of a Fort containing 4 twenty-six pounders, and under a heavy discharge of Musketry,—on the 7th of August, 1807, . . .	50
Lieutenant Richard Head, H.M.S. *Euryalus*, who, in the command of the Boats manned and armed, attacked and carried in the most spirited manner, a Danish Gun-boat	

	Value. £.
and three smaller Vessels, lying under the protection of a Battery, near the entrance of *The Naskon*, on the Coast of Denmark, in the night of the 11th of June, 1808, ...	50
Lieutenant ROBERT HENDERSON, H.M.S. *Osprey, off Martinique*, who, in the Cutter under his command, without waiting for two other Boats, in the most brave and determined manner, and under a heavy fire from the Guns and Musketry of the Schooner, boarded and captured The French Schooner Privateer *La Resource*, mounting four four-pounders, and having on board 43 men,—Mr. HENDERSON was slightly wounded, ...	50
Mr. JAMES HEWITT, Midshipman, of H.M.S. *Inconstant*, for gallant conduct, in cutting out a vessel from under the batteries of *Goree*, under a very heavy fire from the enemy, on the 24th of April, 1804, ...	30
Mr. WILLIAM HILLYAR, Midshipman, of H.M.S. *Niger*, for his gallant conduct, in boarding and carrying the enemy's vessel after the death of the late Lieutenant JONES, on the 17th of August, 1803, ...	30
Captain GEORGE HOPE, *v.* Captain CHARLES BULLEN.	
Captain HENRY HOTHAM, *v.* Captain The Honble. ALAN GARDNER.	
Lieutenant and Commander W. I. HUGHES, H.M. Fire-Brig *Phosphorus*, for his brave and determined resistance against a larger Lugger, pierced for 16 ports, and having on board from 70 to 80 men, which, after an action of one hour and ten minutes, made sail and sheered off,—*off The Isle of Wight*, on the 14th of August, 1806, .	100
Lieutenant CHARLES KERR, H.M.S. *Jason*, severely wounded in action with the enemy at *Aquadilla*, in *Porto Rico*, when he commanded the detachment, that was landed for the purpose of storming The Fort, on the 1st of June, 1806,	50
Captain RICHARD KING, *v.* Captain CHARLES BULLEN.	

	Value.
Lieutenant Samuel King, First of *The Centaur*, then acting in H.M. Sloop *Drake*, for gallant conduct in carrying, with 30 men only, *The Fort* at *Trinité*, Martinique, when opposed to much superior force, on the 24th of February, 1804, ...	£. 50
Captain John Kirkpatrick, *v.* Captain Nathaniel Dance.	
Captain Sir Francis Laforey, Bart., *v.* Captain Charles Bullen.	
Lieutenant John Lake, H.M.S. *Blanche*, for his masterly and gallant conduct, in cutting out a large coppered Cutter, laying close under the guns of *Monte Christe*, on the 5th of November, 1803, ...	50
Mr. Lamb, Midshipman, of H.M.S. *Franchise*, for his brave and gallant conduct, in the capture of the Spanish Monarch's Brig *Raposa*, in The Bay of *Campeachy*, on the 7th of January, 1806, ...	30
Captain Henry Lambert, H.M.S. *St. Fiorenzo*, for his bravery and skill, in the capture of The French National Frigate *La Psyché*, off *Vizagapatam*, on the 14th of February, 1805,	100
Captain Lapenotiere, *v.* Captain Charles Bullen.	
Captain Thomas Larkins, *v.* Captain Nathaniel Dance.	
Captain Sir Robert Lawrie, H.M.S. *Cleopatra*, for his spirited and gallant conduct, in attacking, and so disabling, *The Ville de Milan*, as to render her incapable of either engaging or escaping H.M.S. *Leander*, on the 14th of February, 1805, ...	100
Captain Richard Lee, *v.* Captain The Honble. Alan Gardner.	
Captain James Lind, H.M.S. *Centurion*, for the gallant and spirited conduct displayed by him, in the defence of H.M.S. *Centurion* against so great a superiority of force, in *Vizagapatam* Road, on the 18th of September, 1804, under every advantage on the part of the assailants, with the complete defeat given The French Admiral, *Mons. Linois*, and his Squadron.—Vice-Admiral Rainier adds,	

Value.
"for my own part I do not hesitate to rank this brilliant action with the most famous of the defensive kind, recorded in the glorious Annals of The British Navy," . £. 100

Captain JOHN CHRISTOPHER LOCKNER, *v.* Captain NATHANIEL DANCE.—Captain LOCKNER now fills with the highest integrity and zeal, the responsible situation of a DIRECTOR of The great Company, THE ROYAL EXCHANGE ASSURANCE CORPORATION, in *Lombard Street.*

Lieutenant NICHOLAS LOCKYER, Third of H.M.S *Tartar*, for his bravery and intrepid conduct, in the capture of The French Privateer *Hirondelle*, off *St. Domingo*, carrying 10 four-pounders, and 50 men, on the 31st of July, 1804, 50

Lieutenant JOHN RICHARD LUMLEY, H.M.S. *Seahorse*, for his intrepid conduct, in the destruction of several of the enemy's Coasting Vessels, in *The Bay of Hieres*, on the 10th of July, 1804, when he was severely wounded, and obliged to suffer amputation at the shoulder joint, . 50

Captain CHARLES LYDIARD, H.M.S. *Anson*, for his intrepidity and zeal, in the capture of The Spanish Frigate *Pomona*, and the destruction of 12 Gun-boats, and the explosion of a Castle, near *The Moro Castle*, in the Island of *Cuba*, on the 23d of August, 1806, ... 100

Captain FREDERICK LOUIS MAITLAND, H.M.S. *Loire*, for his gallant conduct, on the 4th of June, 1805, in *Muros Bay*, 100

Lieutenant SAMUEL MALLOCK, Senior Officer of The Royal Marines, of H.M.S. *Loire*, for his bravery, in the capture of a Privateer of the enemy, on the 2d of June,—and was at the head of The Marines, at the storming *Fort Muros*, on the 4th of June, 1805, ... 50

Captain CHARLES JOHN M. MANSFIELD, *v.* Captain CHARLES BULLEN.

Lieutenant WILLIAM HENRY MASTERMAN, of The Royal Marines, for his spirited conduct, in the successful attack of two

Value. £.

Ships and The Spanish Privateer Schooner *Maria*, in the river *St. Mary's*, on the 7th of July, 1805, . . 50

Lieutenant Jas. W. Maurice, First of H.M.S. *Centaur*, off *Martinique*, for his bravery, in assisting in the complete destruction of a Battery in *Petit Ance d'Arlette*, mounting 6 twenty-four pounders, on the 26th of November, 1803, when he was wounded, ... 50

Lieutenant Giles Meech, of The Royal Marines, H.M.S. *Emerald*, who commanded The Marines in the attack made on the Forts in *Vivero Harbour*, on the Coast of Spain, and in the destruction of The French National Corvette *L'Apropos*, on the 13th of March, 1808, . . 30

Lieutenant Mends, of The Royal Marines, H.M.S. *Franchise*, for his highly meritorious conduct, in the cutting out His Catholick Majesty's Brig *Raposa*, from the Bay of *Campeachy*, on the 7th of January, 1806, . . . 50

Lieutenant—(*now*, Lieutenant Colonel,)—Charles Menzies, of The Royal Marines, H.M.S. *Minerva*, for his intrepidity and zeal when commanding The Marines belonging to that Ship, at the storming *Fort Finisterre*, and in capturing *five Spanish Luggers*, on the 22d of June, 1806, ... 50

Captain Henry Meriton, *v.* Captain Nathaniel Dance.

Captain William Moffatt, *v.* Captain Nathaniel Dance.

Lieutenant Colonel Moore, of The 23d Dragoons, for his gallant conduct at the battle of *Maida*, on the 3d of July, 1806, ... 100

Lieutenant Ogle Moore, H.M.S. *Maidstone*, for his gallant conduct and determined bravery, in the destruction of several of the enemy's Coasting Vessels, in *The Bay of Hieres*, on the 10th of July, 1804, ... 50

Captain Robert Moorsom, *v.* Captain Charles Bullen.

Lieutenant W. H. Mulcaster, First of H.M.S. *Minerva*, for his bravery and meritorious conduct, in the capture of

	Value.
five Spanish Luggers and *Chasse Marées*, in *Finisterre* Bay, and *The Fort*, mounting 8 brass guns of twenty-four and twelve pounders, on the 22d of June, 1806, .	£. 50
Lieutenant Henry Mullah, Second of H.M.S. *Tartar*, for his bravery and intrepid conduct, in the capture of The French Privateer *Hirondelle*, off *St. Domingo*, carrying 10 four-pounders, and 50 men, on the 31st of July, 1804, ...	50
Captain George Mundy, H.M.S. *Hydra*, for his gallant achievement, in the capture of three armed *Polaccas* in the harbour of *Begu*, on the Coast of *Catalonia*, and the storming of a *Fort* containing four twenty-six pounders, strongly defended by nature as well as by art, and opposed to *three times their force* for two hours,—on the 7th of August, 1807, ...	100
Lieutenant Henry John Murton, of The Royal Marines, H.M.S. *Renommée*, who commanded The Marines employed in the boats, at the capture of The Spanish Schooner *Giganta*, on the 4th of May, 1806, . .	30
Lieutenant Edward Nichols, of The Royal Marines, H.M.S. *Blanche*, for his gallant attack, in cutting out a large coppered Cutter, laying close under the guns of *Monte Christe*, on the 5th of November, 1803, . . .	30
When,	
Captain Edward Nichols, of The Royal Marines, H.M.S. *Standard*, for his meritorious and gallant conduct on many important occasions,—particularly for his services in burning a Frigate, and destroying a large Battery, in *The Dardanelles*, as recorded in The London Gazette, of the 5th of May, 1807,—and for his distinguished behaviour, when commanding The Royal Marines, in the boats belonging to H.M.S. *Standard*, at the capture of two armed Gun-boats, off The Island of *Corfu*, as re-	

Value. £.

corded in The London Gazette, of the 3d of January, 1809, ... 50

Lieutenant John Nugent, H.M. Gun Brig *Strenuous*, for his bravery, in the capture of The French Frigate *La Salamandre, off Bouche d'Arkie,* on the 12th of October, 1806, ... 50

Major Abraham Augustus Nunn, of The First West India Regiment, wounded, while faithfully executing the orders of General Prevost, "*not to yield to the enemy one inch of ground,*" in the defence of The Island *Dominica*, on the 22d of February, 1805, ... 50

Captain Maurice Charles O'Connell, of The First West India Regiment, whose wound did not induce him to forego the honour of the Command to which he succeeded, on Major Nunn being disabled,—and in which he resisted the repeated charges of the enemy, notwithstanding their superiority in numbers, until he obliged them to retire with great slaughter,—*Dominica*, the 22d of February, 1805, ... 50

Lieutenant James Oliver, H.M.S. *Bacchante*, for his noble conduct, in so gallantly attacking and carrying a Fort in the Island of *Cuba*, which, with the men it contained, ought to have maintained its position against *fifty times* the number that were opposed, on the 5th of April, 1805, 50

Lieutenant Hyde Parker, H.M.S. *Narcissus*, for his determined bravery, in the destruction of several of the enemy's Coasting Vessels, in *The Bay of Hieres*, on the 10th of July, 1804, ... 50

Lieutenant Sir William Parker, H.M.S. *Renommée*, for his gallantry, in boarding, carrying, and bringing out from under the fire of the guns of the Town and *Torre de Vieja*, and also from under the fire of more than 100

	Value.
musketeers, His Catholick Majesty's Schooner *Giganta*, on the 4th of May, 1806, ...	£. 50
Mr. Edward Perkyns, Mate of H.M.S. *Pallas*, for his bravery, in the capture of The National Corvette *La Tapageuse*, in the river of *Bourdeaux*, on the 5th of April, 1806, .	30
Captain Henry Whitmarsh Pearse, H.M. Sloop *Halcyon*, for his spirited and skilful conduct, in the capture of The Spanish Ship *Neptuno Dios de los Mares*, of 14 twelve pounders, and 72 men, *off Cape St. Martin's*, on the 13th of December, 1806, ...	100
Lieutenant Watkin Owen Pell, H.M.S. *Mercury*, for his determined bravery, in cutting out a Venetian Gun-boat, from the harbour of *Rovigno*, in the night of the 1st of April, 1809, ...	50
Captain Israel Pellew, *v.* Captain Charles Bullen.	
Captain Charles Pelly, H.M. Sloop *Beaver*, for his zeal and gallantry, in the capture of the Dutch National Brig *Atalante*, mounting 16 long twelve-pounders, and 76 men, *off The Vlie*, on the 31st of March, 1804, . .	100
Captain James Pendergrass, *v.* Captain Nathaniel Dance.	

Captain Pendergrass appears to have been animated at an early period with the spirit of Enterprise. In his younger years he advanced with approbation through the different stations of an Officer in the Service of The East India Company,—and, during the vicissitudes of the late eventful War, he had the misfortune to be taken prisoner in *The Princess Royal*, when she was captured in *The Straits of Sunda*, in September, 1793,—but being soon set at liberty, and disdaining an indolent life, whilst the Ship "*Hope*" was building, he entered as a Volunteer on board *The Queen Charlotte*, then commanded by Sir Andrew Snape Douglas, and had the satisfaction to be present at the glorious attack of The French Squadron by Admiral Lord Bridport, close in with Port *L'Orient*, on the 23d of September, 1795. While on board *The Queen Charlotte* at Spithead, a Seaman unfortunately fell from the Foretopmast head over-board,—when Mr. PENDERGRASS instantly, with a manly and characteristick alacrity, let himself down the Ship's side, swam to the object of his humanity, and bore him up, until a boat came to their assistance,—but the poor sufferer having struck his head against the Fore-yard, as he fell, the blow occasioned his death. It is to this circumstance that Sir A. S. DOUGLAS so honourably alludes, as well as the regret which he expresses in losing the Services of an Officer, whom he so much regarded, — especially for that steadiness and bravery, which Captain PENDERGRASS had again an opportunity of displaying in defence of The Ship "*Hope,*" when under his command. But there is a feature about these transactions, which throws additional grace upon them,—the noble Testimonial of Sir A. S. DOUGLAS was not revealed by this meritorious Officer, except to a very few friends,—and the remembrance of the preservation of the poor seaman was reserved for the most secret consolation of his own humane bosom. But I hasten to gratify the feelings of my reader, in presenting the following letter,—

"To
Mr. JAMES PENDERGRASS,

Portsmouth,
Sir, 13th of August, 1795.

The moment I received your letter from Plymouth, requesting your discharge, I answered it by saying that I most cheerfully gave it to you, as it appeared to afford an opportunity for advancing your interest,—at the same time I regretted the loss of your services on board of The Queen Charlotte. I now desire to repeat to you my warmest approbation of your conduct during the time you served under my command, particularly for the Humanity you showed in your endeavours to save the poor man who fell over-board, and for your steadiness and bravery in the late Action with the Enemies' Fleet, on the 23d of June last. My best wishes will always attend you. I am sorry you have not received my letter, directed to Plymouth, as it may have led you to imagine that I had overlooked the merit to which you are so justly entitled in testimony of from me.

I am, Sir, &c.

A. S. Douglas."

The brave and gallant Writer of this letter died about two years afterwards.

	Value.
Lieutenant James Robert Phillips, H.M.S. *Centurion*, for his gallant and spirited conduct, in the defence of H.M.S. *Centurion*, against the Squadron of *Mons. Linois*, in *Vizagapatam* Road, on the 18th of September, 1804, .	£. 50
Lieutenant George Pigot, H.M.S. *Cambrian*, for his gallant conduct, in boarding and carrying two Ships and The Spanish Privateer Schooner *Maria*, on the 7th of July, 1805, in the river *St. Mary's*,—when he was wounded,	50
Captain John Pilfold, *v.* Captain Charles Bullen.	
Brigadier General George Prevost, for his distinguished gallantry and military talents, by which the Sovereignty of The Island *Dominica* was preserved to His Majetsy's arms, on the 22d of February, 1805, ...	100
Mr. John Priest, Midshipman, who commanded the third Boat of H.M.S. *La Loire*, on the 27th of June, 1803, in an attack on *Le Venteux*, ...	30
Captain William Prowse, *v.* Captain Charles Bullen.	
Captain Peter Rainier, H.M.S. *Caroline*, for his bravery and determined conduct, in the capture of The Dutch Republican Frigate, *Maria Reygersbergen*, of 36 guns, and 270 men,—besides which he had to encounter, three other Sloops and Brigs, and several Gun-boats, off *Batavia*,—on the 18th of October, 1806, ...	100
Captain Wilson Rathbone, *v.* Captain The Honble. Alan Gardner.	
Captain Robert Redmill, *v.* Captain Charles Bullen.	
Lieutenant Robert Carthew Reynolds, H.M.S. *Centaur*, for his gallant capture of *The Curieux* French Corvette, on	

	Value.
the 4th of February, 1804, laying close under *Fort Edward* at the entrance of *The Carénage*, Fort Royal Harbour, Martinique, — when he received five severe wounds, ...	£. 50
Lieutenant Richards, H.M.S. *Constance*, for his bravery, in the capture of The French Frigate *La Salamandre*, *off Bouche d'Arkie*, on the 12th of October, 1806,—when he was wounded, ...	50
Lieutenant Henry Rowed, of The Armed Cutter *Sheerness*, for attacking, and capturing, with much inferior force, two French *Chasse Marées*, on the 9th of September, 1803,	50
Mr. Alexander Runciman, Midshipman of H.M.S. *Inconstant*, for gallant conduct, in cutting out a vessel from under the batteries of *Goree*, under a very heavy fire from the enemy, on the 24th of April, 1804, ...	30
Captain William G. Rutherford, *v.* Captain Charles Bullen.	
Mr. Barry Sarsfield, Master's Mate, of H.M.S. *Galatea*, who commanded the Launch at the capture of The French Corvette *Lynx*, as recorded in The London Gazette, of the 14th of April, 1807, ...	30
Lieutenant William Shields, H.M.S. *Scorpion*, for his gallant conduct in command of a Boat in the capture of The Dutch National Brig *Atalante*, on the 3d of April, 1804,	50
Captain Conway Shipley, H.M. Sloop *Hippomenes*, for his gallant conduct in capturing The French Frigate *Egyptienne*, mounting 36 guns, and having on board 240 men, on the 27th of March, 1804,—Captain Shipley handsomely attributes the slight resistance she made, *to the fear of being as severely beaten as she had been four days previously by The Osprey*, Captain Younghusband, .	100
Lieutenant Alexander Shippard, of The hired Armed Cutter *Admiral Mitchell*, for attacking *seven* Vessels belonging to the Enemy, and driving on shore a *French Gun Brig*, and an *armed Sloop*, on the 31st of October, 1803, .	50

	Value. £.
Lieutenant E. R. Sibley, First of H.M.S. *Centaur*, for boarding and carrying by the Division of Boats led on by him, "in a style highly honourable to The National character," the French Corvette *Le Cæsar*, in the entrance of the river *Garonne*,—and to whose gallantry, "no words of mine (Capt. Sir Samuel Hood) are equal to do justice, every one speaks of him in terms of the highest commendation,"—on the 15th of July, 1806,—when he was badly wounded in several places, ...	50
Captain John Stockham, *v.* Captain Charles Bullen.	
Mr. James Sutherland, Master of H.M.S. *Pallas*, for his bravery and spirited conduct, in the capture of The National Corvette *La Tapageuse*, in the river of *Bourdeaux*, on the 5th of April, 1806, ...	50
Lieutenant Francis Temple, H.M.S. *La Loire*, who, with Lieut. James Bowen, cut out *The Venteux*, French Gun Brig, from under the Batteries of *The Isle de Bas*, on the 27th of June, 1803, ...	50
Captain John Thicknesse, of H.M. Sloop *Sheldrake*, for his gallantry and bravery, in the capture of *The Salamander*, a French Frigate-built Ship, mounted with twenty-six long twelve and eighteen pounders, and manned with 150 men, off *Bouche d'Arkie*, on the 12th of October, 1806, ...	100
Lieutenant John Thompson, First of H.M.S. *Narcissus*, for his intrepid conduct and determined bravery, in the destruction of several of the enemy's Coasting Vessels, in *The Bay of Hieres*, on the 10th of July, 1804, ...	50
Mr. W. A. Thompson, Midshipman of H.M.S. *Pallas*, for his bravery, in the capture of The National Corvette *La Tapageuse*, in the river of *Bourdeaux*, on the 5th of April, 1806, ...	30
Captain John Fann Timin's, *v.* Captain Nathaniel Danck.	

Value.

" Captain Timins carried '*The Royal George*' into £.
action in the most gallant manner," — *The Royal George* being the leading ship, " bore the brunt of the Action, *and got as near the enemy as he would permit him.*"

Captain Robert Torin, *v.* Captain Nathaniel Dance.

Mr. John (*since* Captain Loch), who was second Officer of The Ship *Coutts*," in the memorable defeat of *The French Squadron*, now fills with his accustomed ability the important office of a Director of The East India Company.

Captain Troubridge, H.M. Sloop *Harrier*, for his spirited and determined conduct, in company with H.M. Ship *Greyhound*, in the capture of The Dutch Republican Frigate *Pallas*, of 36 guns, the *Victoria*, a two-decked ship of about 800 tons, and the *Batavia*, a ship of about 500 tons, both of them armed for the purpose of war, and richly laden with the produce of *The Moluccas*, in The Java Seas,—on the 26th of July, 1806, . . . 100

Captain Charles Tyler, *v.* Captain Charles Bullen.

Lieutenant Thomas Usher, H.M. Armed Brig *Colpoys*, for his boarding, and carrying three Spanish Luggers, in the Port of *Avillas*, notwithstanding a heavy discharge of grape from the enemy's Battery of twenty-four pounders, and the musketry of a party of Soldiers, on the 21st of March, 1806, ... 50

Captain Richard Budd Vincent, H.M. Sloop *Arrow*, Acting as Commodore, for so nobly supporting the honour of The British Flag, against *two French Frigates* of very superior force in *The Mediterranean*, and successfully protecting the Convoy under his care, on the 4th of February, 1805, ... 100

	Value. £.
Lieutenant WILLIAM WALKER, of The Royal Marines, H.M.S. *Centaur, off Martinique,* for his bravery in assisting in the complete destruction of a Battery in *Petit Ance d'Arlette,* mounting six twenty-four pounders, on the 26th of November, 1803,—when he was wounded, . . .	30
Lieutenant JOHN ELLIS WATT, H.M.S. *Ville de Paris,* who, on the 17th of August, 1803, in the Pinnace, cut out *The Messager,* French Lugger, of much superior force, from within the rocks at *Ushant,*	50
Lieutenant GEORGE EDWARD WATTS, H.M.S. *Comus,* for his gallant exploit, in the boarding and cutting out *The St. Pedro* Spanish Packet, in the Port of *Grand Canaria,* under a severe fire of musketry, and an exceeding heavy fire from all the batteries,—when he received several wounds,—on the 8th of May, 1807, ...	50
Lieutenant EDWARD WHITE, H.M.S. *Beaver,* for gallant conduct in command of Boats, in boarding and carrying The Dutch National Brig *Atalante,* on the 3d of April, 1804, ...	50
Captain HENRY WILSON, *v.* Captain NATHANIEL DANCE.	
Captain WILLIAM FURLONG WISE, H.M.S. *Mediator,* for his determined bravery, in company with H. M. Ship *Bacchante,* in the capture of The French National Schooner *Dauphin,* and the carrying of the Fort at *Samana* by storm, by a detachment of Seamen and Marines under his command,—on the 25th of February, 1807, ...	100
Captain JAMES WOOLDRIDGE, H.M.S. *Mediator,* for his gallant and highly meritorious conduct, when employed as the leading Fire-ship in the attack and destruction of the French Fleet, in *Basque Roads,* on the 11th of April, 1809, ...	100
Captain JOHN WORDSWORTH, *Junr., v.* Captain NATHANIEL DANCE.	

	Value.
Lieutenant JAMES LUCAS YEO, H.M.S. *Loire*, for his gallant conduct, on the 4th of June, 1805, in *Muros Bay*, when he was wounded, ...	£. 50
Lieutenant ROBERT YOUNGE, *v*. Captain CHARLES BULLEN.	
Captain GEORGE YOUNGHUSBAND, H.M. Sloop *Osprey*, for his spirited conduct in attacking and silencing The French Frigate *Egyptienne*, mounting 36 guns, and having on board 255 men, on the 23d of March, 1804, . .	100

Honorary Plate.

	Value.
Brigadier-General Sir Samuel Auchmuty, for his able conduct, in the capture of the important Fortress and City of *Monte Video*, on the 3d of February, 1807,—a Vase, with an appropriate inscription, of the value of .	£. 200
Captain Francis William Austen, H.M.S. *Canopus*, for his brave and gallant conduct, in the brilliant and decisive victory, on the 6th of February, 1806, off *St. Domingue*, —a Vase, of the value of	100
Major General Sir David Baird, for the gallantry with which he effected a landing in the face of a superior force of the enemy, and achieved the important conquest of *The Cape of Good Hope*, on the 18th of January, 1806,—a Vase, of the Value of £300, with an appropriate inscription,	300
Captain Thomas Baker, H.M.S. *Phœnix*, who shared in the danger and glory of the engagement with the French Squadron off *Ferrol*,—a Vase, of the value of .	100
Major-General Beresford, for his gallant and disinterested conduct, in the successful and important capture of *Buenos Ayres* and it's Dependencies, in July, 1806,—a Vase, of the value of	200
Captain Sir Edward Berry, H.M.S. *Agamemnon*,—a Vase, of the value of £100, for his distinguished valour, in the memorable victory of *Trafalgar*, ...	100
And, For his brave and gallant conduct, in the brilliant and decisive victory, on the 6th of February, 1806, off *St. Domingue*,—a Vase, of the value of	100

	Value. £.
Lieutenant RICHARD BOURNE, of H.M. Schooner *Felix*, for his gallant conduct, in an action with an enemy's Privateer of superior force, ...	100
Captain CHARLES BRISBANE, H.M.S. *Arethusa*, for his determined courage and judicious zeal, in the capture of *The Island of Curaçoa*, on the 1st of January, 1807,—a Vase, of the value of	200
Major CAMPBELL, of the 40th Regt., for his bravery in the assault and capture of the important Fortress and City of *Monte Video*, on the 3d of February, 1807, when he was wounded,—a Vase, of the value of . . .	100
Captain The Honble. T. B. CAPEL, H.M.S. *Phœbe*, who shared in the danger and glory of the memorable Victory of *Trafalgar*,—a Vase, of the value of . . .	100
Rear-Admiral The Honble. Sir ALEXANDER COCHRANE, K.B., H.M.S. *Northumberland*, who contributed, by his noble support, to the brilliant and decisive Victory, on the 6th of February, 1806, off *St. Domingue*,—a Vase, of the value of £300., with an appropriate inscription, . .	300
Captain N. D. COCHRANE, H.M. Frigate *Kingsfisher*, for his brave and gallant conduct, in the brilliant and decisive Victory, on the 6th of February, 1806, off *St. Domingue*, —a Vase, of the value of	100
Vice-Admiral Lord COLLINGWOOD, H.M.S. *Royal Sovereign*, —RESOLVED, That a Vase, of the value of £500, be presented to His Lordship,—who, after the death of The Commander-in-Chief in the hour of Victory, at *Trafalgar*, so nobly completed the Triumph of the day, . .	500
Captain JOHN COOKE, H.M.S. *Bellerophon*, killed in the action with The Combined Fleets off *Cape Trafalgar*,—his Widow affectionately expressing, that any mark of Honour to her late Husband's memory would be highly gratifying to her feelings,—RESOLVED, That a piece of Plate of the value of £200, with an appropriate inscription, be presented to Mrs. COOKE, ...	200

	Value.
Captain Daly, H.M.S. *Comet*, for his intrepid conduct, in the destruction of The Forts *St. Salvador de Ano*, and *Sedra*, and *the blowing up* of the Magazine, when he was much scorched,—on the 22d of June, 1808,—a Vase, of the value of £100, with an appropriate inscription, .	£. 100
Commodore Nathaniel Dance, for his able and successful conduct, in the memorable repulse of *The French Squadron*, in the China Seas,—a Vase, of the value of, . . .	100
Vice-Admiral Sir John Thomas Duckworth, K.B., H.M.S. *Superb*, for the brilliant and decisive Victory, on the 6th of February, 1806, in the Bay of the Town of *St. Domingue*, —a Vase, of the value of £400, ornamented with emblematical devices, and an appropriate inscription, .	400
Captain George Duff, H.M.S. *Bellerophon*, killed in action with The Combined Fleets, off *Cape Trafalgar*, on the 21st of October, 1805.—Resolved, That a piece of Plate of the value of £100, with a suitable inscription, be presented to Mrs. Duff, his Widow, to descend to his son, then a Midshipman in The Royal Navy, . .	50
Captain R. D. Dunn, H.M. Frigate *Acasta*, for his brave and gallant conduct, in the brilliant and decisive Victory, on the 6th of February, 1806, off *St. Domingue*,—a Vase, of the value of	100
Captain P. C. H. Durham,—*now, Admiral* Sir Philip Charles Henderson Durham, G.C.B., Flag-officer, at Portsmouth,—for his zeal and spirited exertions, when commanding his H.M. Ship *Spitfire*, in 1793, in capturing the enemy's Privateers,—for his re-captures of British Merchantmen,—and also, for having brought into port *the first Tri-coloured flag, captured since the breaking out of hostilities*,—a piece of Plate of the value of . .	300

Subsequently, this gallant Officer was presented with a Sword (*See*, Captain Charles Bullen), in token of the high sense which The Patriotic Fund Institution entertained of the distinguished part, which he took in the ever memorable battle of *Trafalgar*.

His Majesty Louis the Eighteenth, King of France, conferred upon Sir Philip Durham the *Cordon Rouge*, in reward for his exertions in saving *The Island of Martinique* from falling into the hands of the Buonapartists,—for the re-capture of *The Island of Guadaloupe*,—and, finally, for having had the good fortune, *to cause the last Tri-coloured flag captured*, to be struck to *The Venerable*, his own Flag Ship, in August, 1815.

	Value. £.
Captain Birt Dynely, of H.M. Packet *The Duke of Montrose*, for his most gallant conduct and perseverance, in capturing *L'Imperial* Schooner, *off Dominica*, on the 24th of May, 1806,—a Vase, of the value of . . .	50
Mr. Robert Fair, Master of H.M.S. *Beaver*, for his gallant conduct, in the capture of The Dutch National Brig *Atalante*, ...	50
Captain Thomas Francis Fremantle, H. M. S. *Neptune*, a Vase, of the value of £100., for his determined bravery, in the glorious victory of *Trafalgar*, . . .	100
Captain L. W. Halsted, H.M.S. *Namur*, who shared in the danger and glory of the engagement with the French Squadron off *Ferrol*,—a Vase, of the value of . .	100
Major Hammill, of The Royal Regt. of *Malta*, for his intrepidity, at the battle of *Maida*, 4th of July, 1806, when he was wounded,—a Vase, of the value of . .	100
George Hardinge, Esq., H.M. Justice for the Counties of Glamorgan, Brecon, and Radnor, and Attorney General to The Queen, in commemoration of the zeal, gallantry, and judgement displayed by his adopted son, Captain George Nicholas Hardinge, of H.M.S. *St. Fiorenzo*, who was killed in the brilliant action with *La Piedmontaise* French National Frigate, on the 6th, 7th, and 8th of March,	

	Value.
1808,—a Vase, of the value of £100., with an appropriate inscription, ...	£. 100
Captain Thomas Masterman Hardy, H.M.S. *Victory*, who shared in the danger and glory of the memorable Victory of *Trafalgar*,—a Vase, of the value of . . .	100
Captain William Hargood, H.M.S. *Belleisle*, who shared in the danger and glory of the memorable Victory of *Trafalgar*,—a Vase, of the value of	100
Mr. John Heddle, Surgeon to *The African Corps*, who volunteered his services, and was severely wounded in the defence of *Goree*, ...	50
Captain Hennah, H.M.S. *Mars*, who shared in the danger and glory of the memorable victory of *Trafalgar*,—a Vase, of the value of	100
Mr. Robert Hillier, Gunner, of H.M.S. *Pallas*, for having particularly distinguished himself in command of Boats, by landing and destroying the Signal Posts belonging to the Enemy, in 1806,—a piece of Plate, of the value of .	30
Lieutenant-Colonel Robert Honeyman, of the 93d Regt., for his zealous and spirited conduct, in the conquest of *The Cape of Good Hope*, on the 18th of January, 1806, when he was wounded,—a Vase, of the value of . .	100
Commodore Sir Samuel Hood, H.M.S. *Centaur*, for his intrepidity and zeal, in the capture of *four large French Frigates*, on the 25th of September, 1806,—when he received a severe wound in his right arm, which was afterwards amputated,—a Vase, of the value of . .	300
Captain R. G. Keats, H.M.S. *Superb*, for his noble and gallant efforts, in the brilliant and decisive Victory, on the 6th of February, 1806, *off St. Domingue*,—a Vase, of the value of	100
Captain George Lamb, for his gallant action, on the 21st of May, 1805, between *The Doris* armed Transport of 8	

	Value.
guns, under his command, and a *Spanish* Privateer of 18 guns, ...	£. 50
Captain Thomas Lavie, H.M.S. *Blanche*, for his intrepid conduct, in the capture of *La Guerriere* French Frigate, mounting 50 guns, with a complement of 317 men, after a sharp contest of 45 minutes, *off The Feroe Islands*, on the 19th of July, 1806,—a Vase, of the value of . .	100
Rear Admiral Sir Thomas Louis, H.M.S. *Canopus*, for the bravery and judgement displayed by him in the brilliant and decisive Victory, on the 6th of February, 1806, *off St. Domingue*,—A Vase, of the value of £300., with an appropriate inscription, ...	300
Captain Charles Lydiard, H.M.S. *Anson*, for his able and zealous co-operation, in the capture of His Catholick Majesty's Frigate *Pomona*, the destruction of *twelve gun-boats, and the explosion of a Castle*, mounting 16 thirty-six pounders, in *The Island of Cuba*, on the 23d of August, 1806,—a Vase, of the value of . . .	100
Captain Mc Kenzie, H.M. Frigate *Magicienne*, for his brave and gallant conduct, in the brilliant and decisive Victory, on the 6th of February, 1806, *off St. Domingue*,—a Vase, of the value of	100
Lieutenant-Colonel Mc Leod, of the 78th Regt., for his intrepid conduct, at the battle of *Maida*, 4th of July, 1806, —when he was wounded,—a Vase, of the value of .	100
Captain Pulteney Malcolm, H.M.S. *Donegal*, for his brave and gallant conduct, in the brilliant and decisive Victory, on the 6th of February, 1806, *off St. Domingue*,—a Vase, of the value of	100
Mr. John Marks, Boatswain of H.M. hired Armed Cutter *Sheerness*, for exemplary bravery in protecting a *Chasse Maree*, boarded and taken by a boat from the said Cutter, near *The Bec du Raz*, on the 9th of September, 1803,—a *Silver Call* and *Chain*.	

	Value. £.
Captain JAMES NICOLL MORRIS, H.M.S. *Colossus*, who shared in the danger and glory of the memorable Victory of *Trafalgar*,—a Vase, of the value of . . .	100
Captain MORRISON, H.M.S. *Northumberland*, for his noble and gallant conduct, in the brilliant and decisive Victory, on the 6th of February, 1806, *off St. Domingue*,—a Vase, of the value of	100
Lieutenant-Colonel MUNRO, of the 23d Dragoons, for his intrepid conduct, at the battle of *Maida*, 4th of July, 1806, when he was wounded,—a Vase, of the value of . .	100
Lieutenant OGLE MOORE, H. M. S. *Minerva*, for his gallant support, in the capture of *five Spanish Luggers* and *Chasse Marées*, in *Finesterre Bay*, on the 23d of June, 1806,—a Vase, of the value of	50
Lord Viscount NELSON, H.M.S. *Victory*,—RESOLVED, That a VASE, of the value of £500., ornamented with emblematical devices, and appropriate inscriptions, illustrative of the transcendent and heroic achievements of the late Lord Viscount NELSON, be presented to his Relict, Lady Viscountess NELSON, ...	500
Earl NELSON, of *Trafalgar*,—RESOLVED, That a VASE, of the value of £500. be presented to him, to descend as an Heir-Loom with the Title so gloriously acquired .	500
Mr. WILLIAM NESBITT, Commanding *The Queen Charlotte* Berwick Smack, for his gallant defence, against a *French Privateer* of much superior force, in which he was severely wounded,—in February, 1804,—a *Silver Tankard*, of the value of	30
The Right Honble. Rear-Admiral The Earl of *Northesk*, H.M.S. *Britannia*,—RESOLVED, That a VASE, of the value of £300. be presented to His Lordship, for his heroic achievements in the memorable Victory of *Trafalgar* .	300
Major ABRAHAM AUGUSTUS NUNN, for his meritorious conduct, being dangerously wounded, when faithfully executing	

	Value. £.
the orders of General Prevost, in the defence of *Dominica*, ...	100
Captain Maurice Charles O'Connell, for his distinguished gallantry, in the defence of *Dominica*, ...	100
Major Paulett, of the 44th Regt., for his bravery and zealous conduct, at the battle of *Maida*, 4th of July, 1806,—when he was severely wounded,—a Vase, of the value of	200
Lieutenant Charles Pickford, H.M.S. *Inconstant*, whose presence of mind, and address *in negotiation*, induced the French Garrison at *Goree* to capitulate *with himself*, THEN THEIR PRISONER, as recorded in The London Gazette, of the 28th of April, 1804,—a Vase, with an appropriate inscription.	
Lieutenant George Pigot, H.M.S. *Cambrian*, for his gallant conduct, and in consideration of the *wound* which he received in boarding and carrying *two Ships and a Schooner*, in the river *St. Mary's*, ...	100
Commodore Sir Home Popham, for his zealous, able, and spirited co-operation in the conquest of *The Cape of Good Hope*, on the 18th of January, 1806,—a Vase, of the value of £200., with an appropriate inscription .	200
And, For his gallant and disinterested conduct, in the successful and important capture of *Buenos Ayres*, and it's Dependencies, in July, 1806,—a Vase, of the value of	200
Brigadier General George Prevost, for his distinguished gallantry and military talents, by which the Sovereignty of *The Island of Dominica* was preserved to His Majesty's arms, ...	200
Captain William Prowse, H.M.S. *Sirius*, for his exertion and zeal, in an encounter with a *very formidable Flotilla* of the enemy, *off The Tiber*, on the 27th of April, 1806,	

	Value.
in which the Commander of it was captured,—a Vase, of the value of	£. 100
Mr. James Puckinghorn, Master's Mate of H.M.S. *Beaver*, for his steady conduct, in the capture of The Dutch National Brig *Atalante*, ...	30
Captain Samuel Pym, H.M.S. *Atlas*, for his brave and gallant conduct, in the brilliant and decisive Victory, on the 6th of February, 1806, *off St. Domingue*,—a Vase, of the value of	100
Mr. Francis Rennells, Mate of H.M. hired Schooner *Colpoys*, for his spirited conduct, on the 6th of September, 1804, in the attack of *a Signal Post* at *Benthaume Castle*, near Brest, *when the whole of the Signals were captured*, .	30
Mr. Richard Robinson, Master of *The Scipio*, Collier, for his gallant conduct, in beating off *a French Privateer*, on the 26th of April, 1804, in which action he was severely wounded,—a *Silver Tankard*, ...	30
Mr. William Rogers, Acting Captain of H.M. Packet, *Windsor Castle*, for his gallant conduct, in action with *Le Genie*, a French Privateer, of very superior force, *off Barbadoes*, on the 1st of October, 1807,—a Vase, of the value of .	100
Captain Edward Rotherham, H.M.S. *Royal Sovereign*, who shared in the danger and glory of the memorable Victory of *Trafalgar*,—a Vase, of the value of	100

Captain Rutherford, of *The Helen*, Merchant Ship, for the zeal and activity in the cause of his Country, as displayed in the following extract of a letter addressed to him, by The Honble. Rear-Admiral Cochrane,—

"It is to you, that the British Squadron under Sir John Thomas Duckworth, was indebted for the Victory they obtained over the enemy *off St. Domingo*, on the 6th of February, 1806,—Had you not come up to *Tortola in an open boat*, with the intelligence you had received of an Enemy's Fleet being seen in *The Mona Passage*, and which intelligence you delivered to Captain Cochrane of the *The Kingsfisher*, Sir John must have remained in ignorance of their being there."—a Vase, of the value of . . . . £.100

Value.

Mr. Salmon, Master of H.M. Schooner, *L'Eclair*, for gallantly volunteering his services in cutting out *The Rose*, French Privateer, from under the Batteries of *The Hayes*, Guadaloupe, on the 5th of March, 1804, ... £. 50

Captain Thomas Searle, H.M. Sloop *Grasshopper*, for his gallantry, in capturing *two Spanish Vessels* richly laden from South America, under the protection of *four gun-boats*, and under *a battery* close in with *Faro*, on the 23d of April, 1808,—a Vase, of the value of . . . 100

Captain Michael Seymour, H.M.S. *Amethyst*, for his determined bravery and skill, in the capture of The French Frigate *La Thetis*, of 44 guns, and a crew of 330 men, and 106 soldiers, after a long and severe action, off the Point of *Groa*, on the 10th of November, 1808—a Vase, of the value of 100

Lieutenant Simons, H.M.S. *Defiance*, for his gallant conduct, in the action with The Combined Fleets, *off Cape Trafalgar*, on the 21st of October, 1805,—who, after boarding the French Ship *L'Aigle* then lashed to *The Defiance*, wounding her Captain, and hauling down her Colours, was killed on her Poop in the act of hoisting *The British Colours* in their place,—Resolved, That a piece of Plate, of the value of £100, with an appropriate inscription, be presented to Mr. Simons, Father of the late Lieut. Simons, ... 100

Captain John Stewart, H.M.S. *Seahorse*, for his gallantry and skill, in the capture of *two large Turkish* Ships of war, off The Island of *Scopolo*, on the 5th of July, 1808,—a Vase, of the value of 100

Rear-Admiral Stirling, for his zealous conduct, in the capture of the important Fortress and City of *Monte Video*, on the

	Value.
3d of February, 1807,—a Vase, of the value of £200, with an appropriate inscription, ...	£. 200
Captain The Honble. Robert Stopford, H.M.S. *Spencer*, for his brave and gallant conduct, in the brilliant and decisive Victory, on the 6th of February, 1806. *off St. Domingo*, —a Vase, of the value of	100
Rear-Admiral Sir Richard John Strachan, Bart., H.M.S. *Cæsar*,—Resolved, That a Vase, of the value of £300 be presented to him, for his heroic achievements, *off Ferrol*, ...	300
Major Stuart, of the 78th Regt., for his intrepidity, at the battle of *Maida*, 4th of July, 1806,—when he was wounded,—a Vase, of the value of . . .	100
Major-General Sir John Stuart, for his "gallant conduct at the Battle of *Maida*, 4th of July, 1806, in which the pride of the presumptuous enemy was severely humbled, and the superiority of the British Troops most gloriously proved," —a Vase, of the value of £300, with an appropriate inscription	300
Captain John Fann Timins, for the gallant manner in which he led the action, and placed *The Royal George* along-side The French Admiral's Ship, *The Marengo*, of 84 guns, in the memorable repulse of *The French Squadron*, in the *China Seas*,—a Vase, of the value of . . .	100
Mr. Tracey, Secretary to Commodore Hood, for his gallant conduct in cutting out *The Curieux* French Corvette, close under *Fort Edward*, Port Royal Harbour, Martinique, on the 4th of February, 1804,—when he was wounded, ...	30
Major Trotter, of the 83d Regt., for his gallant conduct, in the assault and capture of the important Fortress and City of *Monte Video*, on the 3d of February, 1807,—when he was wounded—a Vase, of the value of . . .	100
Major Tucker, of the 72d Regt., for his gallant conduct, in the assault and capture of the important Fortress and	

	Value. £.
City of *Monte Video*, on the 3d of February, 1807,—when he was wounded,—a Vase, of the value of . .	100
Lieutenant Henry Walker, H.M.S. *Galatea*, killed in action with the enemy, having left no Relative who required pecuniary assistance, — but that any Honorary testimonial to his memory, would be highly gratifying to the feelings of his Mother, then Mrs. Nation, of Exeter, —Resolved, That a piece of plate, of the value of £50, with an appropriate inscription, be presented to Mrs. Nation, ...	50
Brigade-Major Weir, of the 59th Regt., for his zealous and spirited conduct, in the conquest of *The Cape of Good Hope*, on the 18th of January, 1806,—when he was wounded,—a Vase, of the value of . . .	100
Captain James Athol Wood, H. M. S. *Latona*, for his intrepid and zealous co-operation, in the capture of *The Island of Curaçoa*, on the 1st of January, 1807,—a Vase, of the value of	100
Lieutenant James Lucas Yeo, H.M.S. *Loire*, for his gallant conduct, on the 4th of June, 1805, in *Muros Bay*, .	50
Captain Yescombe, for his gallant conduct, in the defence of H.M. Packet *King George*, on the 31st of July, 1803, on his passage from *Lisbon*,—in which Action he was killed, —it was Resolved, that a piece of plate of £50 value, with a suitable inscription, be presented to his Widow,	50

Honorary Medals,

have been granted on the following glorious occasions, viz.,

Battle of *St. Vincent,* 14th of February, 1797.
Battle of *Camperdown,* 11th of October, 1797.
Battle of *The Nile,* 1st of August, 1798.
Capture of "*Hermione,*" 25th of October, 1799.
Battle of *Trafalgar,* 21st of October, 1805.
Action of *Ferrol,* 4th of November, 1805.
Battle of *St. Domingo,* 6th of February, 1806.
Action *off Lissa,* 18th of March, 1811.
Capture of *The Island of Java,* 20th of August, 1811.
Capture of "*Le Rivoli,*" 12th of February, 1812.
Capture of "*The Chesapeake,*" 1st of June, 1813.
Capture of "*The President,*" 15th of January, 1815

Order of Merit,

Of The Fifth Regiment of Foot;

or,

Northumberland Fusiliers.

Early in the year 1767 a system of Honorary Distinction for long-continued good behaviour was introduced into The Fifth Regiment of Foot, which was found to stimulate the indifferent to good conduct, and those already worthy, to perseverance in well-doing, and it produced such a body of Non-commissioned Officers as few Corps could boast of.

These Distinctions consisted of *Three Classes* of Medals to be worn, suspended by a riband, at a button-hole of the left lappel,—

The first, or lowest Class, which was bestowed on such as had served irreproachably for seven years, was of *gilt metal*, bearing on one side The Badge of the Regiment, *St. George and the Dragon*, with the Motto, "*Quo fata vocant*,"—and, on the reverse, "Vth *Foot, Merit*."

The second was of *Silver*, bearing on one side The Badge and Motto, —and, on the other, "*Reward of Fourteen years' Military Merit*."

The third was similar to the second, but was inscribed with the Name of the Individual, whose conduct had earned it,—" A. B., *for twenty-one years' good and faithful service as a Soldier, had received from his Commanding Officers this Honourable testimony of his Merit*."

THESE MEDALS were bestowed only upon Soldiers, who, for the respective periods of seven, fourteen, or twenty-one years, had never incurred the censure of a Court-martial,—they were given by The Commanding Officer at the head of the assembled Battalion,—and if, which rarely happened, the owner of a Medal subsequently forfeited his pretensions to enrolment among the men of Merit, his Medal was cut from his breast by the Drum-Major as publickly as he had been invested with it.

Those who obtained the third, or twenty-one years' Medal, had also an oval Badge of the colour of the facings on the right breast, embroidered round with gold and silver wreaths, and inscribed in the centre with the word "MERIT," in letters of gold.

Previous to the embarkation of THE FIFTH from Ireland, on the 26th of December, 1832, to do duty in the far-famed Fortress of GIBRALTAR, the circumstance of The Regiment having an "*Order of Merit*," attracted the attention, and elicited the representations of the Local Military Authorities. The Commanding Officer, after the arrival at *Gibraltar*, was in consequence called upon by The General Commanding-in-Chief, Lord HILL, to explain under what Regulations and Arrangements THE ORDER was conferred, Candidates selected, Medals provided and distributed, and other particulars connected with this most laudable Institution. The required information was promptly afforded, proved satisfactory, and the following gratifying Letter was the result,—*viz.*,

"Horse Guards, 20th of June, 1832.

"Sir,

I have had the honour to submit to The General Commading-in-Chief, your Letter of the 4th instant, with it's inclosure, on the subject of the "*Order of Merit*" existing in The Fifth Foot' and am directed to acquaint you, that the explanation afforded by Lieutenant Colonel Sutherland, shows that The Order in question is dispensed under the most laudable Regulations, and has been productive of the best effects, during the long period since it's original Establishment in The Regiment.

It is considered highly desirable, however, that both Officer and Soldier should, under all circumstances be taught to expect professional Honours from The Sovereign alone,—and under this impression, Lord Hill has been induced to recommend to The King to give the Royal Authority for the confirmation and continuance of this Regimental Badge of Distinction, an arrangement, which, while it bestows upon it legitimate existence, will, at the same time, no doubt enhance it's value in the estimation of those on whom it is conferred.

You will, therefore, be pleased to communicate this decision to Lieut. Colonel Sutherland, and acquaint him, that he is at liberty to proceed in the distribution of the Medals and Badges as heretofore.

I have, *&c.*,

John Macdonald,

Adjutant General.

Lieut. General
Sir William Houstoun, G.C.B. and G.C.H.
&c. &c. &c,
Commanding at Gibraltar."

The "Order of Merit," which has been so long holden by The Regiment,—consecrated and enhanced by purity and justice of Distribution, and the real worth of the meritorious, though humble Individuals who earn so Hononrable a mark of good conduct, and brightened by the numerous distinguished Services of THE FIFTH REGIMENT since it's foundation, — is thus confirmed by the sanction of THE SOVEREIGN, The legitimate fount of Honour, Distinction, and Reward.

THE GRAND DUCHY OF OLDENBURG.

THE DOMESTICK ORDER AND ORDER OF MERIT OF DUKE PETER FREDERICK LEWIS.

Having been favoured with a copy of THE STATUTES of THE ORDER of Duke PETER FREDERICK LEWIS, by the kindness of H. F. TIARKS, Esq., the obliging and efficient Consul of *Oldenburg*, at London, I avail myself of the opportunity of now giving a more enlarged description of THIS ORDER, than what I was able to do at page 198.

THIS ORDER was instituted by PAUL FREDERICK AUGUSTUS, Grand Duke of OLDENBURG, at *Oldenburg*, on the 27th of November, 1838, as a Mark of Distinction for those who should render themselves meritorious by Services to The State, and by proofs of Attachment to The Royal House,—and also, for distinguished *Foreigners*.

THE ORDER was named, as that of the Father of the actual Founder, it having been the intention of the late

Duke to establish a similar Institution for the purposes mentioned above.

The following are the STATUTES of THE ORDER.—

1. THE GRAND DUKE, for the time being, shall be Grand Master of THE ORDER. In the case of a Minority, the general affairs of THE ORDER shall be managed by The Regency,—but no Members shall be admitted, or changes effected in THE ORDER or it's STATUTES, by The Regency.

2. THE ORDER shall consist of,—1st., ACTUAL MEMBERS,—and 2dly, HONORARY MEMBERS,—the two Divisions being of equal Rank, and comprising, each, four Classes, — viz., GRAND CROSSES, GRAND COMMANDERS, COMMANDERS, and SMALL CROSSES.

3. Subjects of The State, not being in the service of a Foreign Power, can alone be admitted as Actual Members. The Dignity is forfeited by the acceptance of a Foreign Title or Pension, without the consent of The Grand Duke.

4. The *Actual* Member in a lower Class may at the same time be *Honorary* Member in a higher.

5. The Princes of The Grand Ducal House descended, in the male Line, from Duke PETER FREDERICK LEWIS are Honorary Grand Crosses. The Grand Duke apparent shall bear the title of GRAND PRIOR.

6. Within the Grand Duchy, The Grand Cross may be conferred on persons in the first Class of Service,—The Grand Commander's Cross on those, who are in the first two Classes of Service,—The Commanders' Cross on those, who are in the first three Classes of Service, —and The Small Cross on those, who are in the first five Classes of Service. THE ORDER may, however, be conferred upon *Natives*, who are not in the Service. The Small Cross may be conferred on Officers, or those having the Rank of Officers, for acts of bravery in time of War, or other distinction in the Service.

7. No application shall be made for admission into THE ORDER.

8. Admission into THE ORDER, and promotion, shall be conferred by the Diploma of The Grand Master.

9. Relates to the publication of Admissions, *&c.*

10. The Actual Members shall consist of *Two* Grand Crosses, with annual incomes of 500 rix dollars of gold,—*Two* Grand Commanders, with annual incomes of 400 rix dollars of gold.—*Four* Commanders, with annual incomes of 300 rix dollars of gold,—and *Eight* Small Crosses, of which the four Senior shall enjoy annual incomes of 200 rix dollars of gold.

11. The Honorary Members in the Grand Duchy, with the exception of the Princes of The Royal House, shall comprise *four* Grand Crosses, *four* Grand Commanders, *eight* Commanders, and *sixteen* Small Crosses. The number of The Small Crosses may, however, in time of War, be increased as a reward for Military merit.

12. THE DECORATION of THE ORDER consists,—

1. For The Grand Crosses, of a STAR, worn upon the left breast, and a CROSS attached to a broad dark blue, watered riband, with a narrow red stripe on each side, passed over the right shoulder to the left side.

2. For Grand Commanders, of a STAR, worn on the left breast, and a smaller CROSS suspended from a similar riband of the width of two inches passed round the neck.

3. For the Commanders, of a small CROSS, suspended from the neck by a similar riband one and a half inch broad,—and

4. For the Smaller Crosses, of a small CROSS, attached to the left button-hole by a riband of the same pattern and width, as the last preceding.

5. Those Military persons who have gained The Small Cross in war, shall wear on the riband of THE ORDER a round cockade of the same riband.

6. The Actual Members, in addition to the above Decorations, wear a particalur distinction, differing according to the Rank in THE ORDER of the holder.

13. The Decoration of THE ORDER, adorned with *diamonds*, can only be worn by those upon whom it has been so conferred.

14. The Decoration of this ORDER shall not be worn with other ORDERS, but according to the Statutes of this ORDER.

15. Natives must always wear the Decoration of The Order when at Court, in Service, or in large Assemblies.

16. The Decoration of The Order shall be given up in the event of death, or promotion to a higher Class.

17. Authorises Members, not being entitled to wear a Military Uniform, to assume a particular Dress.

18, 19, 20: Regulate the management of the affairs of The Order.

21. The Chapter of The Order shall be holden annually on the 17th of January.

22, 23, 24. Relate to the management of the affairs of The Order.

25. "*The General Mark of Honour*" connected with The Order, consists of a *metal Cross* of the form of the Decoration of the Small Crosses, and bearing the same impression as the medallion of the Decoration of The Order, and is worn at the left breast, attached to the riband of The Order.

26. *This Mark of Honour* comprises three classes.—The first of which wears the Cross, in *gold*,—the second, in *silver*,—and the third, in *iron*. The possessors receive a Diploma, and are not at liberty to wear the riband *without* the Cross.

27. *The General Mark of Honour* can be conferred on all persons, with the exception of Members of this Order. It may, however, be worn, together with the Decoration of The Order, where it has been previously acquired.

28 *to* 33. Appoint similar Rules for the forfeiture of *The Mark of Honour*, and management of it's affairs, as have been declared for The Order, and appoint a Chancellor, and Minor Officers of The Order.

Description of The Decoration of The Order.

The Decoration consists of a white enamelled Latin Cross, set in gold, the four points of which diminish in size towards the centre. In the centre of the Cross is a round medallion, bearing on the obverse, on a dark blue enamelled field the cypher of Duke Peter Frederick Lewis in gold, surmounted by a Crown, surrounded by a dark red enamelled border, with narrow plain gold edges, on white, in letters of gold, is the Motto of The Duke Peter Frederick Lewis, "*One God, one Right, one Truth,*" as the device of The Order. On the reverse, are the Arms of The House of Oldenburg (*Oldenburg and Delmenhorst*), with the mantle and crown enamelled in a white field. On the reverse of the points of the Cross is inscribed in gold enamelled Roman letters and Arabick cyphers,—on the upper point, "17 *Jan.* 1755," the birth-day of The Duke,—on the right point, "6 *Jul.* 1785," the day on which he entered upon the government of The Duchy,—on the left point, "21 *May* 1829," the day on which The Duke died,—and on the lower point, "27 *Nov.* 6. 1838," the day of the Foundation of The Order. The Cross of The Order is surmounted by a Crown. The size of the Decoration of The Order varies according to the Class.

The Grand Crosses and Grand Commanders wear, in addition, an octagon Star of silver, with 48 rays, in the centre of which, on a dark blue enamelled field is the cypher of The Duke Peter Frederick Lewis in gold, with the Crown surrounded by a dark red enamelled border with narrow silver edges, on which, in golden enamelled Roman letters, is the device of The Order.

The Commanders bear, moreover, the Decoration of The Chapter of The Order, consisting of an enamelled Medal, the two sides of which correspond with the two Medallions of the Decoration of The Order, surrounded by a wreath of oak, attached to the riband of The Order of the breadth of an inch and a half.

Grand Crosses wear this Decoration, of gold, surmounted by a golden Crown,—Grand Commanders wear it, of gold, surmounted by a silver Crown.—Commanders wear it, of gold, without the Crown,—and The Small Crosses wear it, of silver, without the Crown.

AFRICA.

The Order of The Palm and Alligator.

I feel a singular satisfaction in being able to record The Institution of an Order of Knighthood by The Independent Kings and Chiefs of the wild Regions of *Sudan,* on the Western Coast of Africa,—which bespeaks an advancement, hitherto little known, in the cultivation of the blessings of Civilization and Humanity,—blessings which must greatly be increased and confirmed by the Justice, Protection, and Philanthropy so wisely administered to them by Major (*now,* Lieutenant Colonel) Henry Dundas Campbell, late Lieutenant Governor of The Colony of *Sierra Leone,* and it's Dependencies.

On the recal of Governor Campbell being made known to The Colonists and the Monarchs of the adjacent Territories, they hastened to present to him the most honourable and affectionate Addresses and Petitions, demonstrating, at once, their extreme regret at his departure and the good feeling which subsisted between him and the People over whom he was deputed to rule.

These Addresses,—the best proofs of Integrity and Benevolence, and the purest solace of reflection in Retirement,—manifest the friendship which he established, and cemented between The Native Kings and Chiefs, in Western Africa, and the Representative of The British Crown, thus making The English character respected, —an object of the highest importance, long sought, and most essential to the Philanthropick views of The British Nation, which it should be the sedulous aim of all, who desire to witness the progressive Civilization of Africa, to encourage and promote,—independent of the incalculable advantages which The United Kingdom must at the same time derive from opening the Interior of Africa, to the introduction of our Manufactures.

These Addresses are not more remarkable for the excellent feelings which they evince personally towards Major Campbell, and for the confidence which the respective Writers cheerfully repose in him, than for the generous and heartfelt Gratitude which they express. Regardless of the Climate, Major Campbell went far into the Interior of the Country three times, during the years 1836 and 1837. On the first occasion, War was raging with great ferocity between the Natives, whom, by his amicable influence, he induced to assemble at *Mabelly* (a large trading Town on the *Rokelle* river), where they entered into a Treaty of Peace, unanimously declared him to be their *sole Arbitrator*, and subsequently signed a Convention, by which permanent Tranquillity was, and it is much to be hoped, will continue to be secured,—provided the moral influence, which THE LIEUTENANT GOVERNOR established, be maintained, and his liberal views be adopted and carried into operation.

The devoted attachment demonstrated towards him by the whole coloured Population of THE COLONY, on embarking for England, was a certain proof that his exertions to serve them and to improve their condition, were sensibly felt by those over whom he had ruled,—and the flattering documents which he received from the Officers of his Government, and men of the first Character in The Colony, showed that his conduct had obtained their respect and esteem.

In a LETTER from THE KINGS and CHIEFS of *Sudan*, to His late Majesty King WILLIAM the Fourth, they beg to inform His Majesty, " what HENRY DUNDAS CAMPBELL has done. He has given liberty to our Country,—for, before he came to *The Timmanee* Country, the Natives used to kill each other,—and sell and buy each other,—they committed all sorts of wickedness,—they disturbed all the Country, so that there was no safe place for quiet People to cultivate. Nobody could travel, for they disturbed every place,—and these excesses were still committed, when THE KINGS of *Sudan* met together, and sent to HENRY DUNDAS CAMPBELL, who came and made a Treaty, and put a stop to the War, and they accepted his good Regulations.

" They thought it better to deliver into his hands, all the Countries which he had governed more than twelve months. The Kings came to CAMPBELL, and went back, and said, '*We never saw any man like this* HENRY DUNDAS CAMPBELL.'—All the Nations were praying for him,—and those who heard them, said, '*It is true, for we never saw any one better than him among The Kings.*'

"All the Nations of one accord said, 'We will obey, and do whatever CAMPBELL orders us to do. CAMPBELL did not tell us any untruth,—he speaks the truth, *like The Koran.*' CAMPBELL prays fervently,—he never mocks any one who prays, whether he be of his Religion or not. Whoever goes to him, he receives kindly. He can laugh, and play with a *good* man,—he cannot play with a *bad* one,—for every one likes him. He went to *The Fences,* where the people were fighting,—he does not care for Sickness, or Ball,—nothing can hurt him. The Kings are sure, that God protects him, and that God has given him power to do good to their Country.

"We, the People of *Sudan,* do not know how to thank CAMPBELL, or reward him. We offered him gold,—he refused, saying, 'I CAME NOT HERE FOR GOLD,'—*we never yet saw any Christian refuse to take gold, except this* CAMPBELL!—The Kings of *Sudan* begged him, saying, 'Accept this from us, to show King WILLIAM and his Ministers, how much THE AFRICANS loved HENRY DUNDAS CAMPBELL, more than any other Governor.'

"They have given him THE ORDER of THE PALM and ALLIGATOR,—*none but Kings can have* THAT ORDER.

"This Country was under CAMPBELL's command, more than Twelve months, during which time the people travelled over it in safety,—The Kings beg King WILLIAM *to let him wear* THAT ORDER, and to give some reward to HENRY DUNDAS CAMPBELL,—and if The King does so, they will thank him, as much as if he gave it to them.

"O! King WILLIAM, the name ABBAS (*Sultan*), which CAMPBELL possesses, was given to him by us, *that he might be honoured by all Nations.*"

THE DEED, making Governor CAMPBELL, *Abbas*, and granting him Territory in *The Timmanee* Country, is, in the following words,—

"All the Kings and Chiefs of *The Timmanee* Country have met together, to send for HENRY DUNDAS CAMPBELL. When they called him, he left his comfortable House, and his soft Bed, and came to them,—he was not afraid of the suffocating and destructive Climate,—he came to *The Desert* in the hot Season, to the Country which God had given him,—when he arrived there, he found War raging terribly, —but CAMPBELL endeavoured to quiet them,—he succeeded, and stopped the War immediately.

"Then all the Natives were certain, that CAMPBELL was the cause of it,—and that he came on purpose to make Peace.

"When he arrived, he found no Farms, but he began to teach them, —when he did that, all *The Sudan* People wished to see him, which they did, and praised him greatly,—and said, *there is no one like him*, he is very wise, he is just, and he was the means of doing us good.

"When *The Sudan* found him so affectionate and so merciful, they delivered up into his hands *all their Country*, and they made Oath to be faithful to him, and give him the Country that he might restore it,— —They then made a treaty of Peace, and put their Country into his hands, after which nobody fought, but all collected together and made him ABBAS,—therefore, The Christians were jealous.

" *The great name of* ABBAS is given to him and to all his Children.

" *The Sudan* made THE ABBAS their *Leader* and *Commander*—we have made him our *Ruler*, like the first ABBASIDE,—we have given him *The Island of Royell*, now called, ' *Ro Campbell*,' and whatever portion of land he wishes to take for himself or for his Children,—*for ever*, until the end of the World."

THE DEED, making Governor CAMPBELL, *Abbas of Sudan*, and conferring upon him THE ORDER of THE PALM and ALLIGATOR, is, as follows,—

" All the Kings of *Sudan* have praised The Christian King, although you do not know it,—but you must know for certain and be sure, that we are *The Timmanee, Mandingo*, and *Foulah* Nations, and HENRY DUNDAS CAMPBELL *is our Sultan*.

" All the Kings of these Countries have given to him THE ORDER of THE PALM and ALLIGATOR, and they have put into his charge all the Country, to restore what was destroyed and ruined.

" All the Kings of *Sudan* gathered together, and said, ' We have seen The Chiefs, Captain ABBOTT, Royal African Corps, and Dr. AITKIN, who assisted our *Sultan* in doing good to our Country."

" All the Kings of the Countries thank Captain ABBOTT and Dr. AITKIN, and ask God to bless them, for helping *The Sultan* to settle our Country.

" All the Kings of *Sudan* have said, ' *Well done!*'—The Princes of other Nations, called *Dakta, Abkero, Kelora, Sager*, will be gathered together of one accord, by his order, to help *The Sultan*, for the good of our Country.

" It is finished.

" All The Mussulmen who inhabit the whole of *Sudan*, with the inhabitants of *Sego*, have united together and said ' *We have made you* SULTAN *over us, like our former* ABBAS, *who was the greatest* SULTAN,—

there can be no SULTAN *higher than* ABBAS, *you are one of* THE ABBASSIDES."

"May God bless you, that you may live long, and keep your body in health,—may the cold Climates not injure you.

"O Lord, who art the lover of Mankind, and whose Throne is glorious, who hast life and death in thy power, and doest whatsoever thou wishest, bless him. We ask thee for the Glory of thy countenance which fills thy Throne, and for the Power which thou hast given to all thy Creatures, and for the Mercy which thou hast bestowed,—there is no God besides thee. Bless him, O Lord,—nourish him, O Lord God."

On the 18th of April, 1837, at *Mabelly*, a grand Procession of The Native Kings and Chiefs waited on Lieutenant Governor CAMPBELL, to conduct him to *The Palaver*, or, *House of Assembly*, where they conferred upon him, in the presence of many Thousands, THE ORDER OF THE PALM and ALLIGATOR,—And, on the 22d of the same month, they again assembled, and invested him with THE TURBAN, declaring him ABBAS *of Sudan, supposed by them to be the highest Rank, that a Mortal can attain,* and shortly after The Lieutenant Governor's return to *Freetown*, the same honours were conferred upon him, by THE KING OF TIMBO, and THE FOULAH NATION.

THE BADGE of THE ORDER is a Cross of pure gold, of seven points, splendidly clustered with diamonds,—in the centre whereof is a round shield, containing a Palm Tree, and an Alligator at the base, enamelled in their proper colours, surrounded by a circle, enamelled green, on which is the MOTTO, "*God is Great*," in Arabick characters, of gold, and surmounted with a Turban. It is worn, pendent to a massy gold COLLAR, composed, alternately, of a Turban richly adorned, and the obverse of The Badge, suitably enamelled.

The Star which is also of pure gold, is similar in shape to that of The Knights of The Garter, and contains in the centre, the obverse of The Badge, encircled with large and costly Brilliants.

Sudan is derived from *Souda* or *Suda,* which signifies in Arabick, " *Black.*"—a name given by The Moors and Arabs to *Negroland,* or *Nigritia,*—and sometimes more particularly to The Empire of *Haussa,* or *Houssa.*

It may not be improper also to add, that the farther a Traveller proceeds into *The Interior* of *Africa,* the purer he finds the *Arabick Language* spoken by The Natives.

THE UNITED STATES OF NORTH AMERICA.

The Society of The Cincinnati.

A Society, denominated "The Society of Cincinnati," was formed in The United States of America towards the close of the year 1783, by The Officers of the Army, just before the disbanding of the troops. It was so called after the Roman Dictator, Cincinnatus, and intended to perpetuate the memory of The Revolution, the mutual Friendship, and the Union of The States,—and also to raise a Fund for the relief of poor Widows and Orphans, whose husbands and fathers had fallen during the War, and for their Descendants.

In the month of October, 1783, General Washington subscribed himself President of this Order.

The General Society, which was to meet at least once in three years, was divided into State Societies, which were to meet annually on the 4th of July, The Anniversary of American Independence.

In order to raise a Fund for the benevolent purposes of The

Society, each Member was to subscribe one month's pay to the general Treasury, and the Fund was to be augmented by Private Donations. The Interest only of the money thus raised, was to be expended in acts of Charity.

The Members of The Institution were to be distinguished by wearing a Medal, emblematical of the design of The Society. The device was a bald Eagle, of gold, and it was suspended by a deep blue Riband, edged with white, descriptive of The Union of America and France. The Emblems borne on the breast of the Eagle, were the following,—the principal figure Cincinnatus, and three Senators presenting him with a Sword, and other military Ensigns,—on a field, in the back ground, his Wife standing at the door of the Cottage, and near it a Plough and other implements of Husbandry,—round the whole, "*Omnia reliquit servare rempublicam.*" On the reverse, the Sun rising, a City with open gates, and Vessels entering the Port,—Fame crowning Cincinnatus with a wreath, inscribed "*Virtutis premium,*"—below, hands joining, supporting a heart, with a Motto, "*Esto perpetua,*"—round the whole, "*Societas Cincinnatorum instituta,* A. D. 1783."

The honours and advantages of this Society were to be *Hereditary* in the Line of the eldest Male heirs, and in default of Male issue, in that of the collateral Male heirs. Honorary Members were to be admitted, but *without* the *Hereditary* advantages of The Society, and provided their number should never exceed the ratio of one to four of the Officers or their Descendants.

The ostensible views of this Society, however honourable and praise-worthy, could not screen it from popular jealousy,—and it was alleged by an able Writer (Ædanus Burke), that the principles on which The Society was formed, would, in process of time, introduce and establish *an Order of Nobility* in the Country, which would be repugnant to the genius of *The Republican* Governments of America, and dangerous to Liberty.

Early in the year 1784, the Provinces of *Pennsylvania* and *Massachusetts* declared THE INSTITUTION unjustifiable, and their resolution to discountenance it. The Province of *Rhode Island* proceeded so far as to annul the Privileges of all the Subjects of it's State, who should be Members of this Society, and to declare them incapable of any Office under Government.

In consequence of this alarm, THE CINCINNATI, in their first General Meeting convened at *Philadelphia*, on the 3d of May, 1784, thought proper to new model the Institution of their Society. They professed to withdraw the claim of *Hereditary* Honour, to disclaim all interference with political subjects, and to place their Funds under the immediate cognizance of the several Legislatures, through the medium of a general Charter. Indeed, they relinquished without hesitation every thing in their new constitution, except their personal friendships, of which they could not be divested, and the acts of beneficence which it was their intention, should flow from them.

With these professions, however, they retained their Funds, their General Meetings, and their Ribands.

THE ARGENTINE REPUBLICK.

The Argentine Republick consists of that part of The South Eastern Coast of South America, which extends from *The River Plate* to *Cape Horn* on the South, to *The Andes* on the West, and to *The State of Bolivia* on the North, formerly known as The Vice-Royalty of Buenos Ayres.

Medals have occasionally been given to Officers, who have distinguished themselves in Actions of War.

In the Heraldick Collections of the late Sir George Nayler, Garter, in The College of Arms, is a Drawing of The Star of The Order of St. Henry, in The Republick of Haiti or St. Domingo,—and in The "*Almanach Royal d'Haiti,*" of 1820, is a List of The Members.—There are also Drawings of Badges from original *Insignia,* for services rendered to The Republick of Chili,—but they are not accompanied, with descriptions.

THE EMPIRE OF THE BRAZILS.

See, also, .

THE KINGDOM OF PORTUGAL.

The Order of Peter.

This Order, founded by Peter the First, is the first of The Brazilian Empire. It comprises but one class, and is conferred upon Sovereigns alone.

The Decoration displays on the white medallion of the centre, a golden Phœnix, the breast of which is covered by an escutcheon of Silver bearing, in gold, the cypher P. 1., and holding in it's claws an antique crown. A deep blue circle round it, contains the words, in gold, "*Founder of the Empire of Brazil.*" The five points of the Cross are of white enamel, and the whole is surmounted by The Imperial Brazilian Crown.

This Decoration is worn attached to a wide green riband, with a white border on each side, passed like a scarf from right to left.

THE ORDER OF THE CROSS OF THE SOUTH.

THIS ORDER, which was founded by PETER the First, and of which THE EMPEROR is Grand Master, is divided into GRAND CROSSES, DIGNITARIES, OFFICERS, and KNIGHTS.

On the obverse of the medallion of the centre is a bust, in relief, of The Emperor PETER the first, upon gold, and on the deep blue circle around, the inscription in letters of gold, "*Peter I., Emperor of the Brazils.*" On the reverse of the medallion, on a deep blue ground, is *The Constellation of the Cross of The South*, composed of white Stars, and, within the circle, the words, "*The reward of the deserving.*" A crown of laurel passes between the five points of the Cross, which is white, and surmounted by a crown.

The Grand Crosses, Dignitaries, and Officers, wear a golden JEWEL surmounted by a Star, displaying the same Constellation, as the reverse of the Decoration.

The Grand Crosses wear the *Insignia* of THE ORDER attached to a sky blue riband, passed from right to left. Dignitaries wear it, pendent from the neck,—and the Officers and Knights, attached to a golden *Buckle*, on the left breast.

THE ORDER OF THE ROSE.

The Emperor PETER the First, by a Decree, dated the 17th of October, 1829, founded THIS ORDER, on the occasion of his Marriage with the Princess AMALIE, of LEUCHTENBERG and EICHSTEDT.

THE EMPEROR is Grand Master of this Order of *Military* and *Civil* Merit. The Prince Royal is always Grand Cross, and Grand Dignitary at the same time. The other Imperial Princes are Grand Crosses.

The ORDER comprises *Eight* GRAND CROSSES, and *Eight* HORORARY GRAND CROSSES,—*Sixteen* GRAND DIGNITARIES,—*Thirty-two* DIGNITARIES,—and an *unlimited* number of COMMANDERS, OFFICERS, and KNIGHTS. The possession of the title of "*Excellency*" is a necessary preliminary to the honour of Grand Cross. All Grand Dignitaries possess this title of right.

In order to become a DIGNITARY, it is necessary to have the honorary distinction of the *Senhoria*, a situation which belongs of right to the class of COMMANDERS. OFFICERS of THE ORDER are entitled to the honours due to *Colonels*, and KNIGHTS to those due to *Captains*.

THE INSIGNIA of THE ORDER consist of a Cross, of six points, of white enamel, between which passes a garland of Roses. On the centre, is the cypher, in gold, P. A. (*Pierre et Amalie*), surrounded by the words, "*Love and Fidelity*,"—the whole surmounted by a crown.

The Dignitaries, Commanders, and Officers, wear the JEWEL, without a crown. Grand Dignitaries and Grand Crosses add this latter ornament.

The Decoration, suspended from a rose coloured riband, with a white border on both sides, is worn by The Grand Crosses, from right to left,—by Grand Dignitaries and Dignitaries, around the neck,—and by Commanders, Officers, and Knights, on the left breast.

The eight actual Grand Crosses wear in addition, on grand Festivals of the Court, a CHAIN of gold, composed of enamelled Roses.

The Royal Order of Saint Elizabeth.

This Order was founded, on the 4th of November, 1801, by John, Prince of Brazil, afterwards King of Portugal,—who, by a Decree, dated the 17th of December of the same year, authorized his Consort, the Princess Charlotte, to frame Statutes for it's regulation and government.

On the 25th of April, 1804, the following Statutes were accordingly published,—

The Insignia of this Order shall consist of a Medal of gold, bearing on one side the representation of *Saint Elizabeth,* and the inscription, "*Pauperum Solatio,*"—and on the other, the initial letters of the name of the Princess Charlotte in a cypher, and, around, the words, "*Royal Order of Saint Elizabeth.*" On the Festival days of The Order, or Gala days, and in all publick functions, this Order shall be worn over the dress, suspended from a rose coloured scarf thrown across the right shoulder,—on other days, it shall be attached to the left breast by a knot of riband of the same colour, but of narrower dimensions.

The Order shall consist of *Twenty-six Ladies,* besides The Royal Family, and the other Royal Personages. This number not to be increased without very sufficient reason.

The Ladies who are admitted into this Order, must be *Twenty-six* years of age, or married.

The reception into The Order shall take place in one of the Halls of The Palace.

On the days of *Saint Isabel* and *Saint Carlos*, the Princess will honour the Ladies of The Order with a particular kissing of hands, and in the form of a Chapter. The precedency of each Lady being regulated by her seniority.

The Ladies of this Order shall visit in turn, once in each week, *The Foundling Hospital*, and shall observe the articles relating to the particular regimen and domestick government of The Hospital, and other Acts of Charity to be put in practice respecting the treatment of *The Expostos*,—which, when completed and reduced to rules, shall be united to these Statutes.

The Festival of this Order shall be celebrated on the day of *Saint Isabel*, it's Patroness, in a Church to be appointed by the Princess, at which all the Ladies shall be present, who are not prevented by distance or illness. On the same day The Princess will visit *The Casa dos Expostos*, accompanied by all the Ladies.

The Secretary of The Order shall be appointed by the Princess, and shall have the custody of it's Archives and all other things belonging to it,—referring himself in all matters relating to The Order to the Secretary of The Princess, from whom he will receive her Royal determinations.

He shall also keep Books, for the purpose of entering Memoranda of the Receptions, and Deaths of the Ladies, and registering correctly all Regulations, Orders, and other Provisions by direction of The Princess,—He shall give the necessary Notices for holding the Acts of The Order,—shall take the notes, and do also whatever may be necessary,—He shall attend to *The Insignia*, collecting those of the Ladies who shall die,—He shall be present at the Receptions and other acts of Ceremony of The Order, and shall perform whatever shall appertain to his office, or be deputed to him by these Statutes.

By the 13th article, The Princess orders the printing and distribution of the Statutes, reserving to herself the right to enlarge or vary them, or make others, *de novo*.

INDEX.

LONDON:

PRINTED BY W. NICOL, 60, PALL-MALL.

www.ingramcontent.com/pod-product-compliance
Ingram Content Group UK Ltd.
Pitfield, Milton Keynes, MK11 3LW, UK
UKHW020129290926
15471UKWH00003B/378

9 781843 426950